Biochemistry and Physiology *of* Polyamines in Plants

Editors
Robert D. Slocum
Department of Botany
Williams College

and

Hector E. Flores
Department of Plant Pathology
Pennsylvania State University

CRC Press
Boca Raton Ann Arbor London

Library of Congress Cataloging-in-Publication Data

Biochemistry and physiology of polyamines in plants / Robert D. Slocum
and Hector E. Flores, editors.
p. cm.
Includes bibilographical references and index.
ISBN 0-8493-6865-0
1. Polyamines—Metabolism. 2. Plants—Metabolism. 3. Plants—
Development. 4. Growth (Plants) I. Slocum, Robert D.
II. Flores, Hector E.
[DNLM: 1. Plants—growth. 2. Polyamines—metabolism. QU 60 B6145]
QK898.P756 1991
581.19'245—dc20
DNLM/DLC
for Library of Congress 91-40655
CIP

Direct all inquiries to CRC Press, Inc., 2000 Corporate Blvd., N. W., Boca Raton, Florida, 33431.

International Standard Book Number 0-8493-6865-0

Library of Congress Card Number 91-40655
Printed in the United States 1 2 3 4 5 6 7 8 9 0

PREFACE

Over the last three decades, an impressive body of evidence has been gathered suggesting that polyamines play essential roles in the growth and development of plants and other organisms. Thus, the old adage: "polyamines are basic", which seems self-evident to any biochemist!

During this period, the number of publications encompassing plant polyamine biology has increased from a handful to more than 100 per year currently. In recognition of this emerging new discipline, two international conferences on plant polyamine research were convened within the past year. *Polyamines and Ethylene: Biochemistry, Physiology and Interactions* was the topic of the 5th Penn State Symposium which attracted over 150 scientists in May, 1990. A second conference, *Polyamines As Modulators of Plant Development*, sponsored by the Juan March Foundation, brought together another group of scientists in Madrid, Spain in February 1991.

In view of the growing literature dealing with plant polyamine biology and the increasing number of new scientists entering this field, it is surprising that a comprehensive treatment of this subject has not been previously published. To be sure, there have been many excellent reviews by individual authors, but each of these was somewhat limited in scope or focused on a particular aspect of polyamine research. Our goal as editors of the present volume, was to bring together chapters written by leading experts from the major areas of plant polyamine research. The authors were asked to present a broad overview of a subject, providing both current information and historical perspectives. They were also charged with identifying important research problems for the future in order to stimulate interest in the field. The contributing authors have done an outstanding job in meeting these goals and we hope that the present volume will be a useful research reference for many years to come.

This volume is introduced by Dr. Terence A. Smith of the Long Ashton Research Station in Bristol, U.K. Dr. Smith is one of the early pioneers in plant polyamine research, and his introductory chapter provides the reader with a personal accounting of the major developments in this discipline. The next eight chapters deal with various aspects of polyamine metabolism in plants, followed by six chapters exploring the functions of polyamines in plant growth and development. Two additional chapters comprise the final section, which examines analytical methods for polyamines and enzymes of polyamine metabolism. This introduction to many of the commonly used methods in polyamine research, including discussions of problems unique to plant studies, should prove invaluable to workers entering the field.

The successful completion of this volume is due not only to the efforts of the contributing authors but also to the many other dedicated scientists whose work is described in these pages. As with any discipline, however, the field of plant polyamine research has grown and thrived because of the efforts of a few individuals who have had both the insight to ask the important questions and the ability to attract students and other scientists to join them in pursuit of the answers.

Terence Smith is one such individual. His work in the early 1960s documented the effects of abiotic stress on polyamine biosynthesis, and in the last two decades, he has characterized many of the enzymes of polyamine metabolism in plants. During this period, Dr. Nello Bagni and his colleagues at the University of Bologna also carried out pioneering studies on the effects of polyamines on cell division and morphogenesis in plants. In the late 1970s, Dr. Arthur W. Galston and his group at Yale University discovered the antisenescence properties of polyamines and, over the last decade, they have investigated the involvement of polyamines in plant growth and development and stress responses. We are especially grateful to Professor Galston, who introduced us to the fascinating world of plant polyamine research, and who continues to be a source of support and inspiration to us both.

The Editors
February, 1991

THE EDITORS

Robert D. Slocum, Ph.D., is Assisstant Professor of Biology at Williams College in Williamstown, Massachusetts. Dr. Slocum graduated in 1974 from the University of Maine, Orono with a B.A. in Zoology. He obtained his M.S. in Botany from The Ohio State University in Columbus in 1977 and a Ph.D. in Botany from the University of Texas, Austin in 1981. He did postdoctoral work at Yale University before joining the faculty at Williams College in 1985. In 1989, he was a Visiting Scientist at the Boyce Thompson Institute at Cornell University, Ithica, New York.

Dr. Slocum is a member of the American Association for the Advancement of Science, the American Society for Gravitational and Space Biology, the American Society of Plant Physiologists, the International Society for Plant Molecular Biology, and Sigma Xi Scientific Research Society. He is presently serving on the Editorial Board of *Plant Physiology*.

He has received fellowships and research grants from the National Aeronautics and Space Administration, Research Corporation and the National Science Foundation. He has published more than 30 research papers and review articles and has been an invited lecturer at national and international research meetings and at colleges and universities. His current research interests include the regulation of polyamine biosynthesis in plants in response to environmental stresses and the involvement of cell wall-localized amine oxidases and peroxidases in the control of plant growth.

Hector E. Flores, Ph.D., is Associate Professor of Plant Pathology at the Pennsylvania State Univeristy, University Park. Dr. Flores also holds a joint appointment with the Penn State Biotechnology Institute, and an adjunct faculty position at the Biology Department. Dr. Flores graduated in 1973 from the Universidad de San Marcos in Lima, Peru, with a B.Sc. in Biology. He obtained his M.Sc. in Horticulture from the University of Puerto Rico (Mayaguez) in 1977, and his M.Phil. and Ph.D. in Biology from Yale University in 1981 and 1983, respectively. He did postdoctoral work at the ARCO Plant Cell Research Institute between 1983 and 1985. He became Assisstant Professor of Plant Pathology and Crop Physiology at the Louisiana State University in 1985. Dr. Flores has been affiliated with the Pennsylvania State University since 1988.

Dr. Flores is a member of the American Society of Plant Physiologists, the Phytochemical Society of North America, Sigma Xi Research Society, and the American Association for the Advancement of Science. He has received research grants and fellowships from the National Science Foundation, the U.S. Department of Agriculture, the Pennsylvania Research Corporation, and the Japan Health Sciences Foundation.

He has served as ad hoc reviewer for the National Science Foundation, the USDA Competitive Grants and BARD programs, USAID, and several plant science journals. Dr. Flores has published over 40 research papers and review articles, edited one book, and given numerous invited lectures at national and international meetings and at colleges and universities. His current research interests include the role of polyamines in response to ozone and other environmental stresses and the biosynthesis of bioactive metabolites in plant roots.

CONTRIBUTERS

Dr. Riccardo Angelini
Dipartimento di Biologia Vegetale
Università "La Sapienza"
Rome, Italy

Dr. Nello Bagni
Department of Biology
University of Bologna
Bologna, Italy

Dr. Erin Bell
Department of Biochemistry and Biophysics
Texas A & M University
College Station, Texas

Dr. Helena Birecka
Biological Sciences Department
Union College
Schenectady, New York

Dr. Peter J. Davies
Section of Plant Biology
Cornell University
Ithaca, New York

Dr. Erwin B. Dumbroff
Biology Department
University of Waterloo
Waterloo, Ontario

Dr. Marcos Egea-Cortines
Department of Life Sciences
Ben-Gurion University of the Negev
Beer-Sheva, Israel

Dr. Rodolfo Federico
Dipartimento di Biologia Vegetale
Università "La Sapienza"
Rome, Italy

Dr. Hector E. Flores
Biotechnology Institute
Pennsylvania State University
University Park, Pennsylvania

Dr. Arthur W. Galston
Department of Biology
Yale Univeristy
New Haven, Connecticut

Dr. Ravindar Kaur-Sawhney
Department of Biology
Yale University
New Haven, Connecticut

Dr. Glenn D. Kuehn
Chemistry Department
New Mexico State University
Las Cruces, New Mexico

Dr. Mosbah M. Kushad
Department of Horticulture
Virginia Polytechnic Institute
Blacksburg, Virginia

Dr. Russell Malmberg
Botany Department
University of Georgia
Athens, Georgia

Dr. Josette Martin-Tanguy
Directeur de Recherches, INRA
Lab de Physiopathologie Vegetale
Dijon, France

Dr. Yosef Mizrahi
Department of Life Sciences and
The Institutes for Applied Research
Ben-Gurion University of the Negev
Beer Sheva, Israel

Dr. Gregory C. Phillips
Department of Agronomy and Horticulture
New Mexico State University
Las Cruces, New Mexico

Dr. Rossella Pistocchi
Department of Biology
University of Bologna
Bologna, Italy

Dr. Rajeev Rastogi
Department of Molecular Biology
and Genetics
University of Guelph
Guelph, Ontario, Canada

Dr. Robert D. Slocum
Department of Biology
Williams College
Williamstown, Massachusetts

Dr. Donatella Serafini-Fracassini
Dipartimento di Biologia (Botanica)
Università di Bologna
Bologna, Italy

Dr. Mary A. Smith
Department of Biology
University of Richmond
Richmond, Virginia

Dr. Terence A. Smith
Long Ashton Research Station
Bristol, United Kingdom

TABLE OF CONTENTS

Introduction

Chapter 1

A HISTORICAL PERSPECTIVE ON RESEARCH IN PLANT POLYAMINE BIOLOGY

Terence A. Smith

TABLE OF CONTENTS

I. THE PIONEERS

The history of polyamine research can be traced back to an observation Antoni van Leeuwenhoek (1632–1723) reported to the Royal Society in London, which was published in the Philosophical Transactions of the Royal Society[1] in 1678. Despite his very simple microscope, his drawing of the spermine phosphate crystals which he found in human semen proved to be a remarkably accurate representation. His description of these crystals, which was originally given in Latin, has been vindicated by recent photographs[2] obtained with the scanning electron microscope. In the next 200 years they were rediscovered several times by workers who did not know about the earlier observations of Leeuwenhoek. Vauquelin[3] in 1791 thought that these crystals were of calcium phosphate, but in 1865 Boettcher[4] considered them to be protein. In 1878 it was established that these crystals were due to an organic base, which was given the name "spermin" [*sic*] by Ladenburg and Abel[5] in 1888. The correct structure of spermine was not finally established until 1926, simultaneously in England and Germany. In the same year the closely related base spermidine was identified in animal tissues by the English group. The early history of the polyamines has been summarized by several authors,[2,6-11] and during the last 30 years many comprehensive reviews of plant polyamine biology have been published.[12-24]

By the late 19th century, spermine had been demonstrated in many animal tissues, although it was not until 1925 that Dudley and Rosenheim[25,26] showed that yeast also contains this base. The diamines, putrescine and cadaverine, were first identified in animal tissues subject to bacterial decay, giving rise to the very descriptive names. At first the amines were thought to be among the poisonous components of decayed matter and to be responsible for food poisoning. At that time they were collectively known as the ptomaines (from the Greek *ptoma* ; a corpse) The structures of the common diamines (putrescine, cadaverine) and polyamines (spermidine, spermine) are shown below.

Putrescine $NH_2CH_2CH_2CH_2CH_2NH_2$

Cadaverine $NH_2CH_2CH_2CH_2CH_2CH_2NH_2$

Spermidine $NH_2CH_2CH_2CH_2NHCH_2CH_2CH_2CH_2NH_2$

Spermine $NH_2CH_2CH_2CH_2NHCH_2CH_2CH_2CH_2NHCH_2CH_2CH_2NH_2$

Much of the early work on the characterization of the amino acids and their decarboxylation products was effected by a laborious fractionation with heavy metal salts, evidence for which may be found in the name given to arginine from the Latin word for mercury (*argentum vivum*), since it was first isolated[27] as the mercury salt. The bases were purified by crystallization of the complex salts formed, for instance, as chloroaurates, chloroplatinates, and picrates. Structures were based on elemental analysis, and purity established by melting point. Many of these techniques are described by Werle.[28] The work was of a remarkably high standard, despite the absence of chromatographic procedures and modern physical techniques.

Putrescine and cadaverine, established as bacterial decarboxylation products of ornithine and lysine respectively, were first found in fungi and only later in the higher plants. The first reference to the occurrence of putrescine in higher plants is probably due to Ciamician and Ravenna,[29] who in 1911 demonstrated that *Datura stramonium* contains this diamine, together with the tropane alkaloids derived from it. Subsequently, putrescine was found in *Atropa belladonna* by Goris and Larsoneau[30] in 1921, in grapefruit by Hiwatari[31] in 1927, and in orange juice by Herbst and Snell[32] in 1948. The latter work is of particular interest, since the putrescine was shown to be a growth factor for the bacterium, *Hemophilus parainfluenzae*. In the last paragraph of their paper, Herbst and Snell add the prophetic statement "to our knowledge, this is the first demonstration of an essential nutritional function for one of the putrefactive amines, and indicates that putrescine and possibly additional compounds of this group play a much more important metabolic role than has been previously indicated."

Agmatine, which is now known to be an important intermediate in putrescine biosynthesis in higher plants, was first discovered by Kossel[33] in 1910 in putrefied herring roe, where it arose by the bacterial decarboxylation of the arginine which forms a large part of the protamine in this tissue. He gave this base the name agmatin ([*sic*] probably from the Greek *agma*; a fragment). The first reference to the occurrence of agmatine in higher plants was by Heyl,[34] who in 1919 found this in ragweed (*Ambrosia artemisifolia*) pollen, then calling it "agmantine".

Agmatine $$NH_2\underset{\underset{\textstyle NH}{\|}}{C}NHCH_2CH_2CH_2CH_2NH_2$$

Much early work on these amines in higher plants concerned the identification of the diamines in the Solanaceae in relation to alkaloid formation. A leader in this field was B.T. Cromwell, though early reviews of the plant amines were mainly contributed by German authors, notably Werle[28] in 1955 and by Guggenheim[12] in 1958.

II. POTASSIUM DEFICIENCY

Richards and Coleman[35] found that "communication is intended to direct attention to the fact that in barley, at least, putrescine is produced and accumulated under certain conditions of potassium deficiency." During the course of a study on the effects of various mineral element deficiencies on the content of amino nitrogen in barley, Richards and Coleman[35] in 1952 were among the first to apply the techniques of paper chromatography to the separation of amino acids in higher plants, using ninhydrin as the chromogenic reagent. In separations of extracts of potassium-deficient barley leaves, they found that the most intense spot on their paper chromatograms was due to putrescine. After purification using the newly-introduced methods of ion exchange chromatography, they identified this on the basis of its chromatographic properties, derivative formation, and also by its X-ray diffraction pattern. Since experiments showed that ^{14}C-labelled ornithine was not a very promising precursor for this amine in barley,[36] Richards suggested that the biosynthetic pathway might occur from arginine via its decarboxylation product, agmatine. Extracts of the leaves of potassium-deficient barley plants were fractionated on a carboxylic ion exchange resin. This eventually enabled the purification and crystallization of agmatine as the sulfate salt, which was characterized by its infrared spectrum.[37] Moreover, arginine was decarboxylated upon incubation with crude extracts of the barley leaves. The agmatine concentration, like that of putrescine, was greatly enhanced by potassium deficiency, and the activity of the arginine decarboxylase was correspondingly increased in these plants.[38] Feeding of agmatine to the excised shoots of barley seedlings yielded putrescine as expected, though an additional product was eventually identified as *N*-carbamoylputrescine, whose structure is shown below. The formation of putrescine from agmatine was therefore a two-stage process, with *N*-carbamoylputrescine as an intermediate. *N*-carbamoylputrescine was readily hydrolyzed to putrescine by leaf extracts.[39,40] This amine was also shown to be formed by citrulline decarboxylation in sugar cane grown in culture,[41] and in *Sesamum*,[42] a plant in which the amine accumulates in potassium deficiency.[43]

N-Carbamoylputrescine $$NH_2\underset{\underset{\textstyle O}{\|}}{C}NHCH_2CH_2CH_2CH_2NH_2$$

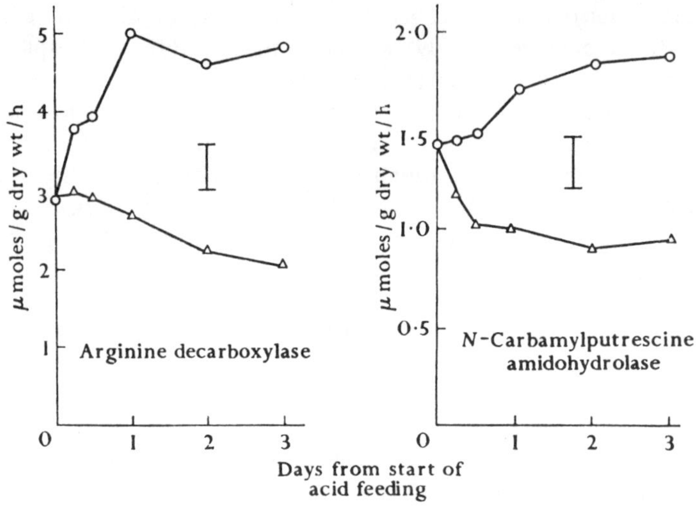

FIGURE 1. Arginine decarboxylase and *N*-carbamoylputrescine amidohydrolase activity in the first leaves of barley seedlings fed 25 m*M* hydrochloric acid (○) or water (Δ) through the roots. Samples were taken at intervals after the 7th day, when the acid feeding commenced. The vertical lines represent the least significant difference at the 5% level of probability.[45]

In 1940 Gale[44] found that certain bacteria will respond to acidification of the growth medium by forming amines, and this appears to be an adaptive response by the bacteria to adjust the pH to a value which is more amenable for growth. By analogy, it was thought possible that plants might be forming the diamines in order to compensate for a decrease in their internal pH. Potassium deficiency, it was argued, might cause a reduction in cytoplasmic pH. To investigate this possibility, barley seedlings were grown in water culture for seven days, and some plants were then transferred to a 25m*M* hydrochloric acid solution. As shown in Figure 1, the activity of arginine decarboxylase was found to be increased significantly by this treatment, on the basis of dry weight or protein.[45] More recent experiments by Young and Galston[46] have served to confirm this finding. Although this simplistic interpretation of these experiments is quite compelling, recent discoveries suggesting that putrescine is formed in response to a variety of stress factors cast some doubt on the involvement of pH alone. For instance, Flores et al.[47,48] have shown that putrescine biosynthesis is increased by exposure to a variety of osmotica (Figure 2). Putrescine synthesis is also a response to cadmium toxicity.[49] The formation of putrescine is apparently of great importance to the plant in potassium deficiency, since in extreme conditions, up to 140 μmol/g dry weight may be accumulated. This represents 1.2% of the dry matter, and the putrescine then accounts for 20% of the total nitrogen[2,50,51] (Figure 3).

III. THE BOLOGNA GROUP

In 1964 Moruzzi and Caldarera[52] were able to demonstrate the presence of the diamines putrescine and cadaverine, and the polyamines spermidine and spermine in the embryos of several species of cereals by the use of paper chromatography and electrophoresis after ion exchange purification. In a paper published in 1967 with these two workers, Bagni[53] studied the distribution of the polyamines in the wheat plant. These workers also showed that spermine acts as a growth substance for *Helianthus tuberosus* explants grown in culture[54] (Figure 4), and in subsequent work, Bagni[55] initiated a major investigation of the function and metabolism of polyamines in this plant, which was reviewed in 1974.[56] Treatment of the *Helianthus* tuber tissue with either polyamines or indole-3-acetic acid (IAA) stimulated growth, and treatment with IAA

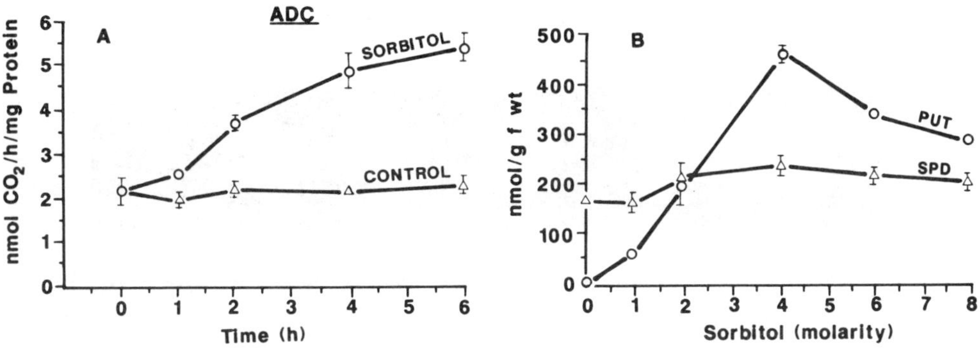

FIGURE 2. (A) Effect of osmotic stress on arginine decarboxylase activity. Leaf segments were floated over buffer (Δ) or buffer + 0.4 *M* sorbitol (○), under light, and samples were taken at the indicated times for enzyme assay. Bars represent the standard error of the mean. (B) Effect of osmotic concentration on polyamine content of peeled oat leaf segments. Polyamines were determined after incubation for 4 h in the buffer with varying concentrations of sorbitol.[47]

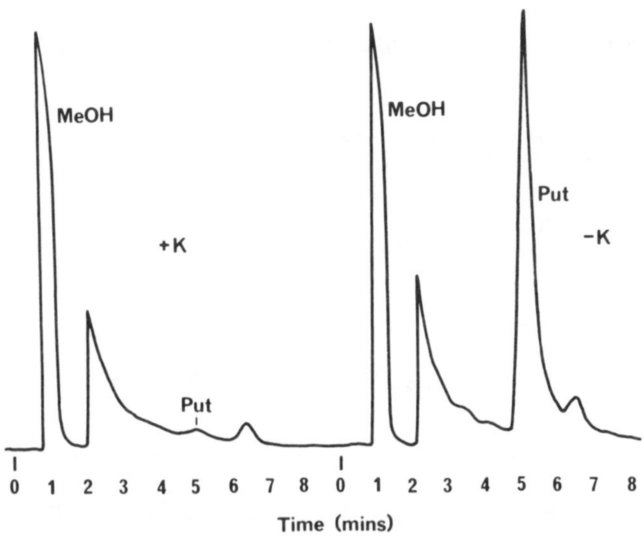

FIGURE 3. Gas chromatograms of the amine fraction of normal (+K) and potassium deficient (-K) barley leaves. (MeOH, methanol solvent peak; Put, putrescine peak.) The sensitivity was increased ×500 2 min after injection.[50]

caused polyamine accumulation in this tissue. It was, therefore, proposed that at least one of the processes by which IAA stimulates growth may be mediated by the polyamines.[57]

Realizing the importance of the polyamines for nucleic acid function, Bagni was among the first to show that polyamines are able to stimulate protein synthesis *in vitro*, using a ribosomal preparation from *Helianthus*.[57] In later work with Serafini-Fracassini and other co-workers, he demonstrated the relationship between polyamines and nucleic acid synthesis during the growth cycle,[58] and studied the binding of the polyamines to tRNA and rRNA.[59] In papers published in 1967 and 1978, reviewing some of the *Helianthus* tissue culture work, he demonstrated polyamine catabolism and studied the biosynthesis of putrescine from citrulline.[60-62] In 1967 and 1968, investigations were made of the occurrence of the polyamines in tomato fruits,[63] in the seeds of various plants,[64] and in bean seedlings.[65] With other workers, Bagni studied the effect of infection with *Agrobacterium tumefaciens* on the polyamines of *Scorzonera*,[66,67] and *Nicotiana*,[68] work which was reviewed in 1979.[69]

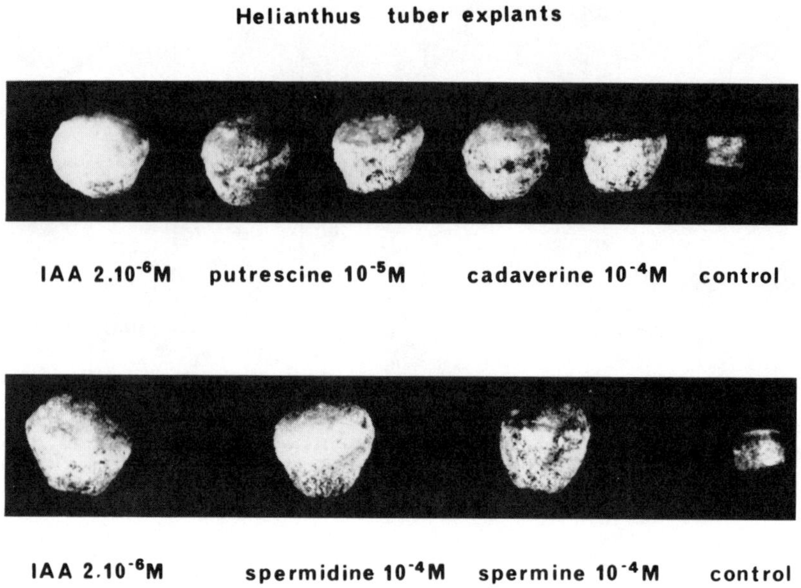

FIGURE 4. Growth stimulation by indoleacetic acid (IAA) and di- and polyamines of *Helianthus tuberosus* (Jerusalem artichoke) *in vitro*. Tuber explants taken in November were incubated on agar containing basal nutrients with IAA, putrescine, cadaverine, spermidine, and spermine at the concentrations stated for 40 days.[16]

Subsequently, the group at Bologna has studied putrescine uptake and conjugation, the relationship between the cell cycle and polyamine metabolism, and the stimulating effect of putrescine sprays on fruit set in commercial crops, among a wide range of other topics. Some of this work has recently been reviewed.[70-72]

IV. BIOSYNTHESIS

The first record of ornithine and lysine decarboxylation in higher plants, studied in extracts of pea seedlings, was published by Hasse et al.[73] in 1967, though the rather tenuous evidence here was based on the detection of pyrroline and piperideine, the oxidation products of putrescine and cadaverine respectively. Yoshida and Mitake[74,75] showed that ornithine and arginine are both incorporated into putrescine in tobacco plants, and Mizusaki et al.[76] detected an ornithine decarboxylase (ODC) activity in 1973 in tobacco roots, where it was thought to be concerned in nicotine biosynthesis. However, recent evidence suggests that arginine decarboxylase (ADC) has a more important role in this pathway.[77] Subsequently, many workers have shown the presence of ODC in higher plants, though in some of these studies an artefactual decarboxylation may have been responsible for part of the activity measured.[78]

Between 1960 and 1970, work with animals had shown that polyamine biosynthesis is greatly accelerated in malignant tissues, and the possibility of controlling cancer by inhibiting the key enzyme ODC was given much attention, especially by the pharmaceutical company Merrell Dow. This research produced some very efficient enzyme-activated inhibitors of ODC,[79] the first of these being DL-α-difluoromethylornithine (DFMO). This and other similar inhibitors are effective against ODC by their ability to form a covalent attachment to the active site on interaction with the enzyme. Although DFMO does reduce the putrescine concentration *in vivo* in many organisms, including plants, its use in cancer chemotherapy has been restricted, due in part to its rapid excretion and also because of the rapid turnover of ODC. Even so, DFMO and related inhibitors are finding use in the control of pathogenic protozoan diseases, including

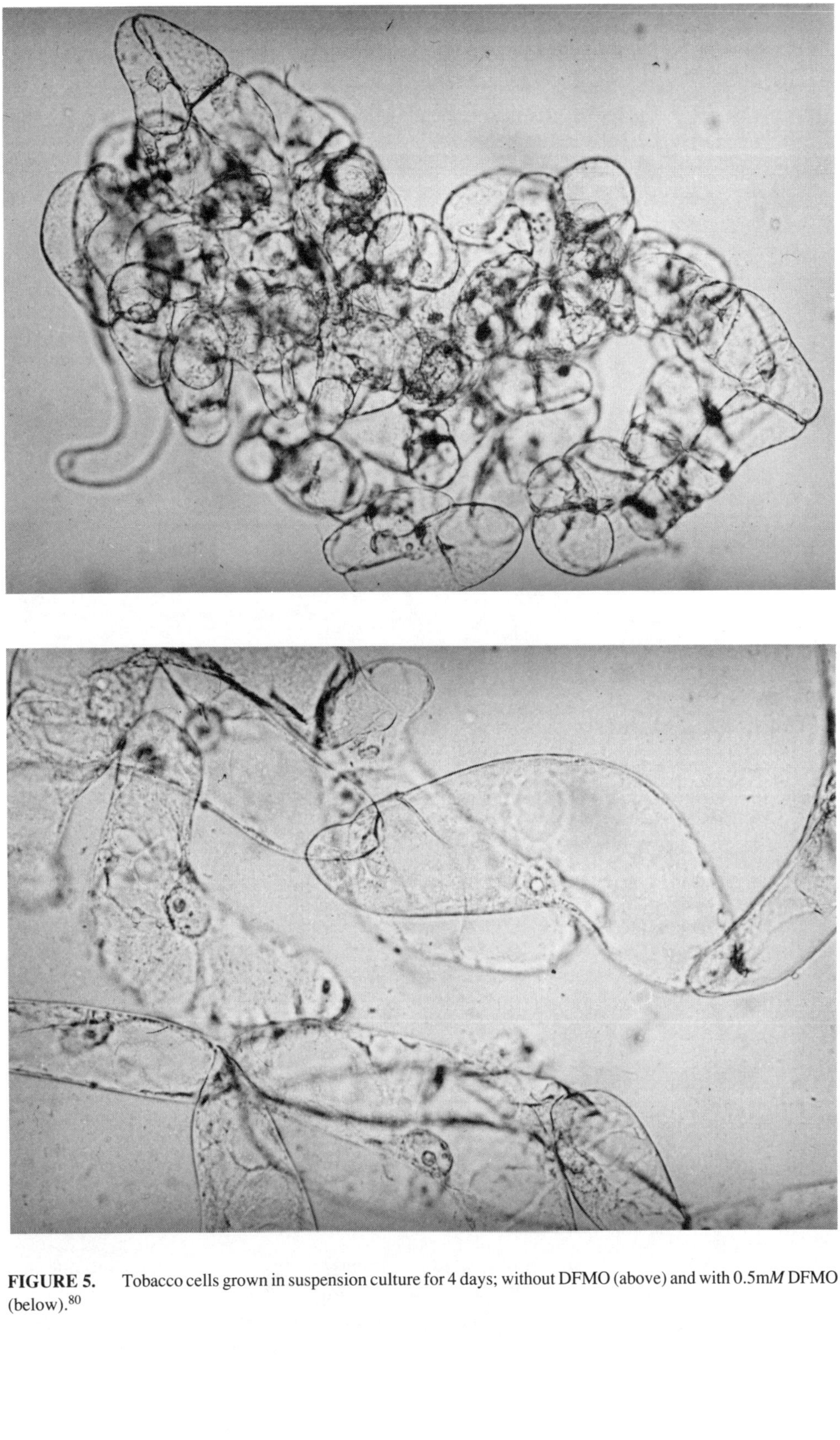

FIGURE 5. Tobacco cells grown in suspension culture for 4 days; without DFMO (above) and with 0.5m*M* DFMO (below).[80]

FIGURE 6. Effect of 1 m*M* DFMA fed to the roots of tall pea seedlings 6 days from sowing, and then grown for an additional 13 days. Left, treated with water only; right, treated with DFMA.[81]

trypanosomiasis and malaria. DL-α-difluoromethylarginine (DFMA), the corresponding arginine analogue of DFMO that is inhibitory to ADC, has been especially helpful with fundamental studies in plant physiology, in view of the importance of ADC in the plant kingdom. Examples[80,81] of the effects on growth which these compounds induce are shown in Figures 5 and 6. The structures of DFMO and DFMA are shown below.

DL-α-Difluoromethylornithine (DFMO)

$$NH_2CH_2CH_2CH_2C(CF_2H)(COOH){-}NH_2$$

DL-a-Difluoromethylarginine (DFMA)

$$NH_2C(=NH)NHCH_2CH_2CH_2C(CF_2H)(COOH){-}NH_2$$

The incorporation into the polyamines of propylamino groups derived from methionine was first demonstrated by Greene in 1957 using *Neurospora* (reviewed by Tabor and Tabor[7]). However, it was not until 1971 that Coppoc et al.[82] found *S*-adenosylmethionine (SAM) decarboxylase activity in mung bean sprout extracts, while Baxter and Coscia[83] detected spermidine synthase activity in *Vinca rosea* two years later.

Cyclohexylamine has been used very widely in the study of polyamine metabolism since it is an effective inhibitor of spermidine synthase in many organisms. However, confusion arose over the structure of the product used in these studies since for several years it was assumed that dicyclohexylamine was the inhibitor, due to an ambiguous name shown in the Sigma Chemical Company (St. Louis, MO) catalogue.[84]

Since *S*-adenosylmethionine is the precursor utilized in both the synthesis of ethylene and of the polyamines, it is of interest that ethylene is a senescence-inducing hormone, while the polyamines tend to delay senescence. Altman and Bachrach[85] noted the potential antagonism between these two pathways, and this switch may constitute an important regulatory mechanism, which is presently the subject of active research[86] (see Chapter 5).

V. AMINE OXIDASES

Although diamine oxidases had been found in animal tissues and bacteria several years earlier, the first report of an amine oxidase in plants was by Cromwell in 1943.[87] Using extracts of *Atropa belladonna*, initial experiments established activity by the decolorization of indigo carmine, which detects the formation of the peroxide[88] that is generated during the oxidative process. Activity was confirmed by analysis of ammonia, and of an aldehyde, detected with 2,4-dinitrophenylhydrazine. He deduced correctly that the enzyme is concerned with the formation of hyoscyamine. In 1948 the first evidence for the remarkable diamine oxidase activity in the legumes was given by Werle and Raub.[89] These workers found that extracts of red clover seedlings rapidly oxidized histamine with a pH optimum of about seven. In subsequent work by Werle and Zabel in 1948,[90] the distribution of this oxidative enzyme, which they termed histaminase, was studied in a wide range of plants. In addition to clover, they found activity in several members of the Rosaceae and the Labiatae. A list of the species in which amine oxidation has been detected is given by Smith.[17] Hydrogen peroxide was established as a product of plant amine oxidase activity by Werle and Pechmann,[91] and oxygen consumption was measured by manometry in the Warburg apparatus. The enzyme was shown to have a broad specificity, though with special affinity for the diamines putrescine and cadaverine.

In 1952, Kenten and Mann[92] observed peroxide generation in crude extracts of pea seedlings, and they eventually showed that this could be ascribed to diamine oxidase. In a very thorough study of this enzyme, Mann and his co-workers confirmed much of the pioneering studies of Werle. They found that the enzyme is especially sensitive to copper chelating agents,[93] and eventually showed that this is a copper-containing enzyme.[94] The sensitivity of the enzyme to carbonyl reagents suggested that this might be due to pyridoxal phosphate, though experimental evidence was unconvincing. In fact, only recently has the carbonyl function been identified tentatively as pyrroloquinoline quinone.[95]

In 1961 Mann[96] and Werle et al.,[97] achieved extensive purification of the pea seedling enzyme, yielding a product with a characteristic spectrum with a pink color. Under anaerobic conditions, the pink color disappeared with the addition of substrate, and the color reappeared with the admission of oxygen (Figure 7).

In a study of the metabolism of the polyamines in barley seedlings, Smith[98] fed spermidine and spermine to the cut shoots, and the amine fraction obtained by an ion-exchange fractionation from these and from unfed seedlings was subjected to analysis by gas chromatography. The resulting separations (Figure 8) showed significant enhancement of the peak attributable to diaminopropane, and also of a peak, at that time unidentified, which eluted more slowly than putrescine. Removal of diaminopropane from spermine indicates oxidation at the secondary amino group, and it was deduced that the resulting aldehyde fragment was aminopropylpyrrolinium, a cyclic amine already identified by Wrede and his co-workers in 1926[99,100] as a major product of the aerobic oxidation of spermine in the presence of an alkaline copper catalyst.

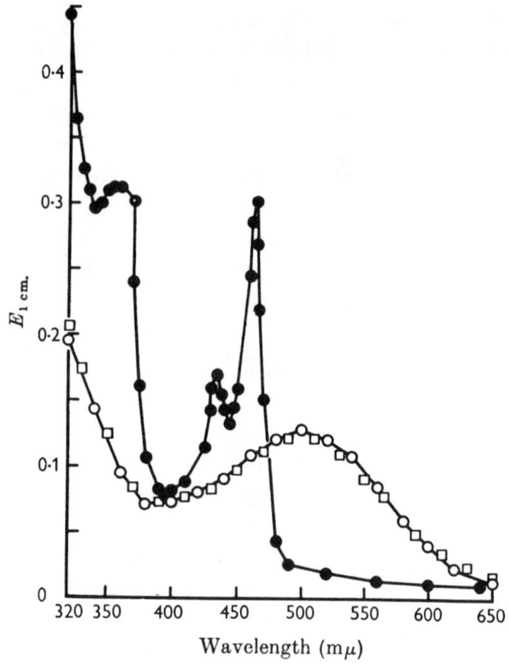

FIGURE 7. Changes in the absorption spectrum of a purified diamine oxidase preparation from pea seedlings upon reaction with putrescine. Before addition of putrescine (○); after addition of putrescine under anaerobic conditions (●); after subsequent oxygenation of the mixture of enzyme and putrescine (□).[96]

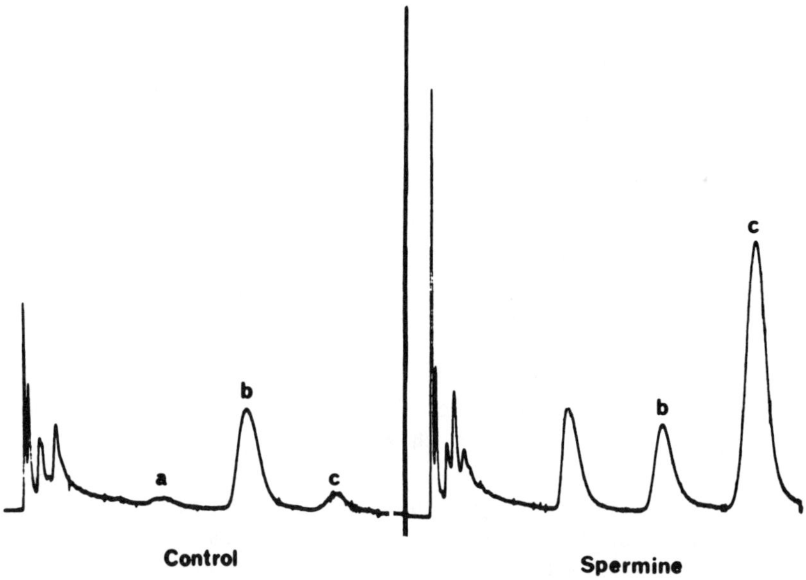

FIGURE 8. Gas chromatograms of the amine fractions of barley seedling leaves fed for 24 h with water alone (left), and 10 m*M* spermine (right). The point of injection is on the left in each case. Peak a, 1,3-diaminopropane; peak b, putrescine; peak c, diazabicyclononane (labeled 1-(3-aminopropyl)pyrroline in the original figure).[98]

The unknown substance and the product synthesized by the method of Wrede proved to be identical. This oxidation product is now known to exist at physiological pH as diazabicyclononane.[101]

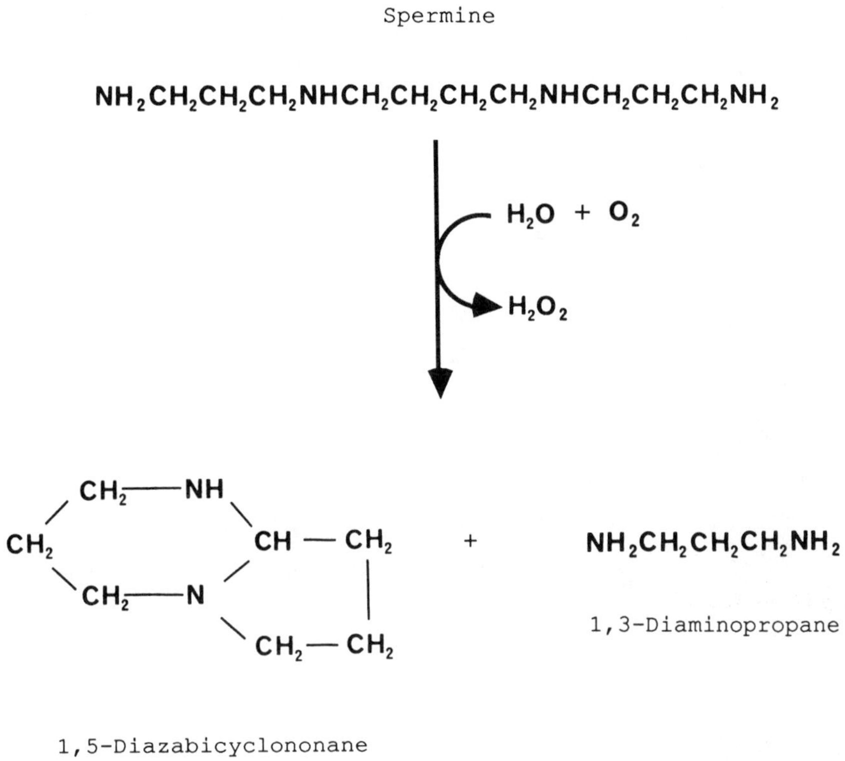

The amine oxidase present in the barley seedlings oxidized polyamines by a mechanism already found in the bacterium *Neisseria perflava* by Weaver and Herbst[102] in 1958. Subsequent work has shown that this enzyme is present in many species within the Gramineae, and it has been recorded sporadically in other monocotyledonous plants. The prosthetic group in the polyamine oxidase from maize seedlings was identified as FAD by Suzuki and Hirasawa in 1973.[103,104] After 260-fold purification, the maize enzyme was light yellow in color, and the apoenzyme could be reactivated by FAD but not by FMN or riboflavin.[103,105] In further work, Suzuki studied amine oxidases from other species and characterized these and several other enzymes involved in polyamine metabolism. Previous work on the plant amine oxidases has been reviewed by Hill and Mann,[106] Smith,[107] Rinaldi et al.,[108] and Smith and Barker[109] (see also Chapter 3).

Early work had suggested that the polyamine oxidase of barley seedlings is associated with the vascular system, and it appeared to be associated with a particulate fraction in barley seedlings, from which it could be removed on washing in 0.5 *M* sodium chloride.[110] Evidence for the association of this enzyme with cell walls was provided by Kaur-Sawhney et al.,[111] who also found high activity of this enzyme in the guard cells of oat leaves (Figure 9). Extensive studies by Federico and his co-workers have conclusively shown the cell wall localization not only of the polyamine oxidase of the cereals but also of the diamine oxidase in the legumes.[112]

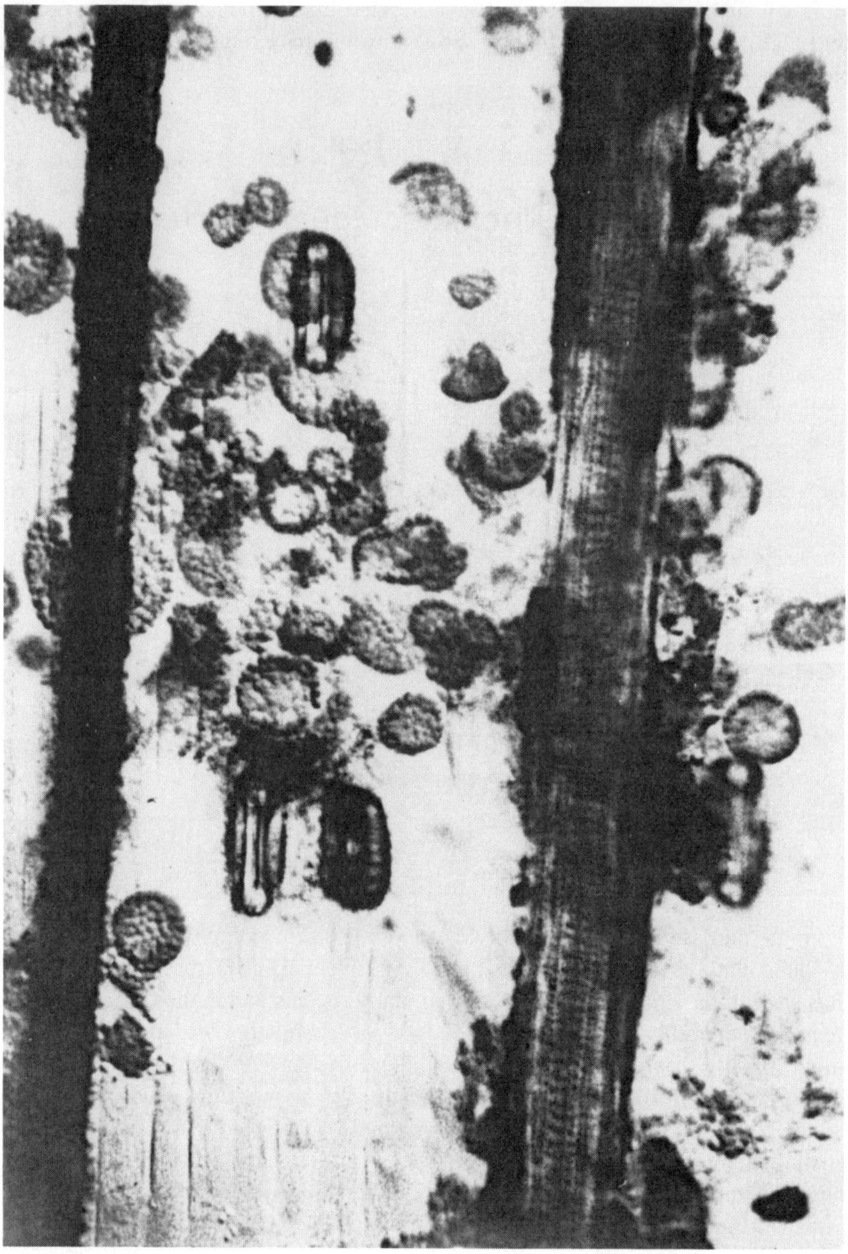

FIGURE 9. Localization of polyamine oxidase activity in oat leaves. The leaf tissues were incubated in a mixture of starch, potassium iodide, and spermine. Activity is reversed by the dark blue color of the starch iodine complex formed when the enzymically generated peroxide oxidizes the potassium iodide to iodine. The enzyme is particularly active in the vascular system and in the guard cells.[111]

They suggest that these enzymes may contribute the hydrogen peroxide needed for lignification in the cell walls.

VI. CONJUGATES

The macrocyclic polyamine alkaloids were first discovered in the horsetail ferns (*Equisetum* spp). These were known to be toxic to cattle and the toxic fraction, named palustrine, was partially characterized by Glet et al.[113] in 1936. On feeding palustrine to cattle, symptoms were produced which were similar to those found on feeding the intact horsetails. In 1960, palustrine was eventually shown by Baumann et al.[114] to contain spermidine, which was isolated after alkaline hydrolysis.

Palustrine

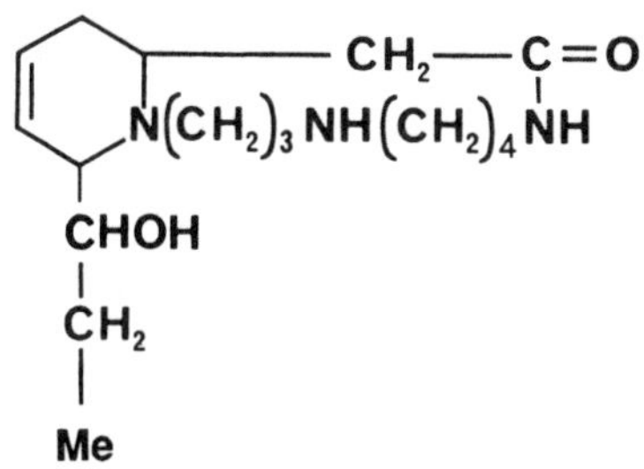

In the same year, Potier et al.[115] demonstrated spermidine in lunarine, which is the most abundant alkaloid found in the seeds of honesty (*Lunaria*). Five additional alkaloids were found in these seeds, all containing spermidine conjugated between two hydroxycinnamic acids.

Lunarine

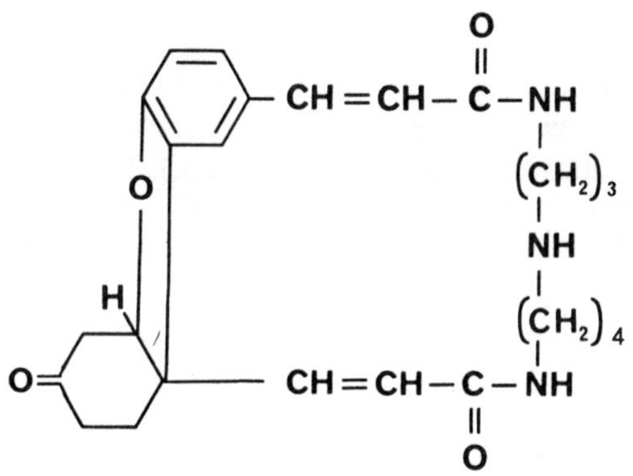

This basic structure occurs in many similar polyamine conjugates found in plants. Other alkaloids having a large loop derived from spermidine or spermine have been found throughout the plant kingdom and are now known as the macrocyclic alkaloids. The simple amides derived from the conjugation of the polyamines with the hydroxycinnamic acids, which appear to be the precursors for these alkaloids have been recognized as plant constituents for many years.

In 1949, Ryabinin and Il'ina[116] discovered feruloylputrescine in *Salsola subaphylla*, a shrub which grows in the sandy deserts of Central Asia. They called this substance subaphylline, identifying the structure from its hydrolysis products and from its empirical formula.

Feruloyl Putrescine

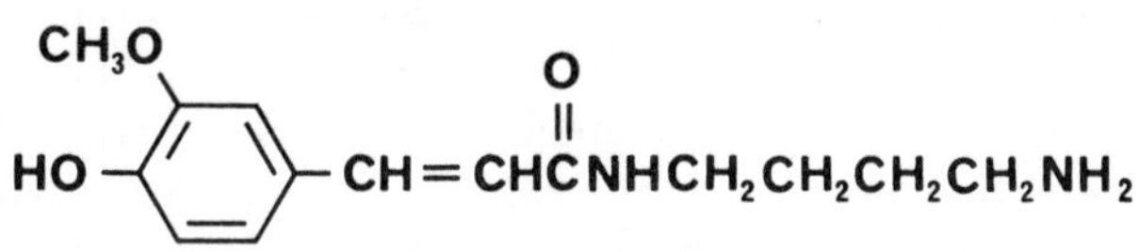

In 1969, Wheaton and Stewart[117] discovered large amounts of feruloylputrescine in grapefruit. In 1970, Ripperger, et al.[118] found methylated putrescine as an amide of 4-hydroxy and 4-methoxy cinnamic acids in the leaves of *Kniphofia*. Also at this time, Stoessl, Samborski, and Rohringer[119] showed that infection of wheat with rust or with a pseudomonad caused accumulation of hydroxycinnamic acid amides formed with the unusual diamine 2-hydroxyputrescine.

Hydroxycinnamic acid amides of putrescine were found in tobacco callus in tissue culture by Mizusaki et al.[120] in 1970, and were subsequently demonstrated in various parts of the tobacco plant by Buta and Izac[121] in 1972. The antifungal hordatines which are found in barley seedlings were characterized in 1970 by Stoessl and his co-workers.[122] These dimers of coumaroylagmatine, occur mainly as the glucosides, together with free coumaroylagmatine. Using extracts of barley seedlings, Smith and Bird showed that this amide is formed via the Coenzyme A derivative of coumaric acid.[123]

Hordatine A

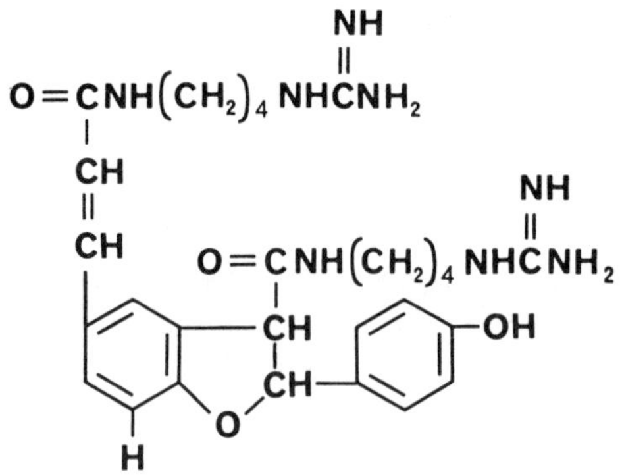

In a series of papers commencing in 1973, Martin and Martin-Tanguy[124] have produced considerable evidence for the physiological importance of these conjugates in virus infection and flowering, and Negrel[125] has now established their biosynthetic route in this tissue, which is similar to that found in barley. In a survey of 20 species in 13 families, Martin-Tanguy et al.[126] showed that these amides comprise the main phenolic fraction in the reproductive organs of these plants. The early history of the di- and polyamine conjugates has been reviewed by Eugster,[127] Badawi et al.,[128] Hesse,[129] Smith,[14,15] and Smith et al.[123] More recent work is surveyed by Martin-Tanguy.[130]

VII. THE INDIAN GROUP

P.R. Adiga and his co-workers have made a particularly thorough investigation of polyamine metabolism in *Lathyrus sativus*, commencing in 1974. In this plant, arginine is about 14-fold more efficient than ornithine as a precursor for putrescine biosynthesis, and it appears that ornithine is first converted to arginine before decarboxylation.[131] Upon 1000-fold purification, the *Lathyrus* arginine decarboxylase was found to be composed of six subunits which could be dissociated by SDS electrophoresis. The enzyme was shown to have a pH optimum of 8.5 and to require pyridoxal phosphate as the cofactor.[132] In *Lathyrus*, agmatine was converted to putrescine by an enzyme having multiple activities termed putrescine synthase. This enzyme hydrolyzed agmatine to *N*-carbamoylputrescine, converted this to putrescine and carbamoylphosphate, and carbamoylated ornithine to citrulline. It was also found to generate ATP from the carbamoylphosphate.[133] A similar multifunctional enzyme in cucumber seedlings, was also shown to possess arginine decarboxylase activity.[134] A thiol-dependent transamidinase with a very broad specificity was found which could convert putrescine to agmatine, agmatine to arcain, ornithine to arginine, cadaverine to homoagmatine, and lysine to homoarginine.[135] Cadaverine was shown to be formed by two alternative routes in *Lathyrus*. Lysine decarboxylation was fourfold faster than the route originating from homoarginine via homoagmatine. Homoarginine, which was found in the seeds, was converted to homoagmatine in the axis.[131] Decarboxylation of lysine and homoarginine in *Lathyrus* seedlings was shown to be due to a single enzyme which was purified 100-fold. It had an absolute requirement for Mn^{2+} or Fe^{2+} and was quite distinct from arginine decarboxylase.[136] Spermidine was formed by two distinct pathways in *Lathyrus*. In animals and bacteria, it is well established that the aminopropyl group is derived from decarboxylated *S*-adenosylmethionine (SAM), the enzyme effecting this being stimulated by putrescine. Indeed, crude extracts of *Lathyrus* seedlings showed an enhanced decarboxylation of SAM in the presence of putrescine. However, this was mainly artefactual, due to the formation of hydrogen peroxide on oxidation of putrescine by diamine oxidase, an enzyme also present in the homogenates. This decarboxylation was inhibited by catalase. The biosynthetic SAM decarboxylase was characterized after removal of the diamine oxidase. This decarboxylase was shown to be unaffected by putrescine but stimulated by Mg^{2+}.[137] Evidence for spermidine synthase was obtained, though this was not purified.[137]

An alternative pathway of spermidine synthesis first described in bacteria was also demonstrated by this group in *Lathyrus* (see Chapter 8). Aspartic acid and/or homoserine are first converted to aspartic semi-aldehyde, which condenses with putrescine to give a Schiff base. This is reduced by an enzyme purified 45-fold and shown to have a requirement for NADPH. The reduced conjugate is then decarboxylated to spermidine by an enzyme which requires pyridoxal phosphate. Spermine did not appear to be formed by this pathway, which is probably the major route of spermidine formation in *Lathyrus*, and which may be used when methionine is limiting.[138]

Homospermidine, the higher homologue of spermidine whose structure is shown below, was discovered in leaves of the sandal tree by Kuttan et al. in 1971,[139,140] where it comprised up to 1.5% of the dry weight. In *Lathyrus*, an enzyme system specifically for the biosynthesis of homospermidine, and which was dependent upon NAD, was demonstrated by Srivenugopal and Adiga.[141] The work on polyamine metabolism in *Lathyrus* has been reviewed by Adiga and Prasad.[142]

Homospermidine $NH_2CH_2CH_2CH_2CH_2NHCH_2CH_2CH_2CH_2NH_2$ Structure N

VIII. THE YALE GROUP

Considerable interest lies in the regeneration of protoplasts, and much attention has been given to protoplasts of cereal leaves in view of the agronomic importance of this group. Cereal protoplasts were found to continue synthesis of nucleic acid and protein for only a short period after isolation; they also divided only rarely in comparison with the protoplasts derived from other plants like tobacco. Arthur W. Galston and his co-workers, while investigating ways of increasing protoplast viability, discovered that the process of senescence could be greatly delayed by incubating the protoplasts with L-lysine, L-arginine, cadaverine, putrescine, or spermidine.[143,144] They showed that the di- and polyamines reduce the rise in ribonuclease activity associated with senescence[145,146] and also induce DNA synthesis and mitosis in these protoplasts.[147,148] They suggested that the dibasic amino acids function by their prior conversion to the amines, and proposed that one of the main mechanisms for protoplast stabilization resides in the ability of the polyamines to interact with membranes. They showed that the polyamines will also retard the loss of chlorophyll in the dark, suggesting that polyamines are potentially useful antisenescent agents for plants.[149] Diaminopropane, which is derived from the polyamines in the cereals by the activity of polyamine oxidase, was found to be especially active in the reduction of senescence.[150]

In the course of this work, a method for plant polyamine analysis, based upon the HPLC separation of benzoylated amine products, was developed by Flores and Galston[151] in 1982. This offered significant advantages over the dansyl method[152] in certain respects, and modifications of this method are now widely used[153] (see Chapter 16). Some of the work produced by this

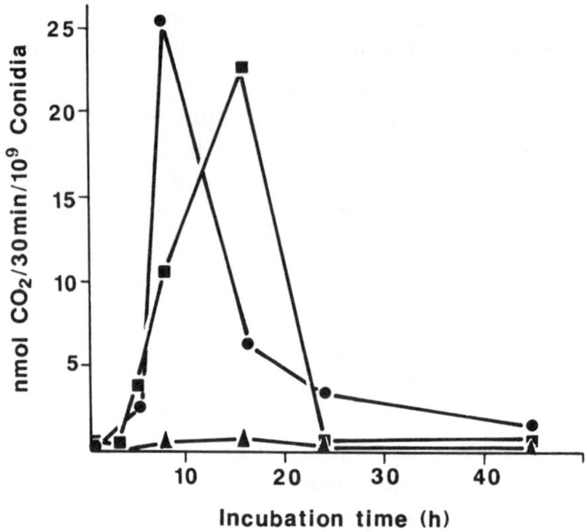

FIGURE 10. Ornithine decarboxylase activity of conidia from strains of *Aspergillus nidulans*. From the prototrophic strain germinating in the absence of putrescine (●); from the prototrophic strain germinating in the presence of putrescine (■); from the auxotrophic strain germinating in the presence of putrescine (▲). Putrescine concentration about 10^{-5} *M*.[165]

outstanding laboratory has been reviewed by Galston[154,155] and by Galston and Kaur-Sawhney.[156] More recently, discoveries from Professor Galston's group concerning the polyamines have ranged over a variety of physiological and biochemical topics, including the role of amides in flower differentiation, regulation of polyamine metabolism by light and hormones, conjugation of polyamines with protein, and the selective control of plant pathogenic fungi by interference with polyamine biosynthesis.

IX. LOWER PLANTS

The occurrence of "unusual" polyamines, generally higher and lower homologues of spermidine and spermine, was demonstrated in various algae by Hegewald and Kneifel[157] and earlier work on this subject was reviewed by Kneifel[158] in 1979. These polyamines were later studied by Villanueva[159] who developed an elegant technique for their separation and estimation based upon classical amino acid analytical methods. These strange polyamines have now been demonstrated in a wide range of autotrophic lower plants.[160,161] Similarly, unusual polyamines have recently been found in other higher plants.[162,163]

In 1955, Sneath[164] showed that putrescine is essential for the growth of a UV-induced mutant of the fungus *Aspergillus nidulans*. The lack of growth in the absence of this diamine appears to be due to the inability of the fungus to synthesize ODC at a critical stage in spore germination[165] (Figure 10). A concentration of putrescine as small as 10^{-5} *M* allowed full growth of the spores. An exhaustive review of polyamine function and metabolism in fungi was published by Stevens and Winther[166] in 1979. Studies by Galston, with Rajam and Weinstein[167,168] in 1985 showed that DFMO will inhibit the growth of certain fungi due to their dependency on the ODC pathway for putrescine biosynthesis and that this inhibition could be reversed by putrescine. The Uredinales appear to be particularly sensitive to DFMO inhibition,[169] and the control of certain rust diseases on crop plants is now the subject of patents.[170,171]

Gutta cavat lapidem - Ovid, Ex Pont. IV, x, 5.

ACKNOWLEDGMENTS

This review is dedicated to the memory of the late Dr. F. J. Richards, FRS, who introduced me to the polyamines and taught me the strategy and techniques of research. I am very grateful to Mrs. Jacqueline Barker for her constructive suggestions and to Dow Elanco Ltd. Letcombe, Oxford, U.K., for full financial support. I greatly regret that it has not been possible to mention all of the workers who have made significant contributions in the field of polyamine research. I offer my sincere apologies to those who have been omitted.

REFERENCES

1. **Leeuwenhoek, A. van,** Observationes D. Anthonii Lewenhoeck de Natis e semine genitali Animalculis, *Phil. Trans. Roy. Soc.*, 12, 1040, 1978.
2. **Smith, T.A.,** The physiology of the polyamines and related compounds, *Endeavour*, 31, 22, 1972.
3. **Vauquelin, M.,** Expériences sur le sperme humain, *Ann. Chem.*, 9, 64, 1791.
4. **Boettcher, A.,** Farblose Krystalle eines eiweissartigen Körpers aus dem menschlichen Sperma dargestellt, *Virchow's Arch. Path. Anat. Physiol. Med.*, 32, 525, 1865.
5. **Ladenburg, A. and Abel, J.,** Ueber das Aethylenimin (Spermin?), *Ber. Dtsch. chem. Ges.*, 21, 758, 1888.
6. **Guggenheim, M.,** *Die biogenen Amine*, Verlag S. Karger, Basel, 1951.
7. **Tabor, H. and Tabor, C.W.,** Spermidine, spermine and related amines, *Pharmacol. Rev.*, 16, 245, 1964.
8. **Mann, T.,** *The Biochemistry of Semen*, Methuen, London, 1964.
9. **Williams-Ashman, H.G., Nicolas Louis Vauquelin,** *Investigative Urol.*, 2, 605, 1965.
10. **Cohen, S.S.,** *Introduction to the Polyamines*, Prentice-Hall Inc., Englewood Cliffs, New Jersey, 1971.
11. **Bachrach, U.,** *Function of Naturally Occurring Polyamines*, Academic Press, New York, 1973.
12. **Guggenheim, M.,** Die biogenen Amine in der Pflanzenwelt, in *Encyclopedia of Plant Physiology*, Vol. 8, Ruhland, W. and Mothes, K., Eds., Springer-Verlag, Berlin, 1958, 889.
13. **Smith, T. A.,** The biosynthesis of putrescine in higher plants and its relation to potassium nutrition, in *Recent Aspects of Nitrogen Metabolism in Plants*, Hewitt, E.J. and Cutting, C.V., Eds., Academic Press, London, 1968, 139.
14. **Smith, T.A.,** The occurrence, metabolism and functions of amines in plants, *Biol. Rev.*, 46, 201, 1971.
15. **Smith, T.A.,** Recent advances in the biochemistry of plant amines, in *Prog. in Phytochem.*, 4, 1977, 27.
16. **Smith, T.A., Bagni, N., and Serafini Fracassini, D.,** The formation of amines and their derivatives in plants, in *Nitrogen Assimilation in Plants*, Hewitt, E.J. and Cutting, C.V., Eds., Academic Press, London, 1979, 557.
17. **Smith, T.A.,** Plant amines, in *Encyclopedia of Plant Physiology*, Vol. 8, Bell, E.A. and Charlwood, B.V., Eds., Springer-Verlag, Berlin, 1980, 433.
18. **Smith, T.A.,** Amines, in *The Biochemistry of Plants*, Vol. 7, Conn, E.E., Ed., Academic Press, New York, 1981, 249.
19. **Waring P.F., Ed.,** *Plant Growth Substances 1982*, Academic Press, London, 1982.
20. **Slocum R.D., Kaur-Sawhney, R., and Galston, A.W.,** The physiology and biochemistry of polyamines in plants, *Arch Biochem. Biophys.*, 235, 283, 1984.
21. **Smith, T.A.,** Polyamines, *Annu. Rev. Plant Physiol.*, 36, 117, 1985.
22. **Smith, T.A.,** Polyamines in plants, in *Polyamines in Plants*, Galston, A.W. and Smith, T.A., Eds, Martinus Nijhoff/ Dr W. Junk, Dordrecht, 1985, vii.
23. **Evans, P.T. and Malmberg, R.L.,** Do polyamines have roles in plant development? *Annu. Rev. Plant Physiol., Plant Mol. Biol.*, 40, 235, 1989.
24. **Flores, H.E., Protacio, C.M., and Signs, M.W.,** Primary and secondary metabolism of polyamines in plants, in Plant Nitrogen Metabolism, *Recent Advances in Phytochemistry*, Vol. 23, Conn, E.E., Ed., Plenum Press, New York, 1989, 329.
25. **Dudley, H.W. and Rosenheim, O.,** Notes on spermine, *Biochem. J.*, 19, 1034, 1925.
26. **Dudley, H.W., Rosenheim, O., and Starling, W.W.,** The chemical constitution of spermine. III. Structure and synthesis. *Biochem. J.*, 20, 1082, 1926.
27. **Schulze, E. and Steiger, E.,** Ueber einen neuen stickstoffhaltigen Bestandteil der Keimlingen von *Lupinus luteus, Berichte*, 19, 1177, 1886.
28. **Werle, E., Amine und Betaine,** in *Modern Methods of Plant Analysis*, Vol. 4, Paech, K. and Tracey, M.V., Eds., Springer-Verlag, Berlin, 1955, 517.

29. **Ciamician, G. and Ravenna, C.,** Ricerche sulla genesi degli alcaloidi nelle piante, *Atti Accad. Lincei Roma*, 20, 1, 614, 1911.
30. **Goris, A. and Larsonneau, A.,** Recherches sur la composition chimique des feuilles de la belladone, *Bull. Sci. Pharm.*, 28, 499, 1921.
31. **Hiwatari, Y.,** Über die stickstoffhaltigen Bestandteile der Früchte von *Citrus grandis* Osbeck: Form. Buntan, Hayat, *J. Biochem.*, 7, 169, 1927.
32. **Herbst, E.J. and Snell, E.E ,** Putrescine as a growth factor for *Hemophilus parainfluenzae*, *J. Biol. Chem.*, 176, 989, 1948.
33. **Kossel A.,** Über das Agmatin, *Hoppe-Seyler's Z. Physiol. Chem.*, 66, 257, 1910.
34. **Heyl, F.W.,** The protein extract of ragweed pollen, *J. Amer. Chem. Soc.*, 41, 670, 1919.
35. **Richards, F.J. and Coleman, R.G.,** Occurrence of putrescine in potassium-deficient barley, *Nature*, 170, 460, 1952.
36. **Coleman, R.G. and Hegarty, M.P.,** Metabolism of DL-ornithine-2-^{14}C in normal and potassium-deficient barley, *Nature*, 179, 376, 1957.
37. **Smith, T.A. and Richards, F.J.,** The biosynthesis of putrescine in higher plants and its relation to potassium nutrition, *Biochem. J.*, 84, 292, 1962.
38. **Smith, T.A.,** Arginine carboxy-lyase of higher plants and its relation to potassium nutrition, *Phytochemistry*, 2, 241, 1963 .
39. **Smith, T.A. and Garraway, J.L.,** *N*-Carbamylputrescine — an intermediate in the formation of putrescine in barley, *Phytochemistry*, 3, 23, 1964.
40. **Smith, T.A.,** *N*-Carbamylputrescine amidohydrolase of higher plants and its relation to potassium nutrition, *Phytochemistry*, 4, 599, 1965.
41. **Maretzki, A., Thom, M., and Nickell, L.G.,** Products of arginine catabolism in growing cells of sugarcane, *Phytochemistry*, 8, 811, 1969.
42. **Crocomo, O.J., Basso, L.C., and Brasil, O.G.,** Formation of *N*-carbamylputrescine from citrulline in *Sesamum*, *Phytochemistry*, 9, 1487, 1970.
43. **Crocomo, O.J. and Basso, L.C.,** Accumulation of putrescine and related amino acids in potassium deficient *Sesamum*, *Phytochemistry*, 13, 2659, 1974.
44. **Gale, E.F.,** The production of amines by bacteria. I. The decarboxylation of amino acids by strains of *Bacterium coli*, *Biochem. J.*, 34, 392, 1940.
45. **Smith, T.A. and Sinclair, C.,** The effect of acid feeding on amine formation in barley, *Ann. Bot.*, 31, 103, 1967.
46. **Young, N.D. and Galston, A.W.,** Putrescine and acid stress, Induction of arginine decarboxylase activity and putrescine accumulation by low pH. *Plant Physiol.*, 71, 767, 1983.
47. **Flores, H.E. and Galston, A.W.,** Polyamines and plant stress, activation of putrescine biosynthesis by osmotic shock, *Science*, 217, 1259, 1982.
48. **Flores, H.E., Young, N.D., and Galston, A.W.,** Polyamine metabolism and plant stress, in *Cellular and Molecular Biology of Plant Stress*, Key, J.L. and Kosuge, T., Eds., Alan R. Liss, New York, 1985, 93.
49. **Weinstein, L.H., Kaur-Sawhney, R., Rajam, M.V., Wettlaufer, S.H., and Galston, A.W.,** Cadmium-induced accumulation of putrescine in oat and bean leaves, *Plant Physiol.*, 82, 641, 1986.
50. **Smith, T.A.,** The biosynthesis and metabolism of putrescine in higher plants, *Ann. N. Y. Acad. Sci.*, 171, 988, 1970.
51. **Smith, T.A.,** Putrescine and inorganic ions, in *Phytochemical Adaptations to Stress*, Vol. 18, Timmermann, B.N., Steelink, C., and Loewus, F.A., Eds., Plenum Press, New York, 1984, 7.
52. **Moruzzi, G. and Caldarera, C.M.,** Occurrence of polyamines in the germs of cereals, *Arch. Biochem. Biophys.*, 105, 209, 1964.
53. **Bagni, N., Caldarera, C.M., and Moruzzi, G.,** Spermine and spermidine distribution during wheat growth, *Experientia*, 23, 139, 1967.
54. **Bertossi, F., Bagni, N., Moruzzi, G., and Caldarera, C.M.,** Spermine as a new growth-promoting substance for *Helianthus tuberosus* (Jerusalem artichoke) *in vitro*, *Experientia*, 21, 80, 1965.
55. **Bagni, N.,** Aliphatic amines and a growth-factor of coconut milk as stimulating proliferation of *Helianthus tuberosus* (Jerusalem artichoke) *in vitro*, *Experientia*, 22, 732, 1966.
56. **Bagni, N. and Serafini Fracassini, D.,** The role of polyamines as growth factors in higher plants and their mechanism of action, in *Plant Growth Substances 1973*, Hirasawa Publ. Co. Tokyo, 1974, 1205.
57. **Cocucci, S. and Bagni, N.,** Polyamine-induced activation of protein synthesis in ribosomal preparation from *Helianthus tuberosus* tissue, *Life Sci.*, 7, 113, 1968.
58. **Serafini Fracassini, D., Bagni, N., Cionini, P.G., and Bennici, A.,** Polyamines and nucleic acids during the first cell cycle of *Helianthus tuberosus* tissue after the dormancy break, *Planta*, 148, 332, 1980.
59. **Bagni, N., Stabellini, G., and Serafini Fracassini, D.,** Polyamines bound to tRNA and rRNA of eukaryotic plant organisms, *Physiol. Plant*, 29, 218, 1973.
60. **Bagni, N.,** Absorption of spermine and spermidine and their degradation in explants of *Helianthus tuberosus in vitro*, *Z. Pflanzenphysiol.*, 57, 22, 1967.

61. **Bagni, N., Calzoni, G.L., and Speranza, A.,** Polyamines as sole nitrogen sources for *Helianthus tuberosus* explants *in vitro*, *New Phytol.*, 80, 317, 1978.
62. **Speranza, A. and Bagni, N.,** Products of L-[^{14}C-carbamoyl] citrulline metabolism in *Helianthus tuberosus* activated tissue, *Z. Pflanzenphysiol.*, 88, 163, 1978.
63. **Bagni, N.,** Identificazione di alcuni fattori di crescita nel succo di pomodoro, *Giorn. Bot. Ital.*, 101, 81, 1967.
64. **Bagni, N.,** Spermina e spermidina nei semi, *Giorn. Bot. Ital.*, 102, 67, 1968.
65. **Bagni, N.,** Metabolic changes of polyamines during the germination of *Phaseolus vulgaris*, *New Phytol.*, 69, 159, 1970.
66. **Bagni, N., Serafini Fracassini, D., and Corsini, E.,** Tumors of *Scorzonera hispanica*, their content in polyamines, *Z. Pfanzenphysiol.*, 67, 19, 1972.
67. **Speranza, A. and Bagni, N.,** Putrescine biosynthesis in *Agrobacterium tumefaciens* and in normal and crown gall tissue of *Scorzonera hispanica* L., *Z. Pflanzenphysiol.*, 81, 226, 1977.
68. **Audisio, S., Bagni, N., and Serafini Fracassini, D.,** Polyamines during the growth *in vitro* of *Nicotiana glauca*, R. Grah. habituated tissue, *Z. Pflanzenphysiol.*, 77, 146, 1976.
69. **Bagni, N. and Serafini Fracassini, D.,** Polyamines and plant tumors, *Ital. J. Biochem.*, 28, 392, 1979.
70. **Bagni, N., Serafini Fracassini, D., and Torrigiani. P.,** Polyamines and growth in higher plants, in *Advances in Polyamine Research*, Vol. 3, Caldarera, C.M., Zappia, V., and Bachrach, U., Eds., Raven Press, New York, 377, 1981.
71. **Bagni, N. and Pistocchi, R.,** Polyamines as growth substances in higher plants, in *Progress in Polyamine Research, Advances in Experimental Medicine and Biology*, Vol. 250, Zappia, V. and Pegg, A.E., Eds., Plenum Press, New York, 1988, 547.
72. **Bagni, N.,** Polyamines in plant growth and development, in *The Physiology of Polyamines*, Vol II, Bachrach, U. and Heimer, Y.M., Eds., CRC Press, Boca Raton, FL, 1989, 107.
73. **Hasse, V. K., Ratych, O.T., and Salnikow, J.,** Transaminierung und Decarboxylierung von Ornithin und Lysin in höheren Pflanzen, *Hoppe-Seyler's Zeit. Physiol. Chem.*, 348, 843, 1967.
74. **Yoshida, D. and Mitake, T.,** Agmatine and *N*-carbamoylputrescine as intermediates in the formation of nicotine by tobacco plants, *Plant Cell Physiol.*, 7, 301, 1966.
75. **Yoshida, D.,** Formation of putrescine from ornithine and arginine in tobacco plants, *Plant Cell Physiol.*, 10, 393, 1969.
76. **Mizusaki, S., Tanabe, Y., Noguchi, M., and Tamaki, E.,** Changes in the activities of ornithine decarboxylase, putrescine *N*-methyltransferase and *N*-methylputrescine oxidase in tobacco roots in relation to nicotine biosynthesis, *Plant Cell Physiol.*, 14, 103, 1973.
77. **Tiburcio, A.F. and Galston, A.W.,** Arginine decarboxylase as the source of putrescine for tobacco alkaloids, *Phytochemistry*, 25, 107, 1986.
78. **Smith, T.A. and Marshall, J.H.A.,** Oxidative decarboxylation of amino acids by plant extracts, *Biochem. Soc. Trans.*, 16, 972, 1988.
79. **Bey, P., Danzin, C., and Jung, M.,** Inhibition of basic amino acid decarboxylases involved in polyamine biosynthesis, in *Inhibition of Polyamine Metabolism*, McCann, P.P., Pegg, A.E., and Sjoerdsma, A., Eds., Academic Press, Orlando, 1987, 1.
80. **Berlin, J. and Forche, E.,** D,L-α-difluoromethylornithine causes enlargement of cultured tobacco cells, *Z. Pflanzenphysiol.*, 101, 277, 1981.
81. **Smith, M.A., Davies, P.J., and Reid, J.B.,** Role of polyamines in gibberellin-induced internode growth in peas, *Plant Physiol.*, 78, 92, 1985.
82. **Coppoc, G.L., Kallio, P., and Williams-Ashman, H.G.,** Characteristics of *S*-adenosyl-L-methionine decarboxylase from various organisms, *Int. J. Biochem.*, 2, 673, 1971.
83. **Baxter, C. and Coscia, C.J.,** *In vitro* synthesis of spermidine in the higher plant *Vinca rosea*, *Biochem. Biophys. Res. Commun.*, 54, 147, 1973.
84. **Batchelor, K.W., Smith, R.A., and Watson, N.S.,** Dicyclohexylamine is not an inhibitor of spermidine synthase, *Biochem. J.*, 233, 307, 1986.
85. **Altman, A. and Bachrach, U.,** Involvement of polyamines in plant growth and senescence, in *Advances in Polyamine Research*, Vol. 3, Caldarera, C.M., Zappia, V., and Bachrach, U., Eds., Raven Press, New York, 1981, 365.
86. **Altman, A.,** Polyamines and plant hormones, in *The Physiology of Polyamines*, Vol. II, Bachrach, U. and Heimer, Y.M., Eds., CRC Press, Boca Raton, FL, 1989, 121.
87. **Cromwell, B.T.,** The role of putrescine in the synthesis of hyoscyamine, *Biochem. J.*, 37, 722, 1943.
88. **Kapeller-Adler, R.,** *Amine Oxidases and Methods For Their Study*, Wiley-Interscience, New York, 1970.
89. **Werle, E. and Raub, A.,** Über Vorkommen, Bildung, und Abbau biogener Amine bei Pflanzen unter besonderer Berücksichtigung des Histamins, *Biochem. Z.*, 318, 538, 1948.
90. **Werle, E. and Zabel, A.,** Über die Verbreitung der Histaminase im Pflanzenreich, *Biochem. Z.*, 318, 554, 1948.
91. **Werle, E. and Pechmann, E.,** Über die Diamin-Oxydase der Pflanzen und ihre adaptive Bildung durch Bakterien, *Liebig's Annalen der Chemie*, 44, 562, 1949.

92. **Kenten, R.H. and Mann, P.J.G.,** The oxidation of amines by extracts of pea seedlings, *Biochem. J.*, 50, 360, 1952.
93. **Mann, P.J.G.,** Purification and properties of the amine oxidase of pea seedlings, *Biochem. J.*, 59, 609, 1955.
94. **Hill, J.M. and Mann, P.J.G.,** The inhibition of pea-seedling diamine oxidase by chelating agents, *Biochem. J.*, 85, 198, 1962.
95. **Glatz, Z., Kovar, J., Macholan, L., and Pec, P.,** Pea (*Pisum sativum*) diamine oxidase contains pyrroloquinoline quinone as a cofactor, *Biochem. J.*, 242, 603, 1987.
96. **Mann, P.J.G.,** Further purification and properties of the amine oxidase of pea seedlings, *Biochem. J.*, 79, 623, 1961.
97. **Werle, E., Trautschold, I., and Aures, D.,** Reinigung und Characterisierung der Diamin-Oxydase aus Erbsen, *Hoppe-Seyler's Z. Physiol. Chem.*, 326, 200, 1961.
98. **Smith, T.A.,** Polyamine oxidase in higher plants, *Biochem. Biophys. Res. Commun.*, 41, 1452, 1970.
99. **Wrede, F.,** Zur Kenntnis des Spermins, IV, *Hoppe Seyler's Z. Physiol. Chem.*, 153, 291, 1926.
100. **Wrede, F., Fanselow, H., and Strack, E.,** Zur Kenntnis des Spermins. V., *Hoppe-Seyler's Z. Physiol. Chem.*, 161, 66, 1926.
101. **Smith, T.A., Croker, S.J., and Loeffler, R.S.T.,** Occurrence in higher plants of 1-(3-aminopropyl)pyrrolinium and pyrroline, products of polyamine oxidation, *Phytochemistry*, 25, 683, 1986.
102. **Weaver, R.H. and Herbst, E.J.,** The oxidation of polyamines by *Neisseria perflava*, *J. Biol. Chem.*, 231, 647, 1958.
103. **Suzuki, Y. and Hirasawa, E.,** Polyamine oxidase from *Zea mays* shoots, *Phytochemistry*, 12, 2863, 1973.
104. **Hirasawa, E. and Suzuki, Y.,** Purification and properties of a polyamine oxidase from *Zea mays*, *Phytochemistry*, 14, 99, 1975.
105. **Suzuki, Y. and Yanagisawa, H.,** Purification and properties of maize polyamine oxidase, a flavoprotein, *Plant Cell Physiol.*, 21, 1085, 1980.
106. **Hill, J.M. and Mann, P.J.G.,** Some properties of plant diamine oxidase, a copper-containing enzyme, in *Recent Aspects of Nitrogen Metabolism in Plants*, Hewitt, E.J. and Cutting, C.V., Eds., Academic Press, London, 1968, 149.
107. **Smith, T.A.,** The di- and polyamine oxidases of higher plants, *Biochem. Soc. Trans.*, 13, 319, 1985.
108. **Rinaldi, A., Floris, G., and Giartosio, A.,** Plant amine oxidases in *Structure and Functions of Amine Oxidases*, Mondovi, B. Ed., CRC Press, Boca Raton, FL, 1986, 51.
109. **Smith, T.A. and Barker, J.H.A.,** The di- and polyamine oxidases of plants, in *Progress in Polyamine Research, Advances in Experimental Medicine and Biology*, Vol. 250, Zappia, V. and Pegg, A.E., Eds., Plenum Press, New York, 1988, 573.
110. **Smith, T.A.,** Purification and properties of the polyamine oxidase of barley plants, *Phytochemistry*, 11, 899, 1972.
111. **Kaur-Sawhney, R., Flores, H.E., and Galston, A.W.,** Polyamine oxidase in oat leaves, a cell wall-localized enzyme, *Plant Physiol.*, 68, 494, 1981.
112. **Angelini, R. and Federico, R.,** Histochemical evidence of polyamine oxidation and generation of hydrogen peroxide in the cell wall, *J. Plant Physiol.*, 135, 212, 1989.
113. **Glet, E., Gutschmidt, J., and Glet, P.,** Das Alkaloid im *Equisetum palustre*, *Hoppe-Seyler's Z. Physiol. Chem.*, 244, 229, 1936.
114. **Baumann, C.G., Dietsche, W., and Eugster, C.H.,** Nachweis des macrozyklischen Natur des Dihydropalustrins. 5. Mitteilung uber Schachtelhalmalkaloide, *Chimia*, 14, 85, 1960.
115. **Potier, P., Le Men, J., Janot, M.-M., and Bladon, P.,** The isolation of spermidine by degradation of lunarine, *Tetrahedron Lett.*, 18, 36, 1960.
116. **Ryabinin, A.A. and Il'ina, E.M.,** Investigation of an alkaloid from the plant *Salsola subaphylla* C.A.M. var. *arenaria* Drob., *Dokl. Akad. Nauk SSSR*, 67, 513, 1949.
117. **Wheaton, T.A. and Stewart, I.,** Feruloylputrescine, isolation and identification from *Citrus* leaves and fruit, *Nature*, 206, 620, 1969.
118. **Ripperger, Von H., Schreiber, K., and Budzikiewicz, H.,** Putresin-Amide aus *Kniphofia*-Arten, *J. Prakt. Chem.*, 312, 449, 1970.
119. **Stoessl, A., Rohringer, R., and Samborski, D.J.,** 2 Hydroxyputrescine amides as abnormal metabolites of wheat, *Tetrahedron Lett.*, 33, 2807, 1969.
120. **Mizusaki, S., Tanabe, Y., and Noguchi, M.,** A new aromatic amide, caffeoylputrescine, from callus tissue culture of *Nicotiana tabacum*, *Agric. Biol. Chem.*, 34, 972, 1970.
121. **Buta, J.G. and Izac, R.R.,** Caffeoylputrescine in *Nicotiana tabacum*, *Phytochemistry*, 11, 1188, 1972.
122. **Stoessl, A. and Unwin, C.H.,** The antifungal factors in barley. V. Antifungal activity of the hordatines, *Can. J. Bot.*, 48, 465, 1970.

123. **Smith, T.A., Negrel, J., and Bird, C.R.,** The cinnamic acid amides of the di- and polyamines, in *Advances in Polyamine Research*, Vol. 4, Bachrach, U., Kaye, A., and Chayen, R. Eds., Raven Press, New York, 1983, 347.
124. **Martin-Tanguy, J., Martin, C., and Gallet, M.,** Présence de composés aromatiques liés à la putrescine dans divers *Nicotiana* virosés, *C.R. Acad. Sci. Paris*, 276, 1433, 1973.
125. **Negrel, J.,** The biosynthesis of cinnamoylputrescines in callus tissue cultures of *Nicotiana tabacum*, *Phytochemistry*, 28, 477, 1989.
126. **Martin-Tanguy, J., Cabanne, F., Perdrizet, E., and Martin, C.,** The distribution of hydroxycinnamic acid amides in flowering plants, *Phytochemistry*, 17, 1927, 1978.
127. **Eugster, C.H.,** Spermidinalkaloide, *Vierteljahrschrift d. Naturf. Ges. Zürich*, 105, 261, 1960.
128. **Badawi, M.M., Bernauer, K., van den Broek, P., Groger, D., Guggisberg, A., Johne, S., Kompis, I., Schneider, F., Veith, H.-J., Hesse, M., and Schmid, H.,** Macrocyclic spermidine and spermine alkaloids, *Pure and Appl. Chem.*, 33, 81, 1973.
129. **Hesse, M.,** Alkaloids - II, in *Biochemical Applications of Mass Spectrometry*, (suppl. vol. 1), Walker, G.R. and Dermer, D.C., Eds., Wiley-Interscience, New York, 1980, 797.
130. **Martin-Tanguy, J.,** The occurrence and possible function of hydroxycinnamoyl acid amides in plants, in *Polyamines in Plants*, Galston, A.W. and Smith T.A., Eds., Martinus Nijhoff, Dordrecht, 1985, 381.
131. **Ramakrishna, S. and Adiga, P.R.,** Amine biosynthesis in *Lathyrus sativus* seedlings, *Phytochemistry*, 13, 2161, 1974.
132. **Ramakrishna, S. and Adiga, P.R.,** Arginine decarboxylase from *Lathyrus sativus* seedlings, *Eur. J. Biochem.*, 59, 377, 1975.
133. **Srivenugopal, K.S. and Adiga, P.R.,** Enzymic conversion of agmatine to putrescine in *Lathyrus sativus* seedlings, purification and properties of a multifunctional enzyme (putrescine synthase), *J. Biol. Chem.*, 256, 9532, 1981.
134. **Prasad, G.L. and Adiga, P.R.,** Arginine decarboxylase is a component activity of the multifunctional enzyme putrescine synthase in cucumber seedlings, *J. Biosci.*, 11, 571, 1987.
135. **Srivenugopal, K.S. and Adiga, P.R.,** Partial purification and properties of a transamidinase from *Lathyrus sativus* seedlings, involvement in homoarginine metabolism and amine interconversions, *Biochem. J.*, 189, 533, 1980.
136. **Ramakrishna, S. and Adiga, P.R.,** Decarboxylation of homoarginine and lysine by an enzyme from *Lathyrus sativus* seedlings, *Phytochemistry*, 15, 83, 1976.
137. **Suresh, M.R. and Adiga, P.R.,** Putrescine sensitive (artifactual) and insensitive (biosynthetic) *S*-adenosylmethionine decarboxylase of *Lathyrus sativus* seedlings, *Eur. J. Biochem.*, 79, 511, 1977.
138. **Srivenugopal, K.S. and Adiga, P.R.,** Coexistence of two pathways of spermidine biosynthesis in *Lathyrus sativus* seedlings, *FEBS Lett.*, 112, 260, 1980.
139. **Kuttan, R., Radhakrishnan, A.N., Spande, T., and Witkop, B.,** *sym*-Homospermidine, a naturally occurring polyamine, *Biochemistry*, 10, 361, 1971.
140. **Kuttan, R. and Radhakrishnan, A.N.,** Studies on the biosynthesis of *sym*-homospermidine in sandal (*Santalum album* L.), *Biochem. J.*, 127, 61, 1972.
141. **Srivenugopal, K.S. and Adiga, P.R.,** Enzymic synthesis of *sym*-homospermidine in *Lathyrus sativus* seedlings, *Biochem. J.*, 190, 461, 1980.
142. **Adiga, P.R. and Prasad, G.L.,** Biosynthesis and regulation of polyamines in higher plants, in *Polyamines in Plants*, Galston, A.W. and Smith, T.A., Eds., Martinus Nijhoff, Dordrecht, 1985, 3.
143. **Altman, A., Kaur-Sawhney, R., and Galston, A.W.,** Stabilization of oat leaf protoplasts through polyamine mediated inhibition of senescence, *Plant Physiol.*, 60, 570, 1977.
144. **Kaur-Sawhney, R., Adams, W.R., Tsang, J., and Galston, A.W.,** Leaf pretreatment with senescence retardants as a basis for oat protoplast improvement, *Plant Cell Physiol.*, 18, 1309, 1977.
145. **Kaur-Sawhney, R., Altman, A., and Galston, A.W.,** Dual mechanisms in polyamine-mediated control of ribonuclease activity in oat leaf protoplasts, *Plant Physiol.*, 62, 158, 1978.
146. **Galston, A.W., Altman, A., and Kaur-Sawhney, R.,** Polyamines, ribonuclease and the improvement of oat leaf protoplasts, *Plant Sci. Lett.*, 11, 69, 1978.
147. **Galston, A.W., Sawhney, R.K., Altman, A., and Flores, H.,** Polyamines, macromolecular synthesis and the problem of cereal protoplast regeneration, in *Advances in Protoplast Research*, Proc. 5th Int. Protoplast Symp., Ferenczy, L. and Farkas, G.L. Eds., Pergamon Press, Oxford, 1980, 485.
148. **Kaur-Sawhney, R., Flores, H.E., and Galston, A.W.,** Polyamine-induced DNA synthesis and mitosis in oat leaf protoplasts, *Plant Physiol.*, 65, 368, 1980.
149. **Kaur-Sawhney, R. and Galston, A.W.,** Interaction of polyamines and light on biochemical processes involved in leaf senescence, *Plant Cell Environ.*, 2, 189, 1979.
150. **Fuhrer, J., Kaur-Sawhney, R., Shih, L.-M., and Galston, A.W.,** Effects of exogenous 1,3-diaminopropane and spermidine on senescence of oat leaves, *Plant Physiol.*, 70, 1597, 1982.

151. **Flores, H.E. and Galston, A.W.,** Analysis of polyamines in higher plants by high performance liquid chromatography, *Plant Physiol.*, 69, 701, 1982.
152. **Smith, T.A. and Best, G.R.,** Polyamines in barley seedlings, *Phytochemistry*, 16, 841, 1977.
153. **Slocum, R.D., Flores, H.E., Galston, A.W., and Weinstein, L.H.,** Improved method for HPLC analysis of polyamines, agmatine and aromatic monoamines in plant tissue, *Plant Physiol.*, 89, 512, 1989.
154. **Galston, A.W.,** Polyamines as modulators of plant development, *BioScience*, 33, 382, 1983.
155. **Galston, A.W.,** Polyamines and plant response to stress, in *The Physiology of Polyamines*, Vol. II, Bachrach, U. and Heimer, Y.M., Eds., CRC Press, Boca Raton, FL, 1989, 99.
156. **Galston, A.W. and Kaur-Sawhney, R.,** Polyamines as endogenous growth regulators, in *Plant Hormones and their Role in Plant Growth and Development*, Davies, P.J., Ed., Martinus Nijhoff, Dordrecht, 1987, 280.
157. **Hegewald, E. and Kneifel, H.,** Amines in algae VI. The occurrence of polyamines in the genus *Chlorella*, *Arch. Hydrobiol.*, 4, Suppl 60, 489, 1982.
158. **Kneifel, H.,** Amines in algae, in *Marine Algae in Pharmaceutical Science*, Hoppe, H.A., Levring, T., and Tanaka, Y., Eds., Walter de Gruyter, Berlin, 1979, 365.
159. **Villanueva, V.,** Polyamines in *Euglena gracilis*, in *Advances in Polyamine Research*, Vol. 3, Caldarera, C. M., Zappia, V., and Bachrach, U., Eds., Raven Press, New York, 1981, 389.
160. **Corbin, J.L., Marsh, B.H., and Peters, G.A.,** An improved method for analysis of polyamines in plant tissue by precolumn derivatization with *o*-phthalaldehyde and separation by high performance liquid chromatography, *Plant Physiol.*, 90, 434, 1989.
161. **Hamana, K., Matsuzaki, S., and Sakakibara, M.,** Distribution of *sym*-homospermidine in eubacteria, cyanobacteria, algae and ferns, *FEMS Microbiol. Lett.*, 50, 11, 1988.
162. **Rodriguez-Garay, B., Phillips, G.C., and Kuehn, G.D.,** Detection of norspermidine and norspermine in *Medicago sativa* L. (alfalfa), *Plant Physiol.*, 89, 525, 1989.
163. **Matsuzaki, S., Hamana, K., Okada, M., Masaru, N., and Samejima, K.,** Novel aliphatic pentaamines found in *Canavalia gladiata*, *Phytochemistry*, 29, 1311, 1990.
164. **Sneath, P.H.A.,** Putrescine as an essential growth factor for a mutant of *Aspergillus nidulans*, *Nature*, 175, 818, 1955.
165. **Stevens, L.,** Ornithine decarboxylase activity in germinating conidia of *Aspergillus nidulans*, *FEBS Lett.*, 59, 80, 1975.
166. **Stevens, L. and Winther, M.D.,** Spermine, spermidine and putrescine in fungal development, *Adv. Microbiol. Physiol.*, 19, 63, 1979.
167. **Rajam, M.V. and Galston, A.W.,** The effects of some polyamine biosynthetic inhibitors on growth and morphology of phytopathogenic fungi, *Plant Cell Physiol.*, 26, 683, 1985.
168. **Rajam, M.V., Weinstein, L.H., and Galston, A.W.,** Prevention of a plant disease by specific inhibition of fungal polyamine biosynthesis, *Proc. Nat. Acad. Sci. USA*, 82, 6874, 1985.
169. **Slocum, R.D. and Galston, A.W.,** Inhibition of polyamine biosynthesis in plants and plant pathogenic fungi, in *Inhibition of Polyamine Metabolism*, McCann, P.P., Pegg, A.E., and Sjoerdsma, A., Eds., Academic Press, Orlando, 1987, 305.
170. **Carson, C. and McCann, P.P.,** Method of controlling phytopathogenic fungus, U.S. Patent 4,760,091, 1988.
171. **Weinstein, L.H. and Galston, A.W.,** Prevention of a plant disease by specific inhibition of fungal polyamine biosynthesis, International Patent WO 88/02986, 1988.

Polyamine Metabolism in Plants

Chapter 2

POLYAMINE BIOSYNTHESIS IN PLANTS

Robert D. Slocum

TABLE OF CONTENTS

I. INTRODUCTION

In recent years, there have been several reviews concentrating on polyamine metabolism in plants.[1-5] In this chapter, various aspects of the biosynthesis of so-called "common" polyamines putrescine (Put), cadaverine (Cad), spermidine (Spd) and spermine (Spm) will be considered in detail. The biosynthesis of uncommon polyamines is examined in Chapter 8, while the incorporation of common polyamines into plant secondary metabolites is reviewed in Chapter 4. Polyamine catabolism in plants is further examined in Chapter 3.

The biosynthesis of polyamines can be conveniently divided into two major steps: the synthesis of diamines (Put, Cad) from basic amino acids and the synthesis of polyamines from the diamine precursors and an aminopropyl moiety derived from L-methionine. The chapter is organized according to this theme.

Wherever possible, polyamine biosynthesis in plants is compared to that in bacteria, animals and fungi, since considerably more is known about polyamine metabolism in these organisms.[6,7] An attempt has been made to: (1) review what is known about the biochemical properties and metabolic regulation of specific enzymes involved in polyamine biosynthesis in plants, and (2) summarize information concerning the effects of various inhibitors on these enzymes. Inhibitors have played an essential role in the elucidation of polyamine metabolic pathways, as well as in helping to establish the basic requirement for polyamines in plant growth and developmental processes. Physiological control of polyamine metabolism is addressed in several chapters in Section C of this volume.

II. THE BIOSYNTHESIS OF PUTRESCINE AND CADAVERINE

In plants, and in bacteria, the diamine Put is synthesized via one of two main pathways (see Figure 1). Put can be formed directly by decarboxylation of L-ornithine (Orn), in a reaction catalyzed by ornithine decarboxylase (ODC; EC 4.1.1.17). This pathway represents the sole route to Put biosynthesis in animals and in most fungi.[6,7] Alternatively, the decarboxylation of L-arginine (Arg) by arginine decarboxylase (ADC; EC 4.1.1.19) leads to the formation of Put, via agmatine (Agm) and *N*-carbamoylputrescine (NCPut) intermediates, in the so-called ADC pathway. The vast majority of studies utilizing inhibitors and radiolabeled precursors and intermediates support the existence of the ODC and ADC pathways, although alternative routes to Put biosynthesis have also been described in some plants (see Sections II.D. and II.F. below).

Cadaverine is another diamine constituent of many plants which appears to be synthesized predominantly as the result of L-lysine (Lys) decarboxylation mediated by lysine decarboxylase (LDC; EC 4.1.1.18). Again, however, other mechanisms of Lys synthesis have been reported (see Section II.E.).

A. ARGININE DECARBOXYLASE (ADC) AND ORNITHINE DECARBOXYLASE (ODC)

ADC and ODC activities have been measured in many different plant species although, in some instances, apparent decarboxylase activities may have been artifactual, depending upon the assay conditions used (see Chapter 17). In most tissues, ADC activity is higher than that of ODC, but this is not always the case and, in some instances, only ODC activity has been measured.[8] Elevated ODC activities appear to be positively correlated with cell division in many plant organs[9] (see also Chapter 10). On the other hand, there is considerable interest in mechanisms regulating ADC activity in plants, since the activity of this enzyme is modulated by light and plant hormones (Chapter 13) and various types of environmental stresses (Chapter 15).

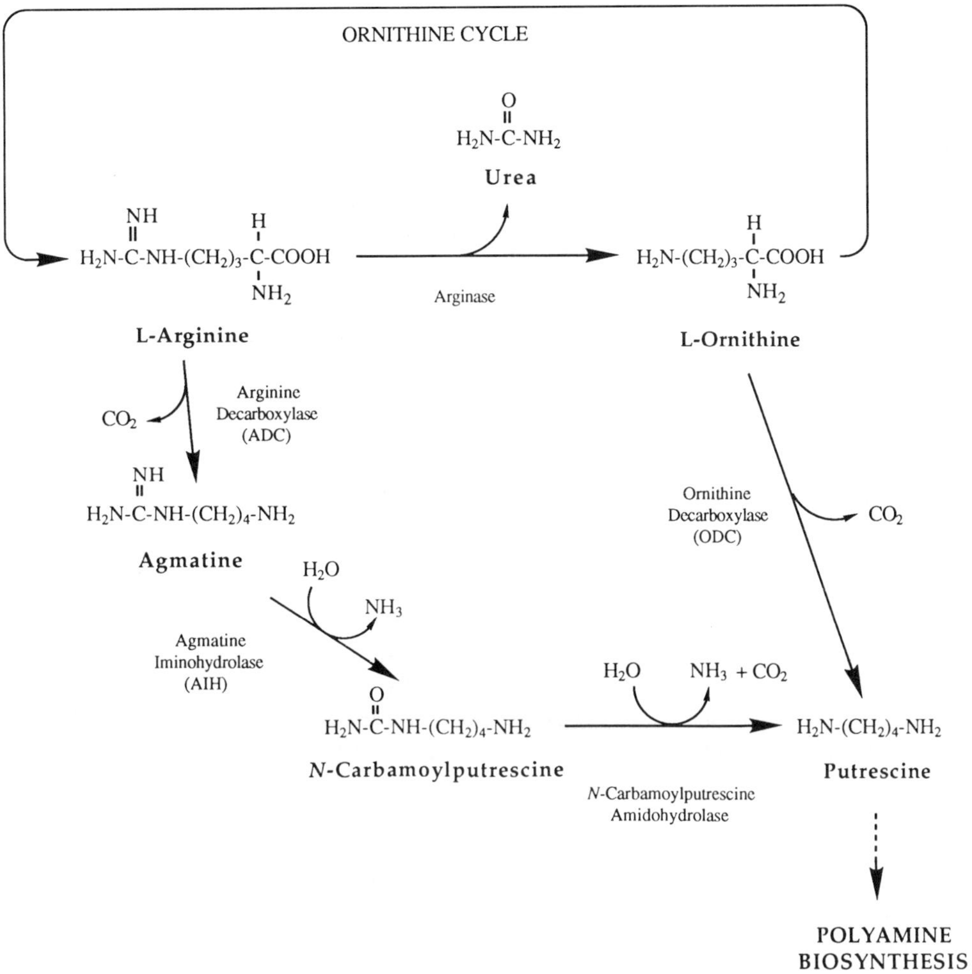

FIGURE 1. Metabolic pathways for the biosynthesis of putrescine in plants. Putrescine formation is initiated by the decarboxylation of L-arginine or L-ornithine via the ADC and ODC pathways.

Arginine decarboxylase has been partially purified and characterized from a number of plants[3] and has been purified to homogeneity from oat,[10] rice,[11] cucumber,[12] and *Lathyrus sativus*.[13] Each of these ADCs requires thiols, has a PLP cofactor, and is specific for Arg, although the cucumber ADC appears to be able to decarboxylate Orn, as well.[12] Oat ADC exhibits a pH 7 to 7.5 optimum and is a hexamer of identical 39,000 Da subunits.[10,14] Similarly, the *Lathyrus* enzyme is a hexamer of 37,000 Da subunits, but displays a pH 8.5 optimum. Rice ADC has a pH 8 optimum and is separated into two fractions (M_r = 88,000 Da and 174,000 Da) by gel filtration chromatography, suggesting that this enzyme is a multimer of somewhat different subunits. Curiously, three nonidentical polypeptides comprise the cucumber enzyme. The purified plant ADCs have K_ms for Arg ranging from 30 μM for the oat enzyme to 1.73 m*M* for the *Lathyrus* ADC. None of the plant ADCs resembles the biosynthetic ADC from *E. coli*, which is a tetramer of 74,000 Da subunits.[15]

Recently, the oat ADC gene has been cloned by Bell and Malmberg[16] (see Chapter 9). Oat ADC cDNA encodes a 66 kDa peptide, although the mature enzyme isolated from leaf tissues exhibits an M_r = 24,000 Da. Antibody against the 24 kDa peptide recognizes both a 34 kDa and the 24 kDa peptide on SDS-PAGE immunoblots. These results suggest that the oat enzyme is synthesized as a precursor protein, which is further processed to one or more mature forms of the enzyme, perhaps during organellar import. Indeed, it has been reported that ADC is a plastid-localized enzyme.[17] A comparison of oat and *E. coli* ADC[18] primary structures shows that there is considerable sequence homology between the plant and bacterial enzymes.

Ornithine decarboxylase has recently been purified from barley[19] and jute.[20] Native barley ODC appears to be a single polypeptide with an $M_r = 55{,}000$, which is similar in size to the subunits of mammalian and fungal ODCs.[21,22] However, the latter enzymes are dimers and tobacco ODC probably is a dimer, since the native enzyme exhibits an $M_r = 107{,}000$ Da.[23] Barley ODC has a pH 8.5 optimum and a $K_m = 0.36$ m*M* for Orn. In contrast, the jute ODC is a 39,000 Da polypeptide with a pH 7.8 optimum and 0.1 m*M* K_m for Orn. Neither plant ODC resembles the biosynthetic ODC from *E. coli*, which is a dimer of identical 81,000 Da subunits with a pH 8.3 optimum and $K_m = 5.6$ m*M* for Orn.[15] In barley, there appear to be distinct cytosolic and chromatin-bound ODC activities that are differentially regulated in germinating seeds and in response to GA_3 treatment.[24]

In *E. coli*, biodegradative ADC and ODC enzymes are induced in cells grown at low pH in the presence of L-Arg or L-Orn, respectively.[15] There does not appear to be any evidence that such enzymes exist in plants.

1. ADC and ODC Inhibitors

All of the ADC enzymes which have been purified and characterized (see above) are strongly inhibited by D-Arg, which is a competitive inhibitor of the L-Arg substrate. Plant ADCs are also inhibited by Agm, by arcain (an Agm analog), and by L-canavanine (a guanidinooxy analog of L-Arg). However, these enzymes are relatively insensitive to Put, Spd, and Spm even at 10 m*M* concentrations.[10-13]

The most widely used ADC inhbitor is DL-α-difluoromethylarginine (DFMA), which is a potent enzyme-activated, irreversible inhibitor.[25] *In vitro*, this inhibitor appears to be highly specific for ADC, with a $K_i = 0.85$ m*M* for the *E. coli* enzyme.[26] However, the specificity of this inhibitor *in vivo* has been questioned in tissues that contain high arginase activities, in view of the finding that radiolabeled DFMA is largely metabolized to labeled DL-α-difluoromethylornithine (DFMO) with the resultant inhibition of ODC, as well as ADC activities.[27] Recently, several other Arg and Agm analog inhibitors have been described with K_i values around 10–50 μ*M* and these compounds strongly inhibit ADC enzymes from barley and oat.[26]

Unlike ADC, plant ODCs are strongly inhibited by Put, Spd, and Spm.[19,20,23] DFMO is also a potent ODC inhibitor[28] but many plant ODCs are not inactivated by this inhibitor. For example, barley ODC activity is inhibited 50% by 0.38 m*M* DFMO,[19] while the jute enzyme is not inhibited by 10 m*M* DFMO.[20] Some bacterial ODCs also appear to be insensitive to DFMO,[29] although they are inactivated by structurally related inhibitors. DFMO, unlike DFMA, does not appear to be metabolized in plant tissues[27] and is considered to be a specific ODC inhibitor *in vivo*.

2. Regulation of ADC and ODC Activities

Apart from the well-documented physiological regulation of ADC activity in plants, little is known about metabolic or molecular mechanisms regulating this enzyme. The apparent lack of ADC inhibition by Agm, Put, and polyamines, except at very high concentrations, would suggest that feedback inhibition of this enzyme might not be significant under physiological conditions. This seems to be the case in many studies. However, Hiatt et al.[30] reported that ADC activity in tobacco cells was fully repressed by exogenous Spd or Spm, while repression of ODC activity was weak. The mechanism of ADC inhibition by polyamines in this study is unknown.

There seems to be no evidence for inducible, proteinaceaous inhibitors of ADC, analogous to ODC "antizyme" (see below), in dialyzed extracts of plants.[31]

There have been a few studies in which changes in ADC activities have been related to changes in synthesis or turnover of the enzyme. For example, stress-induced Put accumulation appears to result primarily from increased ADC activity, which is accompanied by a *de novo* synthesis of ADC protein.[32] Whether this increase in ADC activity is driven by increased Arg synthesis, which is known to occur in response to many types of stress, is an open question.[33] In carrot cell cultures, ADC displayed a half-life of approximately 30 min, suggesting that this enzyme is rapidly degraded in this plant.[34] Since the half-life of carrot ADC was the same in either embryogenic or nonembryogenic cultures, the authors concluded that increased ADC activities in embryogenic cells were not due to differences in turnover rates for this enzyme. The half-life of cucumber ADC was observed to be 3.7 h.[31]

The sensitivity of plant ODC to Put or polyamines suggests that polyamine synthesis via the ODC pathway may be regulated, to some extent, by simple end-product inhibition. In barley, however, it has been found that Put or Spd induce a 16,000 Da proteinaceous inhibitor of ODC, known as ODC "antizyme".[35] ODC-antizyme complexes can be isolated from both cytosolic and chromatin-bound protein fractions[36] and the authors speculate that the antizyme may play an important role in regulating ODC activities in this plant. The role of a similar antizyme in regulating ODC activity in animal cells is controversial[7] and it has been suggested that antizyme may be involved in degradation of ODC protein rather than functioning as an effector molecule. Further studies will be required to clarify the precise functions of the plant ODC antizyme.

Two interesting questions regarding Put biosynthesis in plants are (1) what factors determine the flux rates through the ADC vs. ODC pathways and (2) does inhibition of either pathway result in compensatory increases Put synthesis via the other pathway?

In bacteria, flux through the ADC and ODC pathways appears to be determined by Orn availability, since intracellular Arg levels are always high.[6] Under normal conditions, where Orn levels are not limiting, Put is synthesized primarily by ODC. If the Orn concentration drops, Arg is then utilized in Put synthesis via the ADC pathway. In plants, it appears that the ADC pathway is the major, if not the only, route to Put biosynthesis under a variety of conditions which increase flux through the Arg pathway (i.e., via an Orn intermediate). For example, osmotic stress increases the incorporation of ^{14}C-glutamate label into Put by threefold;[33] significantly, inhibition of ADC with DFMA completely blocks this incorporation, suggesting that Put can not be synthesized from the ^{14}C-Orn intermediate. Such differences may reflect subcellular compartmentalization of intermediates or enzymes in the plant cell that is not a factor in bacterial Put synthesis.

Compensatory mechanisms maintaining Put and polyamine homeostasis have been documented in bacteria,[37] although evidence for such mechanisms in plants is largely circumstantial. In tobacco, it does appear that the ADC and ODC pathways have distinct, indispensible functions in cellular metabolism, as has been reviewed by Malmberg and Hiatt.[38]

B. AGMATINE IMINOHYDROLASE

In the ADC pathway leading to Put synthesis, Agm is converted to NCPut and NH_3 through the activity of agmatine iminohydrolase (AIH; EC 3.4.3.12, also called agmatine deiminase). The participation of AIH in the ADC pathway for Put synthesis in plants is supported by studies demonstrating the incorporation of Agm into NCPut and Put.[39-41]

AIH has also been purified and characterized from several plant species. Sindhu and Desai[42] characterized a partially purified AIH from groundnut, but this enzyme was first purified to homogeneity from corn.[43] Corn AIH exhibited an M_r = 85,000 Da (a dimer of identical 43,000 Da subunits) and appeared to be cytosolic. The enzyme had a pH 6.5 optimum, in agreement with the pH 6.5 to 7.5 optimum reported by Smith,[44] a K_m = 0.19 m*M* for Agm, and was inhibited by the Agm analog arcain and by sulfhydryl reagents. Unlike groundnut AIH,[42] the corn enzyme was not inhibited by Put or polyamines. An AIH enzyme with different properties was purified from rice.[45] The rice enzyme, which was purified 717-fold, displayed an M_r = 183,000 Da and was also a dimer of identical 95,000 Da subunits. Rice AIH had a pH 6.0 optimum, and K_m = 15 m*M* for Agm and was inhibited by Spd and Spm, but not Put.

AIH activity has been reported from numerous plants[41-45] but little is known about the metabolic regulation of this enzyme. Upon feeding soybean seedlings with NH_3, a treatment which is known to increase Put formation via the ADC pathway in this plant, AIH activities increased.[41] In K^+-deficient barley seedlings, in which the activities of ADC and NCPut amidohydrolase are elevated, as compared with control plants, AIH activity could not be detected; similarly, K^+ deficiency did not increase AIH activity in corn seedlings.[46] Thus, enhanced Put synthesis via the ADC pathway is not always accompanied by an increase in AIH activities.

Conflicting reports regarding the effects of Put and polyamines upon AIH activity *in vitro* (see above) suggest that feedback control of AIH activity may or may not play an important role in regulating Put synthesis *in vivo*. Recently, α-monofluoromethylagmatine (FMA) and other agmatine analogs have been reported to be potent enzyme-activated, irreversible inhibitors of bacterial and plant ADC.[26] These compounds appear not to have been tested as inhibitors of AIH and may prove useful in establishing the extent to which AIH is required for Put synthesis in plants, as alternate mechanisms for NCPut synthesis via citrulline decarboxylase and Put carbamoyltransferase have been reported (see Section II.F.).

C. *N*-CARBAMOYLPUTRESCINE AMIDOHYDROLASE

There is evidence that *N*-carbamoylputrescine (NCPut) is an intermediate in the synthesis of Put in several plants that have been studied.[40,47] There are at least two enzyme activities which have been described from plants and bacteria that could carry out Put synthesis via an NCPut intermediate.

The first is *N*-carbamoylputrescine amidohydrolase (EC 3.5.1.-), which catalyzes the hydrolysis of NCPut, releasing Put, NH_3 and CO_2 as products. This enzyme has been characterized in barley and has been detected in extracts of several other plants[48] and has been partially purified from corn.[49] The corn enzyme exhibited a K_m = 90 μ*M* for NCPut, with a pH optimum between 6.5 and 7.0 and an M_r = 125,000 on Sephadex G-200. It was inhibited by Cu^{2+} and Zn^{2+} ions and sulphydryl reagents and exhibited no activity toward citrulline, carbamoyl-aspartate, or urea. NCPut amidohydrolase, like ADC and AIH, displays no significant inhibition by Put or polyamines. Both the corn and barley NCPut amidohydrolases appear to be cytosolic enzymes.[48,49]

A second enzyme, putrescine carbamoyltransferase (EC 2.1.3.6), can also produce Put and carbamoyl-P from NCPut. Putrescine carbamoyltransferase from *Streptoccocus* has been purified and characterized[50] and a similar activity has been reported from pea.[51] Interestingly, putrescine carbamoyltransferase has also been found to be one of four apparently distinct enzyme activities associated with putrescine synthase, a multifunctional enzyme described from *Lathyrus sativus* (see Section II.D. below). Wargnies et al.[50] found that the *Streptococcus* putrescine carbamoyltransferase displayed a broad substrate specificity, carbamoylating not only Put (to form NCPut in the reverse direction), but also Orn (i.e., an ornithine carbamoyltransferase (OCT; EC 2.1.3.3)-like activity) and several other amines, such as 1,3-diaminopropane, Cad, Spd, and Spm. While OCT from this bacterium specifically carbamoylated Orn, Kleczkowski and Wielgat[51] observed that partially purified OTC from pea also carbamoylated Put. However, given the fact that both putrescine and ornithine carbamoyltransferases from *Streptococcus* exhibited many of the same structural features and largely copurified, the partially purified pea enzyme may, in fact, have been Put carbamoyltransferase.

While NCPut amidohydrolase is usually implicated in Put synthesis in reviews of plant polyamine metabolism, evidence for this, based largely upon metabolic labeling studies and the single purification of this enzyme from corn, is limited. In light of the possibility that other enzymes (Put carbamoyltransferase, Put synthase) might carry out Put synthesis from an NCPut intermediate, further studies will be required to firmly establish the route to Put synthesis in plants.

D. PUTRESCINE SYNTHASE

During the course of studies characterizing enzymes involved in Put synthesis from Arg in *Lathyrus satvius*, Srivenugopal and Adiga[52] discovered a single $M_r = 55{,}000$ polypeptide with multiple catalytic activities. Remarkably, this multifunctional enzyme, which they named "putrescine synthase", exhibited AIH (reaction 1), Put carbamoyltransferase (reaction 2), ornithine carbamoyltransferase (reaction 3), and carbamate kinase (EC 2.7.2.2; reaction 4) activities.

1. $\text{Agm} + H_2O \rightarrow \text{NCPut} + NH_3$
2. $\text{NCPut} + \text{Pi} \rightleftharpoons \text{Put} + \text{carbamoyl-P}$
3. $\text{carbamoyl-P} + \text{L-ornithine} \rightleftharpoons \text{L-citrulline}$
4. $\text{carbamoyl-P} + \text{ADP} + H_2O \rightleftharpoons \text{ATP} + CO_2 + NH_3$

In vitro, this enzyme carried out stoichiometric conversion of Agm and Orn to Put and L-citrulline (Cit), in the presence of inorganic phosphate. Label from [*ureido*-^{14}C]-NCPut was transfered intact to Orn, to form Cit. Put synthase activity was also found in crude extracts of pea, cucumber, and corn and appeared to be associated with the cytosolic fraction.

On the basis of these studies, the authors concluded that the putrescine carbamoyltransferase activity associated with Put synthase, rather than the NCPut amidohydrolase activity reported by Smith,[48] was probably involved in the synthesis of Put from NCPut in some plants. The production of $CO_2 + NH_3$ in the Put synthase reaction scheme could be explained in terms of carbamate kinase-mediated hydrolysis of the carbamoyl-P product of the Put carbamoyltransferase reaction, rather than on the basis of NCPut amidohydrolase activity. It should be noted, however, that purified corn NCPut amidohydrolase produces stoichiometric amounts of Put, CO_2 and NH_3 from NCPut *in vitro*.[49]

Putrescine synthase from cucumber has recently been purified and characterized.[53] Interestingly, these authors found that the cucumber enzyme also exhibited ADC activity.[54] At present, it is not clear to what extent multifunctional enzymes of this type may contribute to Put synthesis in other plants. Most experimental evidence supports the scheme for Put synthesis from Arg, via Agm and NCPut intermediates, as is shown in Figure 1.

E. LYSINE DECARBOXYLASE

Lysine decarboxylase (LDC) catalyzes the decarboxylation of L-lysine (Lys) to form cadaverine (Cad). LDC activity has been reported in numerous plants[55-58] and this enzyme has recently been purified from *Hafnia alvei*.[59]

LDC activity appears to be quite low in most plant tissues but in some legumes, its activity is positively correlated with the accumulation of quinolizidine alkaloids, for which Cad is a precursor.[58] In etiolated pea seedlings, exposure to ethylene causes a decrease in ADC activity, Put and polyamine titers, while LDC activity increases as much as threefold and Cad accumulates in the tissue.[60] In these tissues, it is believed that Cad and aminopropylCad serve as Put and Spd analogs, and that increased LDC activity and Cad synthesis is a compensatory response to the loss of polyamines.

In *E. coli*, a similar compensatory response is seen upon depletion of Put and Spd from the cell.[61] Under normal growing conditions, Put and Spd inhibit the activity of a constitutive LDC in the bacterium, but a second inducible LDC activity, which is insensitive to Put and Spd, appears upon polyamine depletion. It is not known whether distinct constitutive and inducible LDC occur in plants or if the activity of plant LDC is modulated by polyamine availability.

A completely different type of compensatory response is seen in an arginase-deficient (*aga*) mutant of *Neurospora crassa*.[62] When these cells are grown in the presence of Arg, Orn synthesis is feedback inhibited and polyamines, which are dervied from Orn via the ODC pathway, are not made. Under these conditions, ODC activity is highly derepresed and this enzyme can serve as an inefficient LDC ($K_m > 20$ m*M* for Lys), synthesizing Cad from Lys. Again, it is not known whether ODC-mediated decarboxylation of Lys, or similar compensatory mechanisms, occur in plants. Indeed, such studies are difficult to carry out in plant systems in which mutants deficient in polyamine biosynthesis are not available and in which significant depletion of tissue polyamines, through the use of inhibitors, is difficult to achieve.[63] An enzyme-activated, irreversible inhibitor of LDC, DL-α-difluoromethyllysine (DFML), has been characterized by Pösö et al.[64] and has recently been shown to specifically inhibit LDC activity and Cad synthesis in pea seedlings.[65] In this study, inhibition of ADC activity and polyamine synthesis with DFMA did not result in a compensatory increase in LDC activity and Cad formation, as was reported for the ethylene-treated pea seedlings (see above).[60]

In *Lathyrus sativus*, Cad can be formed either from Lys or ^{14}C-homoarginine (via homoagmatine) and an enzyme that decarboxylates both substrates has been reported.[66] The partially purified enzyme was distinct from ADC and required Mn^{2+} or Fe^{2+} for activity. In contrast, Fe^{2+} has been reported to inhibit the activity of LDC from pea and *Heimia salicifolia*,[55,57] suggesting that the *Lathyrus* decarboxylase characterized in this study was not LDC.

F. OTHER ENZYMES

With regard to Put biosynthesis, it should be noted that there is no evidence for an agmatine ureohydrolase (also called agmatinase; EC 3.5.3.11) activity in plants.[49] In bacteria, this enzyme catalyzes the hydrolytic cleavage of urea from Agm, to form Put directly.[67] As was discussed previously, Put synthesis from Agm in most plants appears to be carried out in two steps, via AIH and NCPut amidohydrolase.

There is evidence that another enzyme found in bacteria, citrulline decarboxylase (CitDC),[68] may be involved in Put synthesis in some plants. This enzyme catalyzes the decarboxylation of Cit to form NCPut. In sugarcane, [*carbamoyl*-^{14}C]-citrulline was shown to be incorporated primarily into NCPut, with little label detected in Agm, suggesting that assimilation of label into NCPut occurred directly rather than via the Arg pathway.[69] In activated *Helianthus tuberosus* tuber tissues, NCPut and Put were found to be a major products of [*carbamoyl*-^{14}C]-citrulline metabolism, although incorporation into Arg suggested that at least some of the label was shunted to the Put via ADC and AIH.[70] In leaf extracts of K^+-deficient sesame, Crocomo et al.[71] demonstrated the enzymatic decarboxylation of [*carbamoyl*-^{14}C]-citrulline to form [*carbamoyl*-^{14}C]-NCPut. However, as was recently noted by Suzuki et al.[5], there is no direct evidence for CitDC in plants and the finding that *E. coli* ADC may decarboxylate citrulline to form NCPut suggests that plant ADC may play a similar role. In soybean seedlings, NCPut could be formed either from Agm (via AIH) or citrulline (via CitDC?), but when the cotyledons were removed only AIH activity was detected.[41]

The presence of CitDC activity in plant tissues would permit Put synthesis to bypass the ADC pathway (it is assumed that Orn is not appreciably metabolized to Put via the ODC pathway under such circumstances). The physiological significance of this type of metabolism is unknown, but could serve to shunt carbon and nitrogen directly to polyamine biosynthesis while avoiding the energetically costly Arg biosynthetic pathway.

II. SPERMIDINE AND SPERMINE BIOSYNTHESIS

The synthesis of spermidine (Spd) and spermine (Spm) is carried out by addition of an aminopropyl moiety to one or both primary amino groups of Put by spermidine and spermine synthases, respectively (see Section II.B.). It is now well-documented that decarboxylated *S*-adenosylmethionine (dSAM) is the aminopropyl donor in the biosynthesis of these common polyamines. Decarboxylated SAM is derived, via the action of *S*-adenosylmethionine decarboxylase (SAMDC; EC 4.1.1.50), from *S*-adenosylmethionine (SAM). SAM, in turn, is produced by the condensation of the adenosine moiety of ATP and L-methionine in a reaction catalyzed by the enzyme *S*-adenosylmethionine synthase (EC 2.5.1.6).

As can be seen in Figure 2, methionine provides the aminopropyl moiety for polyamine synthesis in this pathway. This has been confirmed in several studies examining the incorporation of [^{14}C]-methionine into SAM, dSAM, Spd, and Spm.[72-74] Similarly, the 1,4-diaminobutane skeleton of Spd and Spm has been shown to be derived from labeled Put (or labeled Arg or Orn incorporated via Put).[74] Incorporation of label into polyamines from methionine or Put is blocked by inhibition of either SAMDC or Spd synthase (see below).

Several studies suggest that incorporation of the aminopropyl moiety into polyamines is regulated at the level of SAMDC and that decarboxylation of SAM by this enzyme commits the aminopropyl moiety to polyamine biosynthesis.[73] For this reason, only SAMDC will be examined with regard to its role in the regulation of polyamine biosynthesis.

Because the plant hormone ethylene is also derived from a SAM precursor, and because ethylene and polyamines promote and inhibit senescence in plant tissues, respectively, there has been much interest in the metabolic interactions between the ethylene and polyamine biosynthetic pathways in recent years. This work is reviewed in Chapters 5 and 14.

A. *S*-ADENOSYLMETHIONINE DECARBOXYLASE (SAMDC)

Suzuki and Hirasawa[75] purified the corn SAMDC 500-fold. This enzyme exhibited an M_r = 25,000 Da, a K_m of 5 μM for SAM and was sensitive to sulfhydryl reagents. It was not stimulated by Put, as are the mammalian and yeast SAMDCs,[76] or by Mg^{2+} ions, as is the bacterial enzyme.[76] The corn SAMDC[75] did not require pyridoxal phosphate as a co-factor, unlike most decarboxylases. However, all SAMDCs that have been characterized from other organisms have a pyruvate, rather than a PLP prosthetic group.[77] Yamanoha and Cohen[78] purified SAMDC 1500-fold to near homogeneity from Chinese cabbage. This enzyme had an apparent molecular weight of 35,000 Da on a Sephacryl S-300 column and displayed a K_m = 38 μM for SAM. As with the corn enzyme, the cabbage SAMDC was neither Put- or Mg^{2+}-stimulated. On the basis of subcellular fractionation studies, it appears that this enzyme is localized primarily in the cytosol.[17,79]

1. SAMDC Inhibitors

Numerous inhibitors of SAMDC have been characterized and these studies have recently been reviewed by Pegg.[77] The most commonly used inhibitor is methyl glyoxal-*bis*-(guanylhydrazone), or MGBG. It has been observed that MGBG is a much more potent inhibitor of the mammalian and yeast SAMDCs, which are Put stimulated, than of bacterial and plant SAMDCs, which are not Put-stimulated.[76] Nonetheless, MGBG strongly inhibits plant SAMDCs, exhibiting a K_i = 0.6 μM for the purified SAMDC from Chinese cabbage.[78] Another compound, 1-aminooxy-3-aminopropane (AOAP), has been reported to be an irreversible inhibitor of SAMDC (K_i = 50 μM).[80] AOAP has recently been shown to be a potent inhibitor of wheat SAMDC.[81]

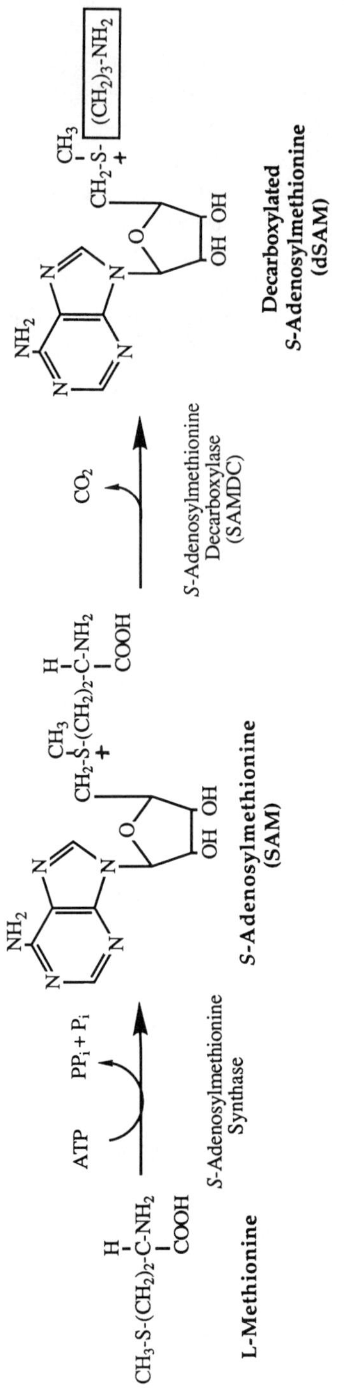

FIGURE 2. Derivation of decarboxylated *S*-adenosylmethionine, the aminopropyl group (box) donor for polyamine biosynthesis, from L-methionine.

The paradoxical enhancement of SAMDC activity in tissues treated with MGBG has been reported. Pegg et al.[82] established that this was due to the stabilization of the animal enzyme, upon binding of MGBG, increasing the half life of SAMDC from 2 to approximately 20 h. A similar phenomenon was reported by Hiatt et al.,[30] who found that the activity of SAMDC from tobacco cells grown in the presence of MGBG increased about 20-fold. They demonstrated that a 35,000 Da protein, which was shown to be SAMDC, was overproduced in the MGBG-treated cultures and that the accumulation of SAMDC protein was due to an increased stability of the enzyme, in the presence of inhibitor, rather than an increase in its rate of synthesis.

MGBG is rapidly taken up by carrot cells in suspension culture.[83] It apparently uses the polyamine transport system, as has been shown for this inhibitor in animal cells,[84] since Spd was a competitive inhibitor of MGBG uptake.[83] In numerous studies, MGBG has been demonstrated to be an effective inhibitor of polyamine biosynthesis *in vivo*. For example, in barley seeds, MGBG treatment resulted in the accumulation of Put and a decrease in Spd and Spm levels, and blocked the incorporation of ^{14}C-arginine or ^{14}C-ornithine into polyamines.[85]

2. Regulation of SAMDC Activity

Cellular polyamine levels play a major role in the regulation of SAMDC activity *in vivo*, as has been reviewed by Pegg.[86] Pajunen et al.[87] reported that DFMO-mediated polyamine depletion in rat prostrate tissue resulted in increased SAMDC protein and mRNA levels. Conversely, it appears that the accumulation of cellular polyamines inhibits SAMDC activity. Hiatt et al.[30] reported that 1 m*M* Spd treatment of tobacco cell cultures resulted in a rapid decrease in SAMDC activity. This decrease in SAMDC activity appeared to be the result of Spd blocking synthesis of the enzyme since, in the presence of Spd, SAMDC activity declined at the same rate as in cyclohexamide-treated cells, with a half-life of 30 to 60 min. The tobacco SAMDC has one of the shortest half-lives reported for any enzyme and the authors speculated that repression of SAMDC synthesis by polyamines might represent an important regulatory mechanism in polyamine biosynthesis.

In addition to polyamine depletion, increased cellular Put levels also stimulate SAMDC activity.[86] In view of this fact, it is curious that an exogenous 50 m*M* Put treatment resulted in a 50% decrease in SAMDC activity in *Saintpaulia* petal tissues after 1 h.[88] In these tissues, however, Spd levels increased fourfold and decreased SAMDC activities may reflect inhibition of enzyme synthesis by Spd, as was found in tobacco.

Decarboxylated SAM is known to be a potent inhibitor of SAMDC activity and displayed a $K_i = 6\ \mu M$ for the purified Chinese cabbage enzyme.[78] However, these authors reported that the 20-fold higher activity of Spd synthase, versus SAMDC, likely explained the observed low tissue levels of dSAM (estimated at 10^{-18} mol/cell). It is not clear whether endogenous dSAM levels exert significant feedback modulation of this enzyme.

3. Artifactual and Biosynthetic SAMDC Activities

Suresh and Adiga[89] documented the occurrence of a Put-sensitive, Mg^{2+}-stimulated SAMDC activity in extracts of *Lathyrus sativus*, suggesting that this plant enzyme was of the mammalian type. It may be recalled, from the discussion above, that those bacterial and plant SAMDCs which have been characterized to date are Put/Mg^{2+}-insensitive. This SAMDC activity, however, was found to be the result of Put oxidation by diamine oxidase; the resulting amine-oxidase-generated H_2O_2 mediated a nonenzymatic decarboxylation of SAM. This artifactual SAMDC activity could be eliminated by adding catalase to the reaction or by removing DAO from the extract. In the absence of DAO, a Put-insensitive SAMDC activity was then detected.

As discussed in detail in Chapter 17, the measurement of artifactual decarboxylase activities in plant extracts has greatly confounded our understanding of polyamine metabolism and extreme caution must be exercised in carrying out such assays.

B. SPERMIDINE AND SPERMINE SYNTHASES

Spermidine synthase (EC 2.5.1.16) and spermine synthase (EC 2.5.1.22), also refered to as Put and Spd aminopropyltransferases, respectively, appear to be present in most plants, since Spd and Spm are common constituents of plant tissues. There have, however, been few studies in which Spd or Spm synthases of plants have been purified or characterized. The biosynthesis of Spd and Spm by these enzymes is shown in Figure 3.

Baxter and Coscia[90] first demonstrated that an enzyme in *Vinca rosea* extracts was capable of utilizing dSAM and Put as substrates, resulting in the production of Spd *in vitro*. Hirasawa and Suzuki[91] purified Spd synthase 115-fold from corn. Significantly, the purified enzyme (pH 7.2 optimum and M_r = 43,000 Da on Sephadex G-200) exhibited no Spm synthase activity. Thus, the plant enzyme appears to be different from the bacterial synthase, which can catalyze aminopropyl addition to either Put or Spd, and similar to the animal enzyme, which is distinct from Spm synthase.[92] Yamanoha and Cohen[77] purified a Spd synthase from Chinese cabbage 1900-fold to near homogeneity. This enzyme exhibited a pH 8.8 optimum, an M_r = 81,000 Da for the native enzyme on Sephacryl S-300, and K_m values of 6.7 μM for dSAM and 32 μM for Put, respectively.[78,93] The effects of Put or polyamines on the activity of Spd synthase, either *in vitro* or *in vivo*, have not been reported. Cohen et al.[94] initially reported that the Chinese cabbage enzyme was associated with the chloroplast, but later concluded that it was most likely localized in the cytoplasm.[95]

Spermine synthase activity has been reported from Chinese cabbage[93] and in Hiproly Barley callus,[74] although this enzyme does not appear to have been isolated and characterized. The Spm synthase in Chinese cabbage appears to utilize dSAM and Spd as substrates.

As is discussed in Chapter 8, aminopropyltransferases generally display a narrow amine substrate specificity (usually Put and Spd, respectively) in bacteria and animals. In some bacteria, however, it seems likely that aminopropyltransferases utilizing dSAM as an aminopropyl group donor and exhibiting broad amine substrate specificities may be involved in the synthesis of uncommon polyamines. Many of these uncommon polyamines occur in plant tissues and there is evidence that broad-specificity aminopropyltransferases may also occur in plants. Villanueva et al.[96] reported the synthesis of *sym*-norspermidine and *sym*-norspermine from labeled SAM and 1,3-diaminopropane in crude extracts of *Euglena gracilis*. The enzymes involved in the synthesis of these polyamines had a pH 7.6 optimum, the same as that of Spd and Spm synthases in this organism. They also exhibited a strong preference for Put as a substrate; 1 m*M* Put inhibited the incorporation of 1,3-diaminopropane into *sym*-norspermidine by 80%.

1. Spermidine and Spermine Synthase Inhibitors

Like other Spd synthases, the plant enzyme is inhibited by cyclohexylamine (CHA), which has commonly been referred to as dicyclohexylamine (DCHA) in the literature, due a labeling mistake in early batches of this compound produced by Sigma Chemical Co., St. Louis (see Batchelor et al.[97]). Sindhu and Cohen[93] reported that the activity of Spd synthase purified from Chinese cabbage was inhibited 50% by 0.54 μM CHA. Biondi et al.[98] have shown that CHA is readily taken up by cells of pine cotyledon tissues, resulting in the accumulation of Put and depletion of Spd and Spm levels.

A transition-state analog inhibitor, *S*-Adenosyl-1,8-diamino-3-thiooctane (AdoDATO), which is specific for Spd synthase and displays no activity toward Spm synthases, inhibited the Chinese cabbage enzyme with an I_{50} = 1.3 μM.[93] The sensitivity of the plant Spd synthase to this inhibitor was similar to that seen for bacterial and mammalian Spd synthases.[99] Recently,

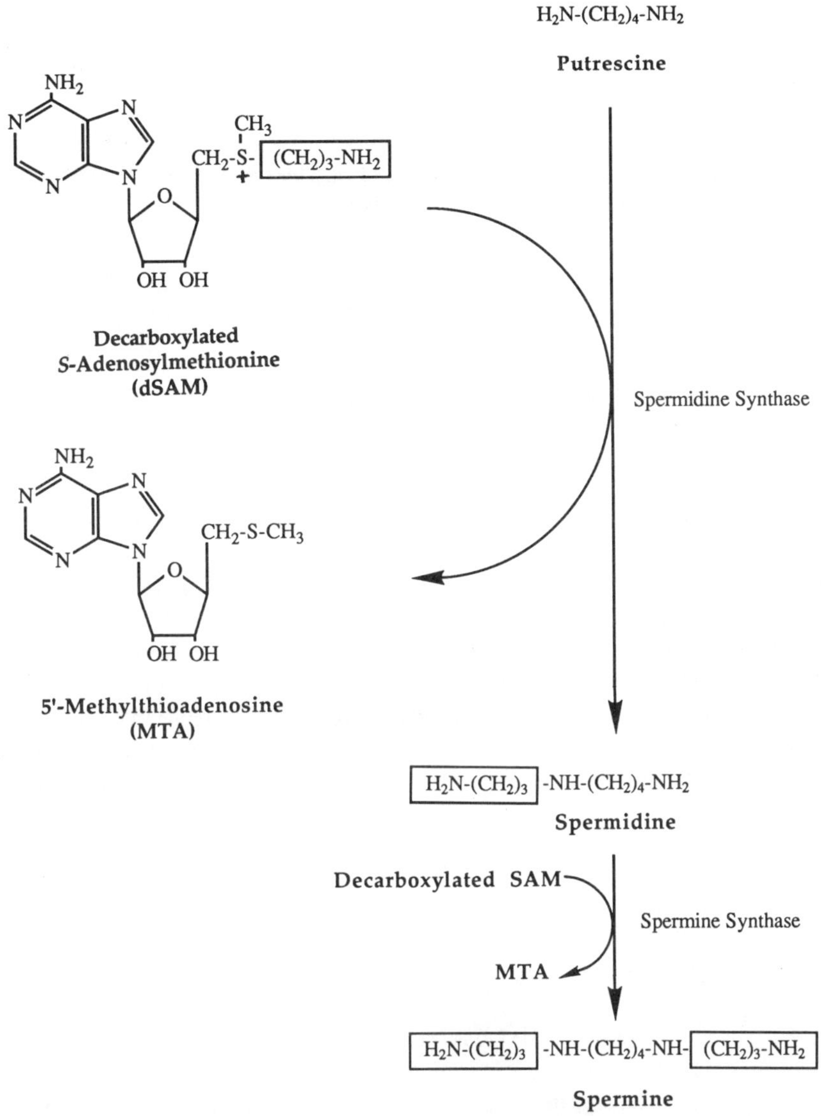

FIGURE 3. Spermidine and spermine biosynthesis via aminopropyl group (boxes) addition to putrescine. The 5′-methylthioadenosine product of this reaction is recycled, regenerating L-methionine.

Woster et al.[92] have described the synthesis and properties of a multisubstrate adduct inhibitor of Spm synthase, *S*-adenosyl-1,12-diamino-3-thio-9-azadodecane (AdoDATAD). AdoDATAD does not inhibit Spd synthase and it may now be possible to use this inhibitor, in conjuction with AdoDATO, to discriminate between Spd and Spm synthase activities in plants and their respective roles in polyamine biosynthesis.

Khomutov et al.[80] have reported that 1-aminooxy-3-aminopropane (AOAP) is a competitive inhibitor of mammalian Spd synthase ($K_i = 2.3\ \mu M$) but AOAP inhibition characteristics for plant Spd synthases have not been reported.

2. Regulation of Spermidine and Spermine Synthases

There is little information regarding the regulation of plant Spd or Spm synthases. Yamanoha and Cohen[78] reported that they were unable to find specific activators of the Chinese cabbage Spd synthase *in vitro*. In this same plant, they found that Spd synthase activity was about 20-fold higher than that of SAMDC, perhaps indicating that dSAM (and/or Put) availability regulates Spd synthesis *in vivo*.

In plant tissues where it has been measured, Spm synthase activity is usually much lower than that of Spd synthase, reflecting the relative levels of Spd and Spm. It has been noted, however, that inhibiton of Spd synthase by CHA *in vivo* leads to a shift in polyamine synthesis, from Spd to Spm. For example, CHA inhibition of Spd synthesis in Chinese cabbage is accompanied by a fourfold increase in Spm and a tenfold increase in dSAM levels, suggesting that dSAM is normally limiting for Spm synthesis.[73] Katoh and Hasegawa[74] found that 10 m*M* CHA decreased the incorporation of ^{14}C-Put or [3,4-^{14}C]-methionine into Spd and inhibited Spd synthase activity while increasing Spm synthase activity tenfold and increasing ^{14}C-Spd or ^{14}C-methionine incorporation into Spm. How this increase in Spm occurs, when Spd synthesis is markedly reduced by CHA, has not been determined but this phenomenon might possibly involve an alternative route to Spd synthesis discussed in the section below.

C. ALTERNATE PATHWAYS FOR SPERMIDINE AND SPERMINE BIOSYNTHESIS

In many bacteria, it appears that the aminopropyl group donor in polyamine biosynthesis is L-aspartate β-semialdehyde, rather than dSAM. As is discussed in Chapter 8 (Section II; reaction scheme 2), the aldehyde forms a Schiff base with Put to form carboxySpd in an NADPH-dependent reaction. CarboxySpd is then decarboxylated, in a step requiring pyridoxal phosphate, to form Spd. Srivenugopal and Adiga[100] provided evidence that a similar pathway was the major route to Spd synthesis in the plant *Latyrus sativus*, although a conventional Spd synthase which utilized dSAM as the aminopropyl donor was also present. They partially purified an NADPH-dependent enzyme activity which synthesized carboxySpd from ^{14}C-Put in extracts of this plant. Recently, a bacterial carboxy*nor*Spd decarboxylase has been purified and characterized[101] and this enzyme also decarboxylated carboxySpd to form Spd. Perhaps a similar, broad-specificity carboxySpd decarboxylase will be found in plants. Interestingly, Spm was not synthesized by this alternate pathway in *Lathyrus* and it is not known whether similar mechanisms may exist for the synthesis of Spm in plant tissues.

IV. CONCLUDING REMARKS

The last few years have witnessed some remarkable advances in our understanding of polyamine biosynthesis in plants. Metabolic studies, aided by the availability of specific inhibitors, have elucidated the major routes for the synthesis of putrescine and the common polyamines, spermidine and spermine. Most of the enzymes of these pathways have been purified and biochemically characterized. Several laboratories are investigating molecular mechanisms regulating these enzymes *in vivo*. There is every reason to believe that the next few years will herald many exciting new discoveries in this area of research.

Nonetheless, reviews of this type always serve to remind the reader how much is *not* known about a particular subject and many questions remain unanswered. For example, to what extent do alternate pathways leading to Put or Spd synthesis play a role in polyamine biosynthesis? What determines the flux of C/N through the ADC vs. ODC pathway? Does Put derived from either pathway have unique and essential metabolic functions or can inhibition of either pathway lead to compensatory increases in the activity of the other in order to maintain intracellular polyamine levels? How is polyamine biosynthesis integrated into plant nitrogen and carbon metabolism as a whole? These and many other questions need to be addressed before a meaningful picture of plant polyamine metabolism will emerge.

ACKNOWLEDGMENTS

This work was supported, in part, by grant DCB-8806240 from the National Science Foundation.

REFERENCES

1. **Slocum, R. D., Kaur-Sawhney, R., and Galston, A. W.,** The physiology and biochemistry of polyamines in plants, *Arch. Biochem. Biophys.*, 235, 283, 1984.
2. **Adiga, P. R. and Prasad, G. L.,** Biosynthesis and regulation of polyamines in higher plants, *Plant Growth Regul.*, 3, 205, 1985.
3. **Smith, T. A.,** Polyamines, *Annu. Rev. Plant Physiol.*, 36, 117, 1985.
4. **Flores, H. E., Protacio, C. M., and Signs, M. W.,** Primary and secondary metabolism of polyamines in plants, in *Plant Nitrogen Metabolism, Recent Advances in Phytochemistry*, Vol. 23, Conn, E. E., Ed., Plenum Press, NY, 1989, 329.
5. **Suzuki, Y., Hirasawa, E., Yanagisawa, H., and Matsuda, H.,** The enzymes of polyamine metabolism in higher plants, in *Polyamines and Ethylene: Biochemistry, Physiology and Interactions*, Flores, H. E., Arteca, R. N., and Shannon, J. C., Eds., Amer. Soc. Plant Physiol., Rockville, MD, 1990, 73.
6. **Tabor, C. W. and Tabor, H.,** Polyamines in microorganisms, *Microbiol. Rev.*, 49, 81, 1985.
7. **Pegg, A. E.,** Recent advances in the biochemistry of polyamines in eukaryotes, *Biochem. J.*, 234, 249, 1986.
8. **Birecka, H., Birecki, M., Cohen, E. J., Bitonti, A. J., and McCann, P. P.,** Ornithine decarboxylase, polyamines, and pyrrolizidine alkaloids in *Senecio* and *Crotalaria*, *Plant Physiol.*, 86, 224, 1988.
9. **Slocum, R. D. and Galston, A. W.,** Changes in polyamine biosynthesis associated with postfertilization growth and development in tobacco ovary tissues, *Plant Physiol.*, 79, 336, 1985.
10. **Smith, T. A.,** Arginine decarboxylase of oat seedlings, *Phytochemistry*, 18, 1447, 1979.
11. **Choudhuri, M. M and Ghosh, B.,** Purification and partial characterization of arginine decarboxylase from rice embryos (*Oryza sativa* L.), *Agric. Biol. Chem.*, 46, 739, 1982.
12. **Prasad, G. L. and Adiga, P. R.,** Decarboxylation of arginine and ornithine by arginine decarboxylase purified from cucumber (*Cucumis sativus*) seedlings, *J. Biosci.*, 10, 203, 1986.
13. **Ramakrishna, S. and Adiga, P. R.,** Arginine decarboxylase from *Lathyrus sativus* seedlings. Purification and properties, *Eur. J. Biochem.*, 59, 377, 1975.
14. **Young, N. D. and Galston, N. D.,** Control of arginine decarboxylase biosynthesis in stressed oat leaf segments, in *Recent Progress in Polyamine Research*, Selmeci, L., Brosnan, M. E., and Seiler, N., Eds., Akademiai Kiadò, Budapest, 1985, 527.
15. **Morris, D. R. and Boeker, E. A.,** Biosynthetic and biodegradative ornithine and arginine decarboxylases in *Escherichia coli*, *Meth. Enz.*, 94, 125, 1983.
16. **Bell, E., and Malmberg, R. L.,** Analysis of a cDNA encoding arginine decarboxylase from oat reveals similarity to the *Escherichia coli* arginine decarboxylase and evidence of protein processing, *Mol. Gen. Genet.*, 224, 431, 1990.
17. **Torrigiani, P., Serafini-Fracassini, D., Biondi, S., and Bagni, N.,** Evidence for the subcellular localization of polyamines and their biosynthetic enzymes in plant cells, *J. Plant Physiol.*, 124, 23, 1986.
18. **Moore, R. C. and Boyle, S. M.,** Nucleotide sequence and analysis of the *speA* gene encoding biosynthetic arginine decarboxylase in *Escherichia coli*, *J. Bacteriol.*, 172, 4631, 1990.
19. **Koromilas, A. E. and Kyriakidis, D. A.,** Purification and properties of ornithine decarboxylase from germinated barley seeds, *Phytochemistry*, 27, 989, 1988.
20. **Pandit, M. and Ghosh, B.,** Ornithine decarboxylase from embryos of jute seeds, *Phytochemistry*, 27, 1609, 1988.
21. **Seely, J. E., Pösö, H., and Pegg, A. E.,** Purification of ornithine decarboxylase from kidneys of androgen-treated mice, *Biochemistry*, 21, 3394, 1982.
22. **Digangi, J. J., Seyfzadeh, M, and Davis, R. H.,** Ornithine decarboxylase from *Neurospora crassa*. Purification, characterization and regulation by inactivation, *J. Biol. Chem.*, 262, 7889, 1987.
23. **Heimer, Y. M. and Mizrahi, Y.,** Characterization of ornithine decarboxylase of tobacco cells and tomato ovaries, *Biochem. J.*, 201, 373, 1982.
24. **Panagiotidis, C. A., Georgatsos, J. G., and Kyriakidis, D. A.,** Superinduction of cytosolic and chromatin-bound ornithine decarboxylase activities of germinating barley seeds by actinomycin D, *FEBS Lett.*, 146, 193, 1982.

25. **Kallio, A., McCann, P. P., and Bey, P.,** DL-α-(difluoromethylarginine: A potent, enzyme-activated irreversible inhibitor of bacterial arginine decarboxylases, *Biochemistry*, 20, 3163, 1981.
26. **Bitonti, A. J., Casara, P. J., McCann, P. P., and Bey, P.,** Catalytic irreversible inhibition of bacterial and plant arginine decarboxylase activities by novel substrate and product analogs, *Biochem. J.*, 242, 69, 1987.
27. **Slocum, R. D., Bitonti, A. J., McCann, P. P. and Feirer, R. P.,** DL-α-Difluoromethyl[3,4-^{3}H]arginine metabolism in tobacco and mammalian cells. Inhibition of ornithine decarboxylase activity after arginase-mediated hydrolysis of DL-α-difluoromethylarginine to DL-α-difluoromethylornithine, *Biochem. J.*, 255, 197, 1988.
28. **Metcalf, B. W., Bey, P., Danzin, C., Jung, M. J., Casara, P., and Vevert, J. P.,** Catalytic irreversible inhibition of mammalian ornithine decarboxylase (EC 4.1.1.17) by substrate and product analogs, *J. Am. Chem. Soc.*, 100, 2551, 1978.
29. **Kallio, A. and McCann, P. P.,** Difluoromethylornithine irreversibly inactivates ornithine decarboxylase of *Pseudomonas aeruginosa*, but does not inhibit the enzymes of *Escherichia coli*, *Biochem. J.*, 200, 69, 1981.
30. **Hiatt, A. C., McIndoo, J., and Malmberg, R. L.,** Regulation of polyamine biosynthesis in tobacco. Effects of inhibitors and exogenous polyamines on arginine decarboxylase, ornithine decarboxylase, and *S*-adenosylmethionine decarboxylase, *J. Biol. Chem.*, 261, 1293, 1986.
31. **Suresh, M. R., Ramakrishna, S., and Adiga, P. R.,** 1978, Regulation of arginine decarboxylase and putrescine levels in *Cucumis sativus* cotyledons, *Phytochemistry*, 17, 57, 1978.
32. **Flores, H. E. and Galston, A. W.,** Osmotic stress-induced polyamine accumulation in cereal leaves. I. Physiological parameters of the response, *Plant Physiol.*, 75, 102, 1984.
33. **Slocum, R. D. and Weinstein, L. H.,** Stress-induced putrescine accumulation as a mechanism of ammonia detoxification in cereal leaves, in *Polyamines and Ethylene: Biochemistry, Physiology and Interactions*, Flores, H. E., Arteca, R. N., and Shannon, J. C., Eds., Am. Soc. Plant Physiol., Rockville, MD, 1990, 157.
34. **Montague, M. J., Armstrong, T. A., and Jaworski, E. G.,** Polyamine metabolism in embryogenic cells of *Daucus carota* II. Changes in arginine decarboxylase activity, *Plant Physiol.*, 63, 341, 1979.
35. **Kyriakidis, D. A.,** Effect of plant growth hormones and polyamines on ornithine decarboxylase activity during the germination of barley seeds, *Physiol. Plant.*, 57, 499, 1983.
36. **Koromilas, A. E. and Kyriakidis, D. A.,** The existence of ornithine decarboxylase-antizyme complex in germinated barley seeds, *Physiol. Plant.*, 72, 718, 1988.
37. **Bitonti, A. J., McCann, P. P., and Sjoerdsma, A.,** Restriction of bacterial growth by inhibition of polyamine biosynthesis by using monofluoromethylornithine, difluoromethylarginine and dicyclohexylammonium sulphate, *Biochem. J.*, 208, 435, 1982.
38. **Malmberg, R. L. and Hiatt, A. C.,** Polyamines in plant mutants, in *The Physiology of Polyamines*, Vol. II, Bachrach, U. and Heimer, Y. M., Eds., CRC Press, Boca Raton, FL, 1989, 147.
39. **Smith, T. A. and Richards, F. J.,** The biosynthesis of putrescine in higher plants and its relation to potassium nutrition, *Biochem. J.*, 84, 292, 1962.
40. **Yoshida, D. and Mitake, T.,** Agmatine and *N*-carbamylputrescine as intermediates in the formation of nicotine by tobacco plants, *Plant Cell Physiol.*, 7, 301, 1966.
41. **LeRudulier, D. and Goas, G.,** Biogenèse de la *N*-carbamoylputrescine et de la putrescine dans les plantules de *Glycine max* (L.) Merr., *Physiol. Vég.*, 18, 609, 1980.
42. **Sindhu, R. K. and Desai, H. V.,** Purification and properties of agmatine iminohydrolase from groundnut cotyledons, *Phytochemistry*, 18, 1937, 1979.
43. **Yanagisawa, H. and Suzuki, Y.,** Corn agmatine iminohydrolase, *Plant Physiol.*, 67, 697, 1981.
44. **Smith, T. A.,** Agmatine iminohydrolase in maize, *Phytochemistry*, 8, 2111, 1969.
45. **Choudhuri, M. M. and Ghosh, B.,** Agmatine deiminase in rice seedlings, *Phytochemistry*, 24, 2433, 1985.
46. **Smith, T. A.,** The biosynthesis and metabolism of putrescine in higher plants, *Ann. N. Y. Acad. Sci.*, 171, 988, 1970.
47. **Smith, T. A. and Garraway, J. L.,** *N*-Carbamylputrescine - an intermediate in the formation of putrescine in barley, *Phytochemistry*, 3, 23, 1964.
48. **Smith, T. A.,** *N*-Carbamylputrescine amidohydrolase of higher plants and its relation to potassium nutrition, *Phytochemistry*, 4, 599, 1965.
49. **Yanagisawa, H. and Suzuki, Y.,** Purification and properties of *N*-carbamylputrescine amidohydrolase from maize shoots, *Phytochemistry*, 21, 2201, 1982.
50. **Wargnies, B, Lauwers, N., and Stalon, V.,** Structure and properties of the putrescine carbamoyltransferase of *Streptococcus faecalis*, *Eur. J. Biochem.*, 101, 143, 1979.
51. **Kleczkowski, K. and Wielgat, B.,** Carbamoylation of putrescine in plant material, *Bull. Acad. Pol. Sci.*, 16, 521, 1968.
52. **Srivenugopal, K. S. and Adiga, P. R.,** Enzymic conversion of agmatine to putrescine in *Latyrus sativus* seedlings. Purification and properties of a multifunctional enzyme (putrescine synthase), *J. Biol. Chem.*, 256, 9532, 1981.

53. **Prasad, G. L. and Adiga, P. R.,** Purification and characterization of putrescine synthase from cucumber seedlings. A multifunctional enzyme involved in putrescine biosynthesis, *J. Biosci.*, 10, 373, 1986.
54. **Prasad, G. L. and Adiga, P. R.,** Arginine decarboxylase is a component activity of the multifunctional enzyme putrescine synthase in cucumber seedlings, *J. Biosci.*, 11, 571, 1987.
55. **Bakhanashvili, M. Icekson, I., and Apelbaum, A.,** Cadaverine formation by specific lysine decarboxylation in *Pisum sativum* seedlings, *Plant Cell Rept.*, 4, 297, 1985.
56. **Bagni, N., Creus, J., and Pistocchi, R.,** Distribution of cadaverine and lysine decarboxylase activity in *Nicotiana glauca* plants, *J. Plant Physiol.*, 125, 9, 1986.
57. **Pelosi, L. R., Rother, A., and Edwards, J. M.,** Lysine decarboxylase activity and alkaloid production in *Heimia salicifolia* cultures, *Phytochemistry*, 25, 2315, 1986.
58. **Schoofs, G., Teichman, S., Hartmann, T. and Wink, M.,** Lysine decarboxylase in plants and its integration in quinolizidine alkaloid biosynthesis, *Phytochemistry*, 22, 65, 1983.
59. **Beier, H., Fecker, L. F., and Berlin, J.,** Lysine decarboxylase from *Hafnia alvei*: purification and molecular data, and preparation of polyclonal antibodies, *Z. Naturforsch.*, 42c, 1307, 1987.
60. **Icekson, I., Bakhanashvili, M., and Apelbaum, A.,** Inhibition by ethylene of polyamine biosynthetic enzymes, enhanced lysine decarboxylase activity and cadaverine accumulation in pea seedlings, *Plant Physiol.*, 82, 607, 1985.
61. **Wertheimer, S. J. and Leifer, Z.,** Putrescine and spermidine sensitivity of lysine decarboxylase in *Escherichia coli*: evidence for a constitutive enzyme and its mode of regulation, *Biochem. Biophys. Res. Commun.*, 114, 882, 1983.
62. **Paulus, T. J., Kiyono, P., and Davis, R. H.,** Polyamine-deficient *Neurospora crassa* mutants and synthesis of cadaverine, *J. Bacteriol.*, 152, 291, 1982.
63. **Slocum, R. D. and Galston, A. W.,** Inhibition of polyamine biosynthesis in plants and plant pathogenic fungi, in *Inhibition of Polyamine Metabolism*, McCann, P. P., Pegg, A. E., and Sjoerdsma, A., Eds., Academic Press, Orlando, FL, 1987, 305.
64. **Pösö, H., McCann, P. P., Tanskanen, R. Bey, P., and Sjoerdsma, A.,** Inhibition of growth of *Mycoplasma dispar* by DL-α-difluoromethyllysine, a selective irreversible inhibitor of lysine decarboxylase, and reversal by cadaverine (1,5-diaminopentane), *Biochem. Biophys. Res. Commun.*, 125, 205, 1984.
65. **Birecka, H., Birecki, M., Bitonti, A. J., and McCann, P. P.,** Inhibition of arginine, ornithine and lysine decarboxylases in pea seedlings, *Phytochemistry*, 30, 99, 1991.
66. **Ramakrishna, S. and Adiga, P. R.,** Decarboxylation of homoarginine and lysine by an enzyme from *Lathyrus sativus* seedlings, *Phytochemistry*, 5, 83, 1976.
67. **Satishchandran, C. and Boyle, S. M.,** Purification and properties of agmatine ureohydrolase, a putrescine biosynthetic enzyme in *Escherichia coli*, *J. Bacteriol.*, 165, 843, 1986.
68. **Akamatsu, N., Oguchi, M., Yajima, Y., and Ohuo, M.,** Formation of *N*-carbamoylputrescine from citrulline in *Escherichia coli*, *J. Bacteriol.*, 133, 409, 1978.
69. **Maretzki, A., Thom, M., and Nickell, L. G.,** Products of arginine catabolism in growing cells of sugarcane, *Phytochemistry*, 8, 811, 1969.
70. **Speranza, A. and Bagni, N.,** Products of L-[^{14}C-carbamoyl]-citrulline metabolism in *Helianthus tuberosus* activated tissue, *Z. Pflanzenphysiol.*, 88, 163, 1978.
71. **Crocomo, O. J., Basso, L. C., and Brasil, O. G.,** Formation of *N*-carbamoylputrescine from citrulline in *Sesamum*, *Phytochemistry*, 9, 1487, 1970.
72. **Even-Chen, Z., Mattoo, A. K., and Goren, R.,** Inhibition of ethylene biosynthesis by aminoethoxyvinylglycine and by polyamines shunts label from 3,4[^{14}C]-methionine into spermidine in aged orange peel discs, *Plant Physiol.*, 69, 385, 1982.
73. **Greenberg, M. L. and Cohen, S. S.,** Dicyclohexylamine-induced shift of biosynthesis from spermidine to spermine in plant protoplasts, *Plant Physiol.*, 78, 568, 1985.
74. **Katoh, Y. and Hasegawa, T.,** Effects of dicyclohexylamine on the changes in spermine levels in Hiproly Barley callus after auxin withdrawal, *Agric. Biol. Chem.*, 53, 1485, 1989.
75. **Suzuki, Y. and Hirasawa, E.,** *S*-Adenosylmethionine decarboxylase of corn seedlings, *Plant Physiol.*, 66, 1091, 1980.
76. **Williams-Ashman, H. G. and Schenone, A.,** Methylglyoxal *bis*(guanylhydrazone) as a potent inhibitor of mammalian and yeast *S*-adenosylmethionine decarboxylases, *Biochem. Biophys. Res. Commun.*, 46, 288, 1972.
77. **Pegg, A. E.,** Inhibitors of ornithine and S-adenosylmethionine decarboxylases, in *The Physiology of Polyamines*, Vol. I, Bachrach, U. and Heimer, Y. M., Eds., CRC Press, Boca Raton, FL, 1989, 303.
78. **Yamanoha, B. and Cohen, S. S.,** *S*-Adenosylmethionine decarboxylase and spermidine synthase from Chinese cabbage, *Plant Physiol.*, 78, 784, 1985.
79. **Walker, M. A., Ellis, B. E., Chapple, C. C. S,. and Dumbroff, E. B.,** Subcellular localization of amines and activities of their biosynthetic enzymes in *p*-fluorophenylalanine resistant and wild type tobacco cell cultures, *Plant Physiol.*, 85, 78, 1987.

80. **Khomutov, R. M, Hyvönen, T., Karvonen, E., Kauppinen, L., Paalanen, T., Paulin, L., Eloranta, T., Pajula, R.-L., Andersson, L. C., and Pösö, H.,** 1-Aminooxy-3-aminopropane, a new and potent inhibitor of polyamine biosynthesis that inhibits ornithine decarboxylase, *S*-adenosylmethionine decarboxylase and spermidine synthase, *Biochem. Biophys. Res. Commun.*, 130, 596, 1985.
81. **Christ, M., Felix, H., and Harr, J.,** Inhibitors influencing plant enzymes of the polyamine biosynthetic pathway, *Z. Naturforsch.*, 44, 49, 1989.
82. **Pegg, A. E., Corti, A., and Williams-Ashman, H. G.,** Paradoxical enhancement of *S*-adenosylmethionine decarboxylase in rat tissue following administration of the specific inhibitor methyl glyoxal *bis*(guanylhydrazone), *Biochem. Biophys. Res. Commun.*, 52, 696, 1973.
83. **Pistocchi, R., Bagni, N., and Creus, J. A.,** Polyamine uptake in carrot cell cultures, *Plant Physiol.*, 84, 374, 1987.
84. **Byers, T. L. and Pegg, A. E.,** Properties and physiological function of the polyamine transport system, *Am. J. Physiol.*, 257, C545, 1989.
85. **Lin, P. P. C.,** Effects of methyl glyoxal-*bis*-(guanylhydrazone) and abscisic acid on polyamine metabolism in embryonectomized barley seeds, *Plant Growth Regul.*, 3, 257, 1985.
86. **Pegg, A. E.,** *S*-Adenosylmethionine decarboxylase: a brief review, *Cell Biochem. Funct.*, 2, 11, 1984.
87. **Pajunen, A., Crozat, A., Janne, O. A., Ihalianen, R., Laitinen, P. H., Stanley, B., Madhubala, R., and Pegg, A. E.,** Structure and regulation of mammalian *S*-adenosylmethionine decarboxylase, *J. Biol. Chem.*, 263, 17,040, 1988.
88. **Bagni, N. and Pistocchi, R.,** Putrescine uptake in *Saintpaulia* petals, *Plant Physiol.*, 77, 398, 1985.
89. **Suresh, M. R. and Adiga, P. R.,** Putrescine-sensitive (artifactual) and insensitive (biosynthetic) *S*-adenosylmethionine decarboxylase activities of *Lathyrus sativus* seedlings, *Eur. J. Biochem.*, 79, 511, 1977.
90. **Baxter, C. and Coscia, C. J.,** *In vitro* synthesis of spermidine in the higher plant *Vinca rosea*, *Biochem. Biophys. Res. Commun.*, 54, 147, 1973.
91. **Hirasawa, E. and Suzuki, Y.,** Biosynthesis of spermidine in maize seedlings, *Phytochemistry*, 22, 103, 1983.
92. **Woster, P. M., Black, A. Y., Duff, K. J., Coward, J. K., and Pegg, A. E.,** Synthesis and biological evaluation of *S*-adenosyl-1, 12-diamino-3-thio-9-azadodecane, a multisubstrate adduct inhibitor of spermine synthase, *J. Med. Chem.*, 32, 1300, 1989.
93. **Sindhu, R. K. and Cohen, S. S.,** Propylamine transferases in Chinese cabbage leaves, *Plant Physiol.*, 74, 645, 1984.
94. **Cohen, S. S., Balint, R., and Sindhu, R. K.,** The synthesis of polyamines from methionine in intact and disrupted leaf protoplasts of virus-infected Chinese cabbage, *Plant Physiol.*, 68, 1150, 1981.
95. **Sindhu, R. K. and Cohen, S. S.,** Subcellular localization of spermidine synthase in the protoplasts of Chinese cabbage leaves, *Plant Physiol.*, 76, 219, 1984.
96. **Villaneuva, V. R., Adlakha, R. C., and Calvayrac, R.,** Biosynthesis of polyamines in *Euglena gracilis*, *Phytochemistry*, 19, 787, 1980.
97. **Batchelor, K. W., Smith, R. A., and Watson, N. S.,** Dicyclohexylamine is not an inhibitor of spermidine synthase, *Biochem. J.*, 233, 307, 1986.
98. **Biondi, S., Bagni, N., and Sansovini, A.,** Dicyclohexylamine uptake and effects on polyamine content in cultured cotyledons of radiata pine, *Physiol. Plant.*, 66, 41, 1986.
99. **Pegg, A. E., Bitonti, A. J., McCann, P. P., and Coward, J. K.,** Inhibition of bacterial aminopropyltransferases by *S*-adenosyl-1,8-diamino-3-thiooctane and by dicyclohexylamine, *FEBS Lett.*, 155, 192, 1983.
100. **Srivenugopal, K. S. and Adiga, P. R.,** Coexistence of two pathways of spermidine biosynthesis in *Lathyrus sativus* seedlings, *FEBS Lett.*, 112, 260, 1980.
101. **Nakao, H., Shinoda, S., and Yamamoto, S.,** Purification and properties of carboxynorspermidine decarboxylase, a novel enzyme involved in norspermidine biosynthesis from *Vibrio alginolyticus*, *J. Gen. Microbiol.*, 136, 1699, 1990.

Chapter 3

POLYAMINE CATABOLISM IN PLANTS

Rodolfo Federico and Riccardo Angelini

TABLE OF CONTENTS

I. INTRODUCTION

The ubiquitous occurrence of aliphatic polyamines in procaryotic and eucaryotic organisms and their implication in growth and key developmental processes has engaged the interest of an increasing number of researchers during the last 20 years.[1-3] Polyamines have been tentatively proposed as a new class of plant growth substances, and a role for these compounds has been postulated in cell division, macromolecular synthesis, stress responses, and senescence.[3,4] In this context, attention has been mostly drawn to the analysis of the regulation of polyamine levels and biosynthetic enzyme activities in a particular physiological event or as a response to external stimuli. In contrast, little attention has been devoted to the study of polyamine catabolism and, only recently, the biochemical characterization and the physiological modulations of enzymes involved in this process have been studied in detail. Far from being only a means of eliminating cellular polyamines, the enzymes involved in polyamine catabolism, and the products deriving from their action, have been demonstrated to be involved in important physiological processes, opening up an exciting new area of biochemical and physiological researches.

II. POLYAMINE CATABOLISM: OXIDATIVE DEAMINATION

Polyamines* are oxidatively deaminated by the action of amine oxidases (AOs), enzymes widespread in bacteria, fungi, animals, and plants.[2,3,5-7] These enzymes are formally grouped as E.C. 1.4.3.6. amine oxidase (copper containing) and E.C. 1.4.3.4. amine oxidase (flavin containing). AOs obtained from several sources exhibit different substrate specificity and pattern of substrate degradation (Figure 1). An alternative grouping would be to divide AOs into those that act at the primary amino group of di- and polyamines and those that act at the secondary amino group of polyamines. As suggested by Morgan,[2] the latter enzyme family should be further subdivided according to whether 1,3-diaminopropane or 3-aminopropionaldehyde are the reaction products (Figure 1).

A specific feature of animal and fungal AOs is their activity towards acetyl derivatives of polyamines which, in some instances, may represent their physiological substrates *in vivo*.[8] In this context, a role in the terminal catabolism of polyamines has been ascribed to copper-containing AOs which catalyze the transformation of their substrates into products which cannot be converted again to other polyamines. Diversely, animal polyamine oxidases transform N^1-acetyl-spermidine and N^1-acetyl-spermine into putrescine and spermidine, respectively, plus 3-acetamidopropionaldehyde, which has been termed an interconversion pathway of polyamine degradation.[1,8] A notable exception to this rule is represented by plant polyamine oxidases which deaminate spermidine and spermine to the corresponding aminoaldehydes with the formation of 1,3-diaminopropane.[5,6]

In the the following sections attention will be focused on copper-containing diamine oxidases (DAOs) and flavin-containing polyamine oxidases (PAOs)** of higher plants which occur at a high specific activity in the Leguminosae and Gramineae, respectively. PAOs are also present in other taxonomic groups, such as the Compositae, Cucurbitaceae, Solanaceae, Euphorbiaceae, Gramineae (DAO), and Pontederaceae (PAO).[5,6]

III. DIAMINE OXIDASE

A. GENERAL AND MOLECULAR FEATURES

DAOs (EC 1.4.3.6.) have been purified and characterized principally from some leguminous

* The general term "polyamine" is hereby used to include both diamines and polyamines.

** This notation, based on substrate specificity, is hereby used in accordance with earlier literature, although this property is not so evident for some plant DAO's.

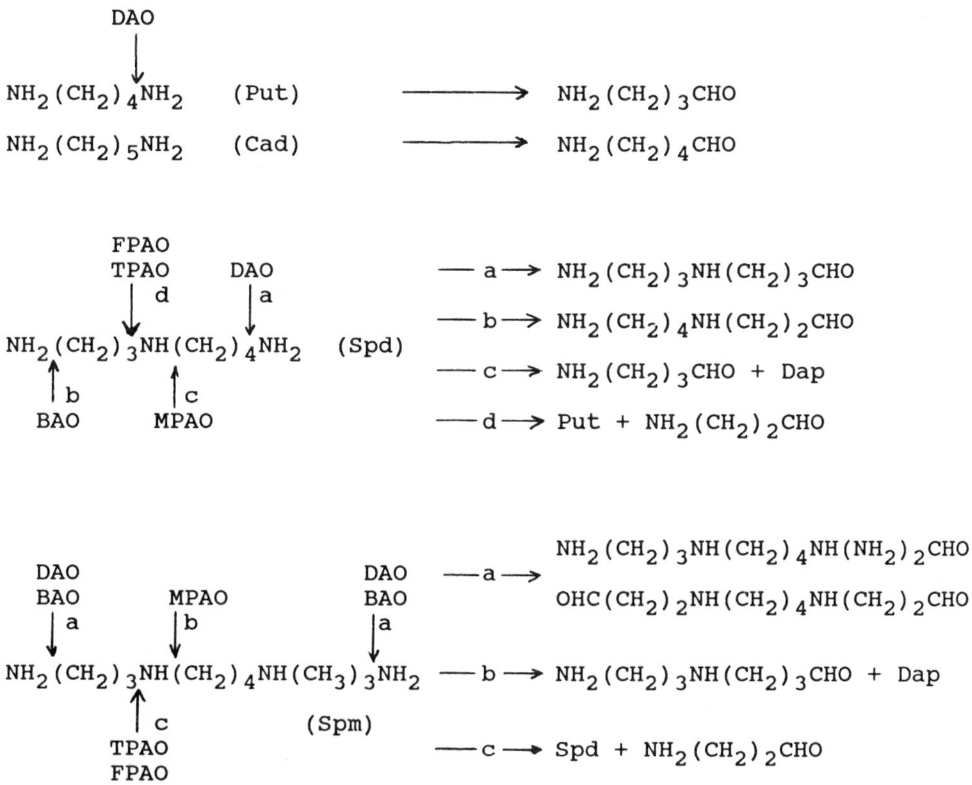

FIGURE 1. Oxidative cleavage of polyamines by amine oxidase from different organisms. Site of attack by the enzymes and aminoaldehydes produced are indicated for putrescine (Put), cadaverine (Cad), spermidine (Spd), and spermine (Spm). NH_3 (in the case of deamination of a primary amino group) and H_2O_2 are co-products of the reactions. DAO, diamine oxidase (lentil seedlings); BAO, bovine serum amine oxidase; TPAO, tissue polyamine oxidase (rat liver); MPAO, maize polyamine oxidase; FPAO, fungal polyamine oxidase (*Aspergillus terreus*, *Penicillium chrysogenum*), for a review see references 2,6.

species in which the enzyme occurs at a high concentration (up to 3% of total protein content for DAO of etiolated pea stem). Specific activity of plant DAOs exceeds that of animal sources by more than 100-fold.[5,9] DAOs oxidize the primary aminogroup of diamines (putrescine, Put; cadaverine, Cad) and spermidine (Spd) with the formation of 4-aminobutyraldehyde (from Put), 5-aminovaleraldehyde (from Cad) and 3-aminopropyl-4-aminobutyraldehyde (from Spd) along with H_2O_2 and ammonia.[3,5,6] Aldehydes produced during the reaction spontaneously cyclize to Δ^1-pyrroline, Δ^1-piperideine and 1,5-diazabicyclononane, respectively (Figure 2).[10,11] It has been recently demonstrated that spermine (Spm) is oxidized by lentil DAO (pH optimum 7.9) at the terminal amino groups producing a dialdehyde.[12] *In vitro* experiments showed that animal and plant DAOs convert the amine substrates to the corresponding aldehydes in absence of oxygen. Ammonia and hydrogen peroxide are released upon reoxidation of the reduced enzyme, this step requiring oxygen or other suitable electron acceptors.[13-15] ESR analysis and low temperature optical spectra showed that any valency changes of the copper are apparent during the catalytic cycle.[13,16]

Plant DAOs have been extracted from several sources: *Pisum sativum*,[9] *Arachis hypogea*,[17] *Glycine max*,[18] *Lathyrus sativus*,[19,102] *Lens culinaris*,[20] *Lupinus luteus*,[21] *Vicia faba*,[22] *Cicer arietinum*,[23] *Trifolium subterraneum*,[24] *Canavalia ensiformis*,[25] *Onobrychis viciifolia*,[26] and *Phaseolus vulgaris*.[103] Among nonleguminous plants, enzymes able to oxidize diamines have been found in: *Euphorbia characias*,[27] *Heliantus tuberosus*,[28] *Cucurbita maxima*,[29] *Ozyza*

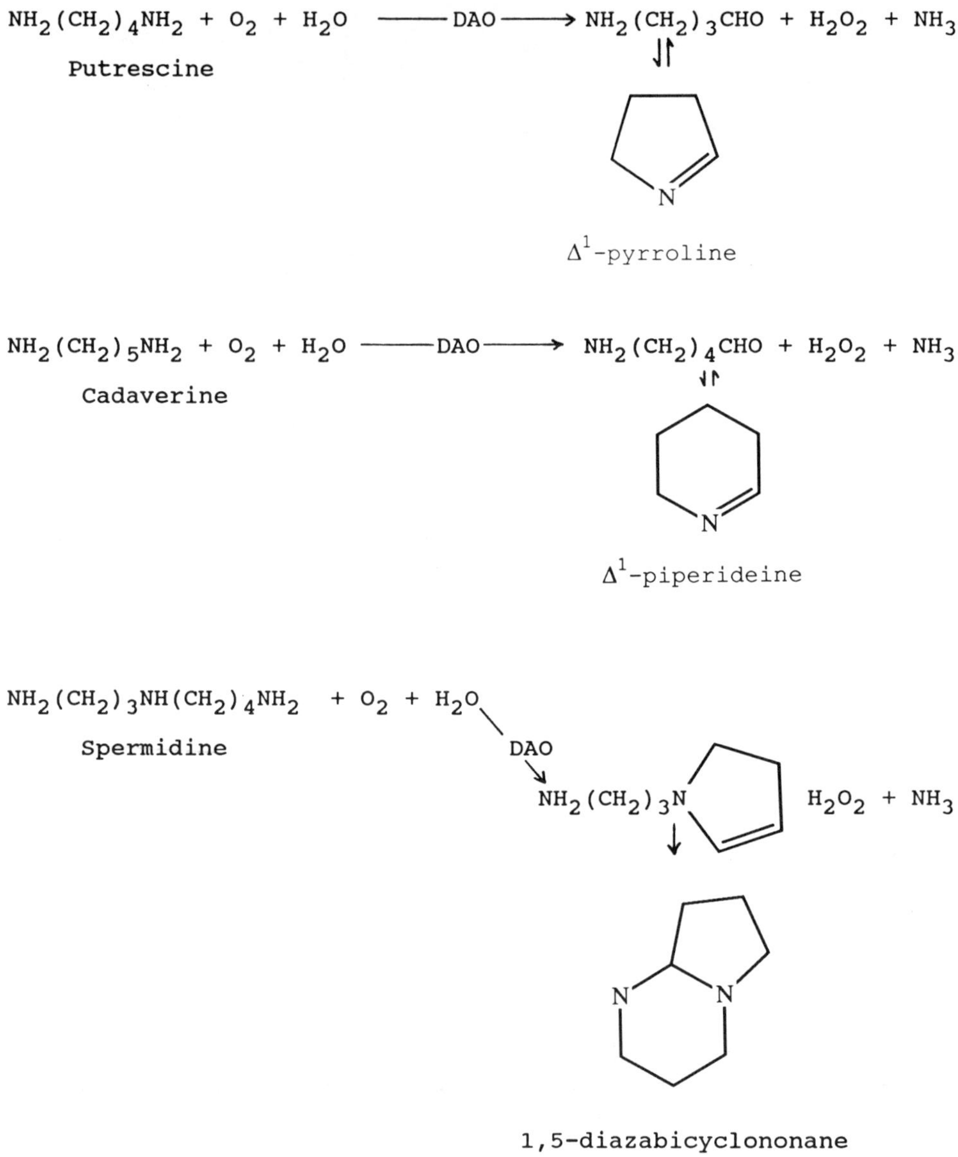

FIGURE 2. Oxidative deamination of polyamines catalized by plant diamine oxidase. Aminoaldehydes produced in the reaction cyclize spontaneously. Spermine is oxidized by lentil DAO at the terminal amino-groups producing a dialdehyde.

sativa,[30] *Triticum aestivum*,[31] *Nicotiana tabacum*,[104,105] *Hordeum vulgare*,[34] and *Hyoscyamus niger*.[106] The enzymes from the stems of *Pisum*,[9] *Lens*,[20] *Cicer*,[23] from the leaves of *Vicia*,[22] *Trifolium*,[24] *Hordeum*,[34] *Oryza*,[30] from the seedlings of *Onobrychis*,[26] *Latyrus sativus*,[19,102] *Lathyrus cicera*,[103] *Phaseolus*,[103] from the roots of *Nicotiana tabacum*,[104,105] *Hyoscyamus*,[106] and from the latex of *Euphorbia*[27] have been purified to homogeneity and partially characterized.[27]

Plant DAOs are generally dimeric glycoproteins formed by similar subunits, containing two copper (Cu II) atoms and one carbonyl residue essential for enzyme activity per dimer. DAO from *Euphorbia characias* latex shows two carbonyl-like groups and two essential SH- groups per dimer.[27,34] DAO from *Oryza sativa* shows unusual characteristics as copper is apparently absent and the enzyme contains four molecules of FAD per dimer.[30] The nature of covalently bound cofactor in eukaryotic copper amine oxidases has been investigated for several years. Initially, pyridoxal phosphate was suggested as cofactor. In subsequent studies, pyrroloquinolinequinone (PQQ), which was previously revealed as cofactor in some bacterial dehydrogenases and bovine serum amine oxidases, has been proposed as the carbonyl cofactor for plant DAOs.[35] More recently Janes et al.[36] have suggested that 6-hydrohydopa and not PQQ is the active site cofactor in mammalian and, perhaps, all eukaryotic DAOs. Electrofocusing techniques have evidenced a marked microheterogeneity of lentil DAO which shows at least six to seven subforms. The electrophoretic pattern is modificable by a glycohydrolase pretreatment of the enzyme.[37] Some molecular and general features of plant DAOs are summarized in Table 1.

B. SPECTROSCOPIC FEATURES

The visible spectrum of plant DAOs is characterized by a broad absorption, peaking between 490 to 525 nm (with a 200 nm width at half height) which confers a typical pink color to purified preparations of the enzyme.[38] Removal of the copper under nonreducing conditions results in the same visible spectrum which has thus been ascribed to the carbonyl cofactor.[14] This shows several electronic transitions as demonstrated by CD analysis of the lentil enzyme.[7]

Spectroscopic studies on animal and plant DAOs have provided useful insight into the structural and functional organization of the prosthetic cofactors within the enzyme. EPR analysis has shown that all the metal is in Cu (II) state with no spin or dipolar interaction between the two atoms (≥14 Å apart).[16] The addition of substrates to lentil and pea DAO bleaches absorption at 500 nm and, if the reaction is performed in anaerobiosis, a stable yellow intermediate absorbing at 350 to 360, 430, and 460 nm is observable.[13,14] Shape and position, but not relative intensities of absorption bands, are independent of the different substrates used.[14] This phenomenon results in a minimum change of the copper EPR spectrum of the native enzyme which, in the case of pea DAO, has been ascribed to a change in the binding situation of copper.[39,40] N_3^- and CN^- strongly modify copper EPR spectrum of pig plasma amine oxidase without any observable changes of the visible spectrum.[41] Copper removal experiments and the above mentioned findings suggest that copper is not associated with the organic factor.[16,38] However copper-free lentil DAO does not form the yellow intermediate upon addition of substrates.[14] On addition of air, pink color is restored in copper-containing but not in copper-free enzyme, thus showing that the metal is essential for the reoxidation step.[14,16,42] Dithionite-cyanide treatment on bovine serum AO results in a copper-free reduced enzyme which on addition of Co (II) or Ni (II) restores its original optical spectrum. Ni removal by treatment with dimethylglyoxime and CN^- does not change the absorption at 500 nm; on the other hand, only copper is suitable for reoxidation of the substrate-bleached enzyme.[42] It has been suggested that metal binding induces a conformational change of the enzyme moiety which results in an easier attack of the dithionite-reduced organic cofactor by oxygen.[16] However, only copper is able to facilitate electron transfer from substrate-bleached cofactor to oxygen.[43]

C. SUBSTRATE SPECIFICITY AND CATALYTIC ACTIVITY PARAMETERS

Plant DAOs are most active with the diamines Put and Cad as substrates, although in some species enzymes with a broad substrate specificity occur, including spermidine, spermine, aliphatic and aromatic monoamines, lysine, ornithine, agmatine, histamine.[7,9,23,24,30] Notably, DAO from *Euphorbia characias* latex shows a very restricted substrate specificity oxidizing mainly Put and Cad.[27] The rate of oxidation of Put, Spd, and Spm by lentil DAO (measured at their respective pH optima) are in the ratio 100:42:20. K_m values for Put of both pea and lentil DAO is in the 10^{-4} to 10^{-5} *M* range with specific activity up to 70 units/mg protein (pea DAO) which is considerably higher than that found for animal AOs.[9] The pH optimum for enzyme activity with Put as substrate is 7 for most plant DAOs and it has been reported to be substrate dependent.[7,9,23,24,30]

TABLE 1
Molecular and Catalytic Properties of Plant DAOs*

Properties	Lens[20] culinaris	Pisum[9] sativum	Vicia[22] faba	Trifolium[24] subterraneum	Cicer[23] arietinum	Onobrychis viciifolia	Lathyrus sativus
Molecular weight (Da)							
Gel chromatography	—	180,000	126,000	150,000	—	—	148,000
SDS-PAGE	78,000	85,000	74,000	80,000	—	25,000	75,000
Cu content (g atom·mol^{-1})	2	2	2	2	—	—	—
Isoelectric point	7.4	7.4	—	—	—	8–9	—
Carbohydrate content (neutral sugar %)	14	13	—	—	—	—	0
pH Optimum	7.1	7.0	7.0	7.5	7.5	7.7	8.4
K_m (*M*)	2.3×10^{-4}	8.2×10^{-5}	3.7×10^{-5}	3.8×10^{-4}	2.4×10^{-4}	7.2×10^{-4}	—
Specific activity (u·mg protein^{-1})	63	70	22	66	59	25	1.9

Properties	Lathyrus cicera	Phaseolus vulgaris[103]	Euphorbia[27] characias	Oryza[30] sativa	Hordeum[33] vulgare	Nicotiana tabacum[104,105]	Hyoscyamus niger[106]
Molecular weight (Da)							
Gel chromatography	150,000	150,000	140,000	123,000	150,000	—	135,000
SDS-PAGE	75,000	75,000	72,000	61,000	75,000	70,000	66,000
Cu content (g atom·mol^{-1})	2	2	2	0	2	—	—
Isoelectric point	—	—	5.2	—	6.8	—	—
Carbohydrate content (neutral sugar %)	14	14	12	—	14	—	—
pH Optimum	7.0	7.5	7.1	7.5	7.5	8.0	7.5–8.0*
K_m (*M*)	3.2×10^{-4}	7×10^{-3}	2×10^{-4}	10^{-5}	—	4.5×10^{-4}* (3.3×10^{-4})*	2.85×10^{-3}
Specific activity (u·mg protein^{-1})	51	6.7	27	0.017	7.4	0.16	0.33*

Note: Catalytic parameters were determined at 37°C in 0.1 *M* potassium phosphate buffer with Put as substrate, except for the *Vicia*, *Oryza*, *Trifolium*, *Onobrychis*, *Hyoscyamus*, and *Nicotiana* enzymes whose catalytic parameters were determined at 30°C.

* *N*-Methylputrescine as substrate.

D. INHIBITORS

Plant and animal AOs are all inhibited by metal chelators and carbonyl-directed reagents as well as by other classes of inhibitors with characteristic features depending on the specific compound and enzyme sensitivity.[7,44] Hydrazines and hydrazides form irreversible adducts with all AOs and have represented a powerful tool in establishing the carbonyl nature of the prostethic group of these ezymes. In particular phenylhydrazine irreversibly and completely inhibits beef plasma and lentil DAO with a $K_i = 1.5 \times 10^{-9}$ *M* forming an adduct with characteristic spectroscopic features ($A_{max} = 430$–440 nm with $\varepsilon = 38$–$44 \times 10^3 \cdot M^{-1} \cdot cm^{-1}$).[45] These properties have been exploited in titration studies for the stoichiometry determination of the catalyzed reaction (see below). Noncompetitive inhibition of lentil DAO has been observed for sodium cyanide ($K_i = 3 \times 10^{-3}$ *M*), diethyldithiocarbamate (DDC; $K_i = 4.5 \times 10^{-5}$ *M*), and sodium azide ($K_i = 5 \times 10^{-2}$ *M*).[20] While DDC seems to behave as true copper chelator and an insoluble DDC-Cu complex has been separated from pea DAO with complete loss of enzyme activity, N_3^- and CN^- act as ligand substitution reagents probably displacing water at the coordination sites.[13,16,44]

In titration experiments with 4,4′-dithiopyridine, DAO from *Euphorbia characias* latex was inhibited only if amine was added anaerobically or if the enzyme was denatured in 8 *M* urea. In this experimental condition, two sulfhydryl groups per dimer were titrated with 98% activity inhibition. *N*-ethylmaleimide gives a similar behavior but results in a 45% activity inhibition. Native enzyme is unreactive showing that -SH groups are not available for thiol reagents under normal conditions.[34]

A number of suicide inhibitors have been reported for animal DAOs typically containing unsaturated bonds such as 1,4-diamino-2-butyne, 1-amino-2-propyne, 2-chloroallylamine, cis-imidazolylallylamine. These compounds lead to the formation of reactive allenes on the active site which irreversibly react with unknown nucleophile groups of the enzyme.[44] The use of these compounds with plant enzymes is currently under study.[46] Macholan and Minar reported that 1,4-diamino-2-butanone and 1,5-diamino-3-pentanone, which probably act as reversible inhibitors ($K_i = 1.5 \times 10^{-8}$ *M*), are able to inhibit DAO activity *in vivo* in pea seedlings.[47]

Aminoguanidine and related compounds are effective inhibitors of AO activity although their marked effect on membrane permeability makes these reagents unsuitable for *in vivo* studies.[44,48] Imidazole derivatives such as carnosine, anserine, and impromidine, with both imidazole ring and aliphatic structure resembling Put, act as noncompetitive inhibitors of pea DAO.[49] Methylgliossal-*bis*-guanylhydrazone, which was originally used as an inhibitor of spermine biosynthesis in animal systems, acts as a powerful DAO ihibitor.[9]

Plant and animal DAOs are inactivated by one of their products, H_2O_2, and this effect may be prevented by the action of catalase. Interestingly, the inactivation of DAO from *Euphorbia characias* latex by hydrogen peroxide parallels the modification of the reactive -SH groups.[34] The occurrence of natural DAO inhibitors has been reported in *Arachis hypogea* and *Sorghum vulgare*.[50,51]

E. STOICHIOMETRY AND MECHANISM OF THE DAO CATALYZED REACTION

Titration of lentil DAO with substrate under anaerobic conditions (only one turnover act, as the enzyme cannot be reoxidized) results in two active sites per dimer. Discrepancy exists on titration with phenyhydrazine which results in one active site per dimer.[45] All DAOs follow a double displacement mechanism in which aldehyde is released in absence of air while ammonia and H_2O_2 are released upon reaction with O_2. Performing the reaction with plant DAOs at subzero temperatures or in anaerobiosis, it is possible to observe the yellow intermediate (pseudo-reduced enzyme), of unknown chemical nature.[13,14,52] Stopped-flow experiments performed with *p*-dimethylaminomethylbenzilamine as a substrate for lentil DAO have shown that the enzyme is rapidly bleached (decrease in absorbance at 496 nm) by the substrate in the absence of oxygen, with a complex kinetic process which appears bimolecular

($K = 2x10^5 \cdot M^{-1} \cdot s^{-1}$) in the initial phase and independent of substrate concentration in the subsequent slower phase. The latter has a very similar rate to that measured for the formation of the yellow intermediate. The aldehyde appears to be released after the formation of the yellow intermediate. The reoxidation step is very fast with a second-order rate constant of $2x10^7 \cdot M^{-1} \cdot s^{-1}$ and is accompained by the release of ammonia and H_2O_2.[52] Surprisingly, aldehyde is not released under anaerobic conditions by lentil DAO with Put as substrate.[14] Copper-free lentil DAO reacts with substrate in the presence of oxygen-producing aldehyde and ammonia indicating that the copper-free enzyme still binds oxygen but that the metal cofactor is essential for the release of H_2O_2 in the reoxidation step.[14] In this context, it has been suggested that copper catalyzes oxygen reduction in animal DAO by a mechanism in which a hydroxyl coordinated to copper acts as a nucleophile to facilitate hydride ion transfer to oxygen.[43]

F. PHYSIOLOGICAL ASPECTS

With few exceptions, physiological studies concerning changes of DAO activity in developmental processes have been neglected in earlier research on polyamine functions in plant systems. Werle and Pechmann[53] and Kenten and Mann[54] first reported that resting seeds of leguminous plants were unable to oxidize diamines, while DAO activity appeared during germination; lower activity was found in the adult plant. Werle et al.[55] reported that DAO was unevenly distributed along the pea stem, the enzyme activity being high, on fresh weight basis, in the basal and apical internodes and very low in the intermediate ones. The finding that DAO attacked tryptamine, and the occurrence of aldehyde oxidase activity in extracts of pea seedlings, suggested that the two enzymes could be involved in IAA biosynthesis.[54] This hypothesis was later abandoned on the basis of the kinetic parameters of DAO-catalyzed tryptamine oxidation.[56] Suzuki and Yamasaki[57] have studied the effect of nucleic acid and protein synthesis inhibitors on DAO levels during the first phases of germination and development of pea seedlings. Actinomycin D and 6-methylpurine strongly inhibit (90%) the increase of DAO activity in the cotyledons if these are soaked and germinated in the presence of the inhibitor. Noticeably, if actinomycin D is added after 24 h imbibition, the effect is less dramatic (20% DAO activity inhibition in the following two days of development). On the basis of these results, the authors hypothesized that mRNAs for DAO are synthesized over a limited period of time during germination and then remain operating for several days.[57] DAO activity is highly influenced by many endogenous and environmental factors such as hormones, natural inhibitors and substrates, and light. On soaking pea seeds with IAA or 2,4-D, enzyme activity is reduced. This finding was originally interpreted as a feedback mechanism for the hormone biosynthesis.[58] As previously mentioned, this role seems to be untenable. Gibberellins supplied during seed imbibition increase DAO activity in pea cotyledons. The same effect is observed upon seed imbibition with putrescine, spermidine, and ornithine.[58] Studies on the effect of hormones on DAO activity should be seriously re-investigated in consideration of the marked effect of light on both the distribution and the absolute amount of the enzyme activity.[59,60] During the first synchronous cell cycle induced *in vitro* by 2,4-D in *Heliantus tuberosus* tuber slices, DAO activity exibits a peak 15 h after the addition of auxin, while Put levels peak at 12 h. Long term experiments have shown that DAO activity is high during the first phases of tuber formation, decreases until plant flowering and increases again in the phase preceeding entry of the tuber into dormancy. Put levels parallel the changes of the enzyme activity.[28] Between the effectors inducing changes of DAO activity during seedling development, light is no doubt one of the most effective. Leguminous seedlings show a similar time course of DAO activity appearance. Enzyme activity, absent in the first phases of germination, increases rapidly, on an organ basis, in etiolated pea or lentil stems, reaching a plateau after the first week of development.[47,55,60] Total DAO activity, after an initial rise, decreases in light-grown lentil stems but remains constant in pea.[47,60] Upon exposure of etiolated plants to white light, the increase in enzyme activity is blocked in pea stems while a decrease is detected in lentil stems.[47,60] These phenomena have been interpreted in "nutritional" terms, ascribing to polyamine catabolizing enzymes a key role in the utilization of seed reserves in the nonphotosynthesizing etiolated seedlings.[47] In this context, it has been shown that upon feeding ^{14}C-Put to pea seedlings, the diamine is converted to 4-aminobutyric acid (GABA); which may be further converted to succinic acid. This catabolism of Put is the result of the combined action of DAO, pyrroline dehydrogenase, GABA:pyruvate transaminase, and succinic semialdehyde dehydrogenase; it represents a way to incorporate products of polyamine catabolism into the Krebs cycle.[61] By the use of far-red and red light it has been successively demonstrated that the photomodulation of DAO activity in lentil stems is essentially unrelated to photosynthetic capacity of the seedling. Photoreversibility experiments have shown that the changes of enzyme activity are mediated by the action of phytochrome.[59] Federico et al.[62] performed a radioimmunoassay for lentil DAO using ^{125}I-DAO and affinity-purified ant-DAO IgG. DAO activity was determined as a function of time in etiolated and in light-grown seedlings either by the usual biochemical assay and by the radioimmunological assay obtaining superimposable curves.[62] This finding demonstrates that the photomodulation of DAO activity in lentil seedlings is the result of changes in the absolute amount of the enzyme and is not due to the modification of enzyme specific activity owing, for instance, to the occurrence of endogenous inhibitors.[62]

Evidence has been reported that an urate-degrading DAO-POD system exists in soybean radicles with a reaction rate controlled by tissue concentration of cadaverine.[63] This finding has attracted much attention, as uricase activity has a key role in allantoin accumulation in nitrogen-fixing leguminous plants. An artifactual, putrescine-sensitive, *S*-adenosyl-L-methionine decarboxylase activity has been reported for DAO in crude extracts of *Lathyrus sativus* seedlings.[64] This activity has been ascribed to hydrogen peroxide generated by putrescine degradation catalyzed by DAO.[64]

On addition of copper to Cu-deficient clover plants, DAO activity is restored only in leaves which have not emerged at the time of copper supply, suggesting that the enzyme can only be synthesized in young leaves, as is also demonstrated by immunoprecipitation of labeled DAO after pulse-chase experiments with $^{14}CO_2$.[65] Long term copper-deficiency experiments showed that DAO activity is inhibited by 50% in the first and 80% in the second generation of pea plants.[66]

Immunohistochemical studies performed with affinity-purified antilentil-DAO rabbit antibodies and fluoresceine-labeled goat antirabbit-IgG have provided evidence for an association of DAO with the cell walls and membranes of different tissues of etiolated lentil seedlings.[67] The occurrence of DAO in the apoplast has been further confirmed by the analysis of specific activity in the extracellular fluids obtained by vacuum infiltration of pea and lentil stem segments. This fraction contained less than 1% of the total proteins and 80% of total DAO present in the crude homogenate of whole segments.[68,69] The enzyme is loosely bound to the cell wall as it can be eluted from pea stem segments with low ionic-strength buffers.[69] A high peroxidase (POD) activity has also been detected in the extracellular fluids; POD is more tightly bound to the cell wall, as is shown by the higher salt concentration needed for enzyme elution.[69] On the basis of the above mentioned findings, a coupled function has been suggested for DAO and POD in the apoplast: the H_2O_2 produced by the oxidation of polyamines in the apoplast could be utilized by POD isoenzymes in crucial physiological events occurring in this compartment, i.e., the polymerization of lignin and suberin precursors, IAA catabolism, cross-linking of extensin, and polysaccharide-bound phenols.[69,70] Polyamines have been found in the cell walls of mung bean hypocotyls as well in xylem exudates of several species.[71,72] The significance of this hypothesis has been recently demonstrated by a histochemical approach using syringaldazine and 4-chloro-1-naphthol as peroxidase substrates. DAO and POD apparently are localized in the apoplast of lignifying tissues as well as in the epidermal cell walls of light-grown chickpea stems.[73] DAO distribution in other leguminous species has been studied in young soybean and bean seedlings giving conflicting results. While in soybean the enzyme activity is detectable in the epicotyl and hypocotyl, as well as in root and leaves, in bean, DAO activity appears to be restricted to the cotyledons, shoot apex, and first leaves.[74] In this context, Cogoni et al.[103] reported that amine oxidase activity can be detected in bean extracts only after three chromatographic steps, thus explaining the apparent absence of DAO activity in crude homogenates of this species. A search for inhibitors of the enzyme was unsuccessful.[103] Moreover, preliminary data suggest that histochemical DAO activity appears in bean hypocotyls in parallel to lignification activity.[75]

Using a histochemical method, Scorer et al. [25] have detected amine oxidase activity in all regions of *Canavalia ensiformis*, except roots and the embryo, during early growth. Slocum and Furey have recently demonstrated the exclusive apoplastic localization of DAO in pea seedlings by a cytochemical approach exploiting the detection of H_2O_2 arising by the *in situ* oxidation of polyamines.[76] In an attempt to find functional correlations between DAO and POD activities, Angelini et al.[77] have studied the tissue and organ distribution of the two enzymes in the chickpea stem and their response to external stimuli such as light and mechanical injury. DAO and POD activities decrease acropetally along the stele and the epidermis. The distribution of histochemical enzyme activities are closely related to those of lignin depositions in the sclerenchyma and xylem tissues and with cell-wall stiffening of epidermal cells along the stem. De-etiolation results in a rise of DAO and POD activities in epidermal peels of the fourth internode of chickpea stems. This phenomenon has been tentatively interpreted as DAO-POD system being involved in the stiffening processes occurring in the epidermal cell walls, which follow the light-mediated growth inhibition of lower internodes.

Both DAO and POD activities and Cad levels increase up to seven days after wounding of chickpea stems with superimposable trends. The activity increase is substantial in both soluble and cell wall fractions. Apparent histochemical activities of both enzymes can be detected in the very cell walls where lignosuberized depositions occurr as a result of the wound-healing process.[77,78] Diamine oxidase activity associated with the cell wall in bean is inhibited *in vivo* by exposure to ozone while extracellular peroxidase activity is stimulated. This phenomenon could profoundly alter the movements of polyamines from cell to cell and needs further investigation in view of the crucial role that this gaseous pollutant plays in forest decline.[79] Put (5 mM) elicites a wound response in corn roots in the presence of calcium. It has been proposed that the mechanism of putrescine-induced wounding involves the catabolism of the diamine by a cell wall diamine oxidase, with the production of hydrogen peroxide and free radicals, which results in peroxidative damage to the plasmalemma.[80] In this context, it has been reported that the degradation products of polyamines are responsible for the inhibition of K^+ uptake occurring as a result of injury in pea roots.[81]

IV. POLYAMINE OXIDASE

A. GENERAL AND MOLECULAR FEATURES

Plant polyamine oxidases (PAO; EC 1.4.3.4.) oxidize at the secondary amino groups of spermidine and spermine, with the production of 4-aminobutyraldehyde and 3-aminopropyl-4-aminobutyraldehyde, which spontaneously cyclize to Δ^1-pyrroline and 1,5-diazabicyclononane, respectively (Figure 3). H_2O_2 and 1,3-diaminopropane are coproducts of the reaction.[5,6] These enzymes, which apparently occur only in the monocots, have been purified and partially characterized from *Avena sativa*,[82,83] *Zea mays*,[84-86] *Hordeum vulgare*,[82] *Setaria italica*,[87] and from *Eichornia stricta* (water hyacynth).[88]

PAOs purified to homogeneity show an absorption spectrum with A_{max} at 280, 380, and 450 to 460 nm which confer a yellow color to purified preparations of the enzyme. The flavoprotein nature of PAOs was originally suggested by inhibition experiments with acridine compounds and by their reversal with exogenous FAD or FMN.[89] Maize and barley PAO do not oxidize artificial electron donors such as 2,6-dichlorophenolindophenol, phenazine methosulfate, or

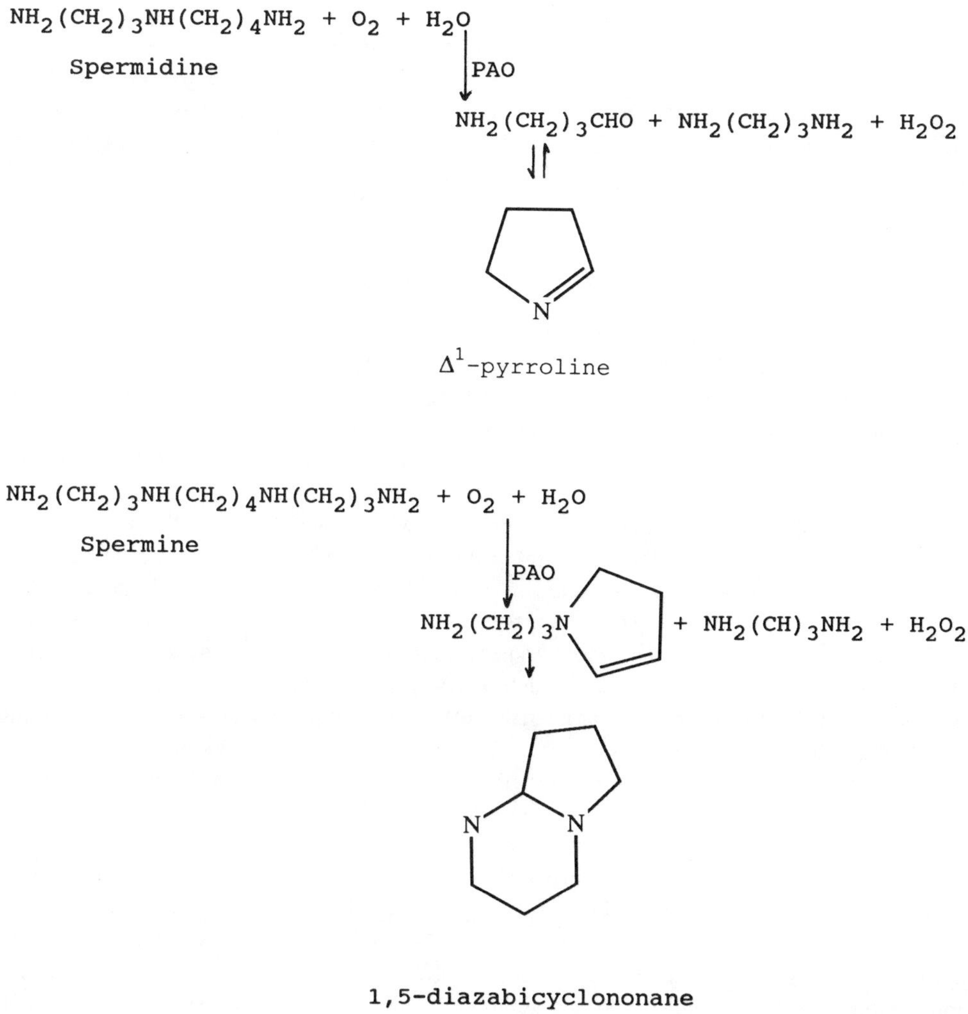

FIGURE 3. Oxidative deamination of polyamines catalyzed by plant polyamine oxidase.

ferricyanide either in anaerobiosis or in the presence of oxygen. Maize PAO is able to utilize *p*-benzoquinone as elecron donor in aerobiosis.[90] Maize PAO, which is the most studied enzyme, is a monomeric glycoprotein with a M_r of 53,000 Da containing one molecule of FAD. The sugar content of the enzyme is 2.5%, mainly represented by arabinose.[85] The enzyme can be purified to apparent homogeneity by only two chromatographic steps from a cell wall preparation of maize seedlings or by a one-step immunoaffinity procedure.[85,91] The enzyme has alanine as N-terminal aminoacid and its N-terminal sequence shows significant degree of homology with tryptophan 2-monooxygenase from *Pseudomonas syringae* pv. *savastanoi*.[85] The addition in anaerobiosis of equimolar amounts of spermidine to oat or maize PAO results in the oxidation of the cofactor as shown by the decrease of absorption at 380 and 456 nm. Upon addition of air the optical spectrum is restored confirming the involvement of flavin in the catalytic cycle.[83,85]

All plant PAOs show a very restricted substrate specificity oxidizing only Spd and Spm.[5,6] The pH optimum for oxidation of the two substrates varies among different species, being in the range 5.0 to 6.6 for the oat and maize enzyme and 4.8 (Spm) and 8.0 (Spd) for the barley enzyme.[82,85,92] Calorimetric analysis on the maize enzyme revealed a single two-state transition at pH 6 with $T_m = 49.8°C$. At pH 5, the thermal stability is increased by more than 14°C.[86] In addition to acridine compounds PAO are inhibited by guanidine analogues.[5] Notably, the oat and maize enzymes are inhibited noncompetitively by N^1-acetylspermine.[86] The products arising from amine oxidation, Δ^1-pyrroline and 1,5-diazabicyclononane, are competitive inhibitors of maize PAO with an apparent $K_i = 400$ and 100 μM, respectively. Moreover these compound improve the thermal stability of the enzyme.[86]

B. PHYSIOLOGICAL ASPECTS

Earlier work on barley PAO showed that the enzyme was associated with a particulate fraction sedimenting at low centrifugal fields (1000 × g) from which it could be eluted by 0.5 *M* NaCl washing but not by incubation with nucleases, thus making a nuclear localization of the enzyme unlikely.[82,92] In the absence of PAO activity in oat leaf protoplasts, and its re-appearance during cell-wall regeneration, Kaur-Sawhney et al.[93] suggested that the enzyme is apoplastic. Histochemical studies utilizing the iodide/starch method for H_2O_2 detection, revealed the association of the enzyme with the vascular system and stomata.[93,94] Specific PAO activity is noticeably higher in extracellular fluids obtained from maize mesocotyls, exceedings by 100 times that found in a crude homogenate of the whole organ.[95] Unlike the oat and maize enzyme, PAO is apparently symplastic in barley primary leaves.[96] The apoplastic nature of maize PAO has been recently confirmed by an electron microscopic cytochemical method which detects H_2O_2 generated by the *in situ* polyamine oxidation.[76] PAO activity is very low in resting seeds and can be detected as soon as germination begins.[97,98] During the first phases of growth, a continuous increase of enzyme activity is observable in the shoot.[98] Subcellular and tissue distribution of PAO activity has been recently re-investigated in maize seedlings, taking into account the high percentage of PAO tightly bound to cell walls. This represents 30 to 90% of total enzyme activity in the shoot and root, respectively.[99] Time course and organ distribution of PAO activity in maize seedlings should be re-investigated taking into account this finding.[97,98] A pathway involving Spd conversion to GABA through the action of PAO and a pyrroline dehydrogenase has been demonstrated in oat. The latter enzyme has been partially characterized.[61] Tissue and organ distribution of PAO in the maize seedling, studied by a histochemical approach similar to that used for DAO, shows an apparent PAO localization mainly in the cell walls of xylem (coleptile), xylem and xylem parenchyma (mesocotyl, root), rhyzodermis, hypodermis and endodermis (root), and epidermis (coleoptile, mesocotyl).[99] Apparent histochemical peroxidase activity shows the same tissue and organ distribution indicating the preferential association of the two enzymes with the cell walls of tissues where lignification, suberization, and wall stiffening occur.[99] Evidence has been reported that polyamines and their oxidation products are responsible for the inhibition of K^+ uptake resulting from cutting injury on maize roots.[81] The available evidence suggests a considerable similarity for the physiological roles of DAO and PAO in the apoplast in particular as regard the functional correlation with peroxidases.

V. CONCLUDING REMARKS

Plant diamine and polyamine oxidases exhibit a high degree of molecular similarities, respectively, in all taxa suggesting an important role in the regulation of intracellular polyamine levels. The situation in higher plants is quite puzzling because these enzymes occur at a high concentration only in some taxa. Two explanations are possible: some species have alternative mechanisms of polyamine catabolism, or the apparent lack of amine oxidase activity is due to a very low tissue concentration of the enzyme and/or to its inhibition by unknown factors, as in the case of barley DAO.[33]

In addition to polyamine catabolism, AOs and their reaction products may have other important functions in some species. For example, DAO is a key enzyme in the biosynthesis of alkaloids containing a pyrrolidine or piperidine ring.[100] The potential roles for amine oxidase-generated H_2O_2 in lignin biosynthesis and cell wall crosslinking reactions which may regulate growth have been discussed. It has also been shown that aminoaldehydes are toxic to a variety of cell types and induce metabolic effects opposite to those of polyamines.[101] In this light, we believe that the study of polyamine catabolism merits a renewed interest and further experimental effort.

ACKNOWLEDGMENTS

This work was supported by a grant of Ministero della Pubblica Istruzione, Italy.

REFERENCES

1. **Tabor, C.W. and Tabor, H.,** Polyamines, *Annu. Rev. Biochem.*, 53, 749, 1984.
2. **Morgan, D.M.L.,** Polyamines, *Essays in Biochem.*, 23, 82, 1987.
3. **Smith, T.A.,** Polyamines, *Annu. Rev Plant Physiol.*, 36, 117, 1985a.
4. **Evans P.T. and Malmberg R.L.,** Do polyamines have roles in plant development?, *Annu. Rev. Plant Physiol.*, 40, 235, 1989.
5. **Smith, T.A.,** The di- and polyamine oxidases of higher plants, *Biochem. Soc. Trans.*, 13, 319, 1985b.
6. **Smith, T.A. and Barker J.H.A.,** The di- and polyamine oxidases of plants, in *Progress in Polyamine Research: Novel Biochemical, Pharmacological and Clinical Aspects*, Zappia, V. and Pegg, A.E., Eds., Plenum Press, New York, 1988, 573.
7. **Rinaldi A., Floris, G., and Giartosio, A.,** Plant amine oxidases, in *Structure and Function of Amine Oxidases*, Mondovì, B., Ed., CRC Press, Boca Raton, 1986, 51.
8. **Seiler N.,** Acetylpolyamines as substrates of amine oxidases, in *Structure and Function of Amine Oxidases*, Mondovì, B., Ed., CRC Press, Boca Raton, 1986, 21.
9. **Yanagisawa, H, Hirasawa, E., and Suzuki, Y.,** Purification and properties of diamine oxidase from pea epicotyls, *Phytochemistry*, 20, 2105, 1981.
10. **Smith, T.A., Croker, S.J., and Loeffler, R.S.T.,** Occurrence in higher plants of 1-(3-aminopropyl) pyrrolinium and pyrroline: products of polyamine oxidation, *Phytochemistry*, 25, 683, 1986.,
11. **Brandange, S., Eriksson, L.H., and Rodriguez, B.,** Ring-chain tautomerism of an N,N-acetal formed by enzymatic oxidation of spemine or spermidine, *Acta Chem. Scand.*, 38, 526, 1984.
12. **Cogoni, A., Padiglia, A., Medda, R., Segni, P., and Floris, G.,** Oxidation of spermine by an amine oxidase from lentil seedlings, *Plant Physiol.*, 95, 477, 1991.
13. **Kluetz, M.D., Adamsons, K., and Flynn, J.E., Jr.,** Cryoenzymology and spectrophotometry of pea seedling diamine oxidase, *Biochemistry*, 19, 1617, 1980.
14. **Rinaldi, A., Giartosio, A., Floris, G., Medda, R., and Finazzi-Agrò, A.,** Lentil seedling diamine oxidase: preparation and properties of the copper-free enzyme, *Biochem. Biophys. Res. Commun.*, 120, 242, 1984.
15. **Rotilio, G., Calabrese, L., Finazzi Agrò, A. and Mondovì B.,** Indirect evidence for the production of superoxide anion radicals by pig kidney diamine oxidase, *Biochim. Biophys. Acta*, 198, 618, 1970.
16. **Rotilio, G.,** Spectroscopic and chemical properties of the copper amine oxidase, in *Structure and Function of Amine Oxidases*, Mondovì, B., Ed., CRC Press, Boca Raton, FL, 1986, 127.
17. **Sindhu, R.K. and Desai, H.V.,** Partial purification and characterization of diamine oxidase from groundnut seedlings, *Ind. J. Biochem. Biophys.*, 17, 194, 1980.
18. **Suzuki, Y.,** Some properties of the amine oxidase of soybean seedlings, *Plant Cell Physiol.*, 14, 413, 1973.
19. **Suresh, M.R., Ramakrishna, S., and Adiga, P.R.,** Diamine oxidase of *Lathyrus sativus* seedling, *Phytochemistry*, 15, 483, 1976.
20. **Floris, G., Giartosio, A., and Rinaldi, A.,** Diamine oxidase from *Lens esculenta* seedlings: purification and properties, *Phytochemistry*, 22, 1871, 1983.
21. **Shutte, H.R., Knofel, D., and Heyer, O.,** Biosynthese von Lupinenalkaloiden 19. Anreicherung und Carakterisierung der Diaminoxydase aus Keimlingen von *Lupinus luteus*, *Z. Pflanzenphysiol.*, 55, 110, 1966.
22. **Matsuda, H. and Suzuki, Y.,** Some properties of the amine oxidase in *Vicia faba* seedlings, *Plant Cell Physiol.*, 22, 737, 1981.
23. **Angelini, R., Di Lisi, F., and Federico, R.,** Immunoaffinity purification and characterization of diamine oxidase from *Cicer*, *Phytochemistry*, 24, 2511, 1985.

24. **Delahize, E. and Webb, J.**, Purification and characterization of diamine oxidase from clover leaves, *Phytochemistry*, 26, 641, 1987.
25. **Scorer, K.N., Caamal, R.M., Oropeza, C., and Loyola-Vargas, V.M.**, Amine oxidase activity in *Canavalia ensiformis* during early growth: a histochemical study, *J. Plant Physiol.*, 135, 346, 1989.
26. **Pec, P., Zajoncova, L., and Jilek, M.**, Diamine oxidase from sainfoin (*Onobrychis viciifolia*). Purification and some kinetic properties, *Biologia (Bratislava)*, 44, 1177, 1989.
27. **Rinaldi, A., Floris, G., and Finazzi-Agrò, A.**, Purification and properties of diamine oxidase from *Euphorbia* latex, *Eur. J. Biochem.*, 127, 417, 1982.
28. **Torrigiani, P., Serafini-Fracassini, D., and Fara, A.**, Diamine oxidase activity in different physiological stages of *Helianthus tuberosus* tuber, *Plant Physiol.*, 89, 69, 1989.
29. **Percival, F.W. and Purves, W.K.**, Multiple amine oxidases in cucumber seedlings, *Plant Physiol.*, 54, 601, 1974
30. **Chaudhuri, M.M. and Gosh B.**, Purification and characterization of diamine oxidase from rice embryos, *Phytochemistry*, 23, 241, 1984.
31. **Nadeau, P., Delaney, S., and Chouinard, L.**, Effect of cold hardening on the regulation of polyamine levels in wheat (*Triticum aestivum* L.) and alfalfa (*Medicago sativa* L.), *Plant Physiol.*, 84, 73, 1987.
32. **Haslam, S.C. and Young, T.W.**, Partial purification of a diamine oxidase from *Nicotiana rustica*, in Amines in Plants, Phytochemical Society of Europe Symposium, University College, London, (Abstract), 1987.
33. **Cogoni, A., Piras, C., Farci, R., Melis, A., and Floris, G.**, *Hordeum vulgare* seedling amine oxidase: purification and properties, *Plant Physiol.*, 93, 818, 1990.
34. **Floris, G., Giartosio, A., and Rinaldi, A.**, Essential sulfhydryl groups in diamine oxidase from *Euphorbia characias* latex, *Arch. Biochem. Biophys.*, 220, 623, 1980.
35. **Glatz, Z., Kovar, J., Macholan, L., and Pec, P.**, Pea (*Pisum sativum*) diamine oxidase contains pyrroloquinoline quinone as a cofactor, *Biochem. J.*, 242, 603, 1987.
36. **Janes, S.M., Mu, D., Wemmer, D., Smith, A.J., Kaur, S., Maltby, D., Burlingame, A.L., and Klinman, J.P.**, A new redox cofactor in eucaryotic enzymes: 6-hydroxydopa at the active site of bovine serum amine oxidase, *Science*, 248, 981, 1990.
37. **Oratore, A., D'Andrea, G., Floris, G., Rinaldi, A., and Finazzi-Agrò, A.**, Microheterogeneity of lentil seedling (*Lens culinaris*) amine oxidase, *Ital. J. Biochem.*, 36, 92, 1987.
38. **Finazzi-Agrò, A.**, Optical spectoscopic properties of copper-containing amine oxidases, in *Structure and Function of Amine Oxidases*, Mondovì, B., Ed., CRC Press, Boca Raton, FL, 1986, 121.
39. **Mondovì, B., Rotilio, G, Costa, M.T., Finazzi-Agrò, A., Chiancone, E., Hansen, R.E., and Beinert, H.**, Diamine oxidase from pig kidney; improved purification and properties, *J. Biol. Chem.*, 242, 1160, 1967.
40. **Nylen, U. and Szybek, P.**, Kinetics and other characteristics of diamine oxidase of pea seedlings, *Acta Chem. Scand.*, 28, 1153, 1974.
41. **Lindstrom, A., Olsson, B., and Pettersson, G.**, Effect of azide on some spectral and kinetic properties of pig plasma benzylamine oxidase, *Eur. J. Biochem.*, 48, 237, 1974.
42. **Suzuki, S., Sakurai, T., Nakahara, Manabe, T. and Okuyama, T.**, Effect of metal substitution on the chromophore of bovine serum amine oxidase, *Biochemistry*, 22, 1630, 1983.
43. **Yadav, K.D.S. and Knowles, P.F.**, A catalytic mechanism for benzylamine oxidase from pig plasma - stopped-flow kinetic studies, *Eur. J. Biochem.*, 114, 139, 1981.
44. **Bardsley, W.G.**, Inhibitors of copper amine oxidases, in *Structure and Function of Amine Oxidases*, Mondovì, B., Ed., CRC Press, Boca Raton, FL, 1986, 135.
45. **Rinaldi, A., Floris, G., Sabatini, S., Finazzi-Agrò, A., Giartosio, A., Rotilio, G., and Mondovì, B.**, Reaction of beef plasma and lentil seedlings Cu-amine oxidases with phenylhydrazine, *Biochem. Biophys. Res. Commun.*, 115, 841, 1983.
46. **Pec, P.**, Personal communication, 1990.
47. **Macholan, L. and Minar, J.**, The depression of the synthesis of pea diamine oxidase due to light and the verification of its participation in growth processes using competitive inhibitors, *Biol. Plant. (Praha)*, 16, 86, 1974.
48. **Srivastava, S.K. and Smith, T.A.**, The effect of some oligoamines and guanidines on membrane permeability in higher plants, *Phytochemistry*, 21, 997, 1982.
49. **Bieganski, T., Osinska, Z., and Maslinski, C.**, Inhibition of plant and mammalian diamine oxidases by hydrazine and guanidine compounds, *Int. J. Biochem.*, 14, 949, 1982.
50. **Sindhu, R.K. and Desai, H.V.**, Inhibitor of diamine oxidase in cotyledons of groundnut seedlings, *Phytochemistry*, 19, 317, 1980.
51. **Joshi, B.H. and Prakash, V.**, Naturally occurring pea diamine oxidase inhibitors obtained from *Sorghum vulgare* during germination, *Experientia*, 38, 315, 1982.
52. **Bellelli, A., Brunori, M., Finazzi-Agrò, A., Floris, G., Giartosio, A., and Rinaldi, A.**, Transient kinetics of copper-containing lentil (*Lens culinaris*) seedling amine oxidase, *Biochem. J.*, 232, 923, 1985.

53. **Werle, E. and Pechmann, E.,** Uber die Diaminoxydase der Pflanzen und ihre adaptative Bildung durch Bakterien, *Justus Liebigs Ann. Chem.*, 562, 44, 1949.
54. **Kenten, R.H. and Mann, P.J.G.,** The oxidation of amines by extracts of pea seedlings, *Biochem. J.*, 50, 360, 1952.
55. **Werle, E, Beaucamp, G. and Schirren,** Uber die Aminoxydase von Erbsenkeimlingen und ihre Bedeutung fur die Keimung, *Planta*, 53, 125, 1959.
56. **McGowan, R.E. and Muir, R.M.,** Purification and properties of amine oxidase from epicotyls of *Pisum sativum*, *Plant Physiol.*, 47, 644, 1971.
57. **Suzuki, Y. and Yamasaki, K.,** Effects of some inhibitors of nucleic acid and protein synthesis on the development of amine oxidase in germinating pea cotyledons, *Physiol. Plant.*, 24, 441, 1971.
58. **Srivastava, S.K., Prakash, V., and Naik, B.I.,** Regulation of diamine oxidase activity in germinating pea seeds, *Phytochemistry*, 16, 185, 1977.
59. **Angelini, R., Federico, R., and Mancinelli, A.,** Phytochrome-mediated control of diamine oxidase level in etiolated lentil (*Lens culinaris* Medicus) seedlings, *Plant Physiol.*, 88, 1207, 1988.
60. **Cona, A, Angelini, R., and Federico, R.,** A preliminary study of hormonal regulation on diamine oxidase activity in lentil seedlings, *Annali di Botanica* Roma, 45, 135, 1987.
61. **Flores, H.E. and Filner, P.,** Polyamine catabolism in higher plants: characterization of pyrroline dehydrogenase, *Plant Growth Regul.*, 3, 277, 1985.
62. **Federico, R., Angelini, R., Cesta, A., and Pini, C.,** Determination of diamine oxidase in lentil seedlings by enzymic activity and immunoreactivity, *Plant Physiol.*, 79, 62, 1985.
63. **Tajima, S., Kanazawa, T., Takeuchi, E., and Yamamoto, Y.,** Characteristics of a urate-degrading diamine oxidase-peroxidase enzyme system in soybean radicles, *Plant Cell Physiol.*, 26, 787, 1985.
64. **Suresh, M.R. and Adiga, P.R.,** Putrescine-sensitive (artifactual) and insensitive (biosynthetic) S-adenosyl-L-methionine decarboxylase activities of *Lathyrus sativus* seedlings, *Eur. J. Biochem.*, 79, 511, 1977.
65. **Delhaize, E., Dilworth, M.S., and Webb, J.,** The effect of copper nutrition and developmental state on the biosynthesis of diamine oxidase in clover leaves, *Plant Physiol.*, 82, 1126, 1986.
66. **Baron Ayala, M. and Sandmann, G.,** Activities of Cu-containing proteins in Cu-depleted pea leaves, *Physiol. Plant.*, 72, 801, 1988.
67. **Federico, R., Angelini, R., Argento-Cerù, M.P., and Manes, F.,** Immunohistochemical demonstration of lentil diamine oxidase, *Cell Mol. Biol.*, 31, 171, 1985.
68. **Federico, R. and Angelini, R.,** Occurrence of diamine oxidase in the apoplast of pea epicotyls, *Planta*, 167, 300, 1986.
69. **Angelini, R. and Federico, R.,** Occurrence of diamine oxidase in the apoplast of Leguminosae seedlings, in *Biochemical Studies of Natural Polyamines*, Caldarera, C.M., Clò, C., and Guarnieri, C., Eds., CLUEB, Bologna, 1985, 183.
70. **Federico, R. and Angelini, R.,** Distribution of polyamines and their related catabolic enzymes in etiolated and light-grown Leguminosae seedlings, *Planta*, 173, 317, 1988.
71. **Goldberg, R. and Perdrizet, E.,** Ratio of free to bound polyamines during maturation in mung-bean hypocotyl cells, *Planta*, 161, 531, 1984.
72. **Friedman, R., Levin, N. and Altman A.,** Presence and identification of polyamines in xylem and phloem exudates of plants, *Plant Physiol.*, 82, 1154, 1986.
73. **Angelini, R. and Federico, R.,** Histochemical demonstration of polyamine oxidation and hydrogen peroxide generation in the apoplast, *J. Plant Physiol.*, 135, 212, 1989.
74. **Scoccianti, V., Torrigiani, P., and Bagni, N.,** Distribution of diamine oxidase activity and polyamine pattern in bean and soybean seedlings at different stages of germination, *Physiol. Plant.*, 80, 515, 1990.
75. **Angelini, R. and Federico, R.,** unpublished data, 1988.
76. **Slocum, R.D. and Furrey, M.J.,** Electron microscopic cytochemical localization of diamine and polyamine oxidase in peas and maize tissues, *Planta*, 183, 443, 1991.
77. **Angelini, R., Manes, F., and Federico, R.,** Spatial and functional correlation between diamine oxidase and peroxidase activities and their dependence upon de-etiolation and wounding in chick-pea stems, *Planta*, 182, 89, 1990.
78. **Scalet, M., Federico, R., and Angelini, R.,** Time courses of diamine oxidase and peroxidase activities and polyamine changes after mechanical injury in chick-pea seedlings, *J. Plant Physiol.*, in press, 1991.
79. **Peters, J.L., Castillo, F.J., and Heath, R.L.,** Alteration of extracellular enzymes in pinto bean leaves upon exposure to air pollutants, ozone, and sulfur dioxide, *Plant Physiol.*, 89, 159, 1989.
80. **DiTomaso, J.M., Shaff, J.E., and Kochian, L.,** Putrescine-induced wounding and its effects on membrane integrity and ion transport processes in roots of intact corn seedlings, *Plant Physiol.*, 90, 988, 1989.
81. **De Agazio, M. Federico, R., and Grego, S.,** Involvement of polyamines in the ihibiting effect of injury caused by cutting on K^+ uptake through the plasma membrane. *Planta*, 177, 388, 1989.
82. **Smith, T.A.,** Polyamine oxidase from barley and oats, *Phytochemistry*, 15, 633, 1976.

83. **Federico, R., Alisi, C., Forlani, F., and Angelini, R.,** Purification and characterization of oat polyamine oxidase, *Phytochemistry*, 28, 2045, 1989.
84. **Suzuki, Y. and Yanagisawa, H.,** Purification and properties of maize polyamine oxidase: a flavoprotein, *Plant Cell Physiol.*, 21, 1085, 1980.
85. **Federico, R., Alisi, C., and Forlani, F.,** Properties of the polyamine oxidase from the cell wall of maize seedlings, *Phytochemistry*, 28, 45, 1989.
86. **Federico, R., Cona, A., Angelini, R., Schininà, M.E., and Giartosio, A.,** Characterization of maize polyamine oxidase, *Phytochemistry*, 29, 2411, 1990.
87. **Hirasawa, E., Watanabe, H., and Suzuki, Y.,** Polyamine oxidase from millet shoots, *Phytochemistry*, 25, 1739, 1986.
88. **Yanagisawa, H., Kato, A., Hoshiai, S., Kamiya, A., and Torii, N.,** Polyamine oxidase from water hyacinth, *Plant Physiol.*, 85, 906, 1987.
89. **Suzuki, Y. and Hirasawa, E.,** Polyamine oxidase from *Zea mays* shoots, *Phytochemistry*, 12, 2863, 1973.
90. **Hirasawa, E. and Suzuki, Y.,** Purification and properties of a polyamine oxidase from *Zea mays*, *Phytochemistry*, 14, 99, 1975.
91. **Federico, R., Alisi, C., Cona, A., and Angelini, R.,** Purification of polyamine oxidase from maize seedlings by immunoadsorbent column, in *Progress in Polyamine Research: Novel Biochemical, Pharmacological and Clinical Aspects*, Zappia, V. and Pegg, A.E., Eds, Plenum Press, New York, 1988, 617.
92. **Smith, T.A.,** Polyamine oxidation by enzymes from *Hordeum vulgare* and *Pisum sativum* seedlings, *Phytochemistry*, 13, 1075, 1974.
93. **Kaur-Sawhney, R., Flores, H.E., and Galston, A.W.,** Polyamine oxidase in oat leaves: a cell wall-localized enzyme, *Plant Physiol.*, 68, 494, 1981.
94. **Smith, T.A.,** Polyamine oxidase in higher plants, *Biochem. Biophys. Res. Commun.*, 41, 1452, 1980.
95. **Federico, R., Alisi, C., and Angelini, R.,** On the occurrence of oxidoreductases in the apoplast of Leguminosae and Gramineae and their significance in in the study of plasmamembrane-bound redox activities, in *Plasma Membrane Oxidoreductases in Control of Animal and Plant Growth*, Crane, F.L., Morrè, D.J., and Low, H., Eds., Plenum Press, New York, 1988.
96. **Li, Z.C. and McClure, J.W.,** Polyamine oxidase of primary leaves is apoplastic in oats but symplastic in barley, *Phytochemistry*, 28, 2255, 1989.
97. **Torrigiani, P., Scoccianti, V., and Bagni, N.,** Polyamine oxidase activity and polyamine content in maize during seed germination, *Physiol. Plant.*, 74, 427, 1988.
98. **Forlani, F., Angelini, R., Cona, A., and Federico, R.,** Distribution of polyamines and polyamine oxidase in etiolated and light-grown maize seedlings, *Annali di Botanica* (Roma), 45, 117, 1987.
99. **Augeri, M.I., Angelini, R., and Federico, R.,** Sub-cellular localization and tissue distribution of polyamine oxidase in maize (*Zea mays* L.) seedlings, *J. Plant Physiol.*, 136, 690, 1990.
100. **Smith, T.A. and Wilshire, G.,** Distribution of cadaverine and other amines in higher plants, *Phytochemistry*, 14, 2341, 1975.
101. **Tabor, H. and Tabor, C.W.,** Spermidine, spermine and related amines, *Pharmacol. Rev.*, 16, 245, 1964.
102. **Suresh, M.R. and Adiga, P.R.,** Diamine oxidase of *Lathyrus sativus* seedlings. Purification and properties, *J. Bioscience*, 1, 109, 1979.
103. **Cogoni, A., Farci, R., Medda, R., Rinaldi, A., and Floris, G.,** Amine oxidase from *Lathyrus cicera* and *Phaseolus vulgaris*: purification and properties, *Prep. Biochem.*, 19, 95, 1989.
104. **Mizusaki, S., Tanabe, Y., Noguchi, M., and Tamaki, E.,** *N*-Methylputrescine oxidase from tobacco roots, *Phytochemistry*, 11, 2757, 1972.
105. **Davies, H.M., Hawkins, D.J., and Smith, L.A.,** Quinoprotein characteristics of *N*-methylputrescine oxidase from tobacco roots, *Phytochemistry*, 28, 1573, 1989.
106. **Hashimoto, T., Mitani, A., and Yamada, Y.,** Diamine oxidase from cultured roots of *Hyoscyamus niger*: its function in tropane alkaloid biosynthesis, *Plant Physiol.*, 93, 216, 1990.

Chapter 4

POLYAMINES AND PLANT SECONDARY METABOLITES

H. E. Flores and J. Martin-Tanguy

TABLE OF CONTENTS

I. INTRODUCTION

Research over the last three decades has resulted in an adequate picture of primary polyamine metabolism in plant cells. The unique features of plant polyamine metabolism include the existence of alternate pathways of putrescine biosynthesis, the sharing of a common precursor (*S*-adenosylmethionine, AdoMet) between the ethylene and the polyamine pathways, and the synthesis of uncommon polyamines in plant cells which have been exposed to drought and/or heat stress. These aspects of primary polyamine metabolism in plant cells are covered in Chapters 2, 5, and 8. Reflecting this emphasis on primary metabolism, most studies which probe the physiology and biochemistry of plant polyamines have focused, until recently, almost exclusively on the pools of "free" polyamines. These are usually defined by the methods used for extraction and analysis, namely homogenization of plant tissues in 5% perchloric acid (PCA) or trichloroacetic acid (TCA), centrifugation of the crude extract, and HPLC/TLC of the supernatant after derivatization. Presumably, polyamines are present inside the cell (at "physiological pH") in ionized form, which is readily extractable in PCA or TCA. It is now apparent, however, that polyamines can also exist in so-called "bound" forms, and that in many cases these can represent a substantial portion of the metabolic pool. For example, putrescine (Put) is present in *Amaranthus* seeds almost exclusively in the bound form;[1] in the same species, 89% of Spd and 60% of Spm are bound. These bound polyamine fractions are presumably covalently linked to low molecular weight compounds, and are usually analyzed after acid (6N HCl) hydrolysis of the PCA/TCA supernatant. However, polyamines can also be released after hydrolysis of he PCA/TCA insoluble fraction, and this may represent covalent linkages with high molecular weight compounds.

Analysis of polyamine fractions before and after hydrolysis provides, however, only a general estimate of the free and bound polyamine pools. Furthermore, many such bound forms are labile to acid extraction, and therefore the quantification of specific compounds using the above methods may be inaccurate. A full understanding of the physiology and biochemistry of bound polyamines can be possible only after determination of their chemical structures, and development of suitable methods of analysis. In animal systems, the major forms of bound polyamines have been characterized as the acetylated derivatives of Spd and Spm.[2] These are physiologically important as the preferred substrates for polyamine oxidases.[3] Acetylated polyamines have not been reported in plant systems, and we still know little about the chemical identity of bound polyamines in plant cells. Over the last two decades, however, it has become apparent that bound polyamines may not only represent a significant portion of the total pool, but may have intriguing physiological roles as well. In addition, it has been known for some time that di- and polyamines can be the precursors of many alkaloids which are unique to higher plants, and which have ecological significance. This review is a brief summary of the current status of research on polyamine-derived secondary metabolites. This aspect of plant polyamine metabolism has been discussed in recent reviews.[4,5]

II. HYDROXYCINNAMIC ACID AMIDES

As mentioned above, polyamines can occur in plant cells not only as free bases but also as bound forms, commonly called conjugates. In animal and bacterial systems, polyamines conjugated to low molecular weight moieties can be acetylated derivatives, form part of a peptide (glutathionylspermidine), amino acids (putreamine), or antibiotics (bleomycin, tallysomycin). Polyamines can also be linked to proteins via a transglutaminase-catalyzed reaction. Although previously reported in mammalian cells,[6] recent evidence indicates that polyamine conjugation to proteins can also occur in plant cells and be mediated by an enzymatic activity.[7] In fact, it is now known that the large polymorphic subunits of 1,5-bisphosphate carboxylase/oxygenase in alfalfa are major substrates for transglutaminases which catalyze the covalent conjugation of ^{14}C-Put.[8] In this review, we will restrict our discussion to low molecular weight conjugated polyamines. In plant cells, the only well-characterized compounds in this category are the hydroxycinnamic acid amides. The metabolism and function of these compounds have been recently reviewed.[9,10]

A. OCCURRENCE AND DISTRIBUTION

Di- and polyamines can be linked with hydroxycinnamic acids by an amide bond, forming the so-called hydroxycinnamic amides (HCAs). There are two major types of HCAs; the water-soluble amides, which have at least one free primary amino function (basic HCAs), and the water-insoluble amides, in which all primary and secondary amino groups have been derivatized (neutral HCAs). The amines of the water-soluble HCAs include only aliphatic di- and polyamines (Put, Spd, Spm), whereas the water-insoluble HCAs include both aliphatic and aromatic amines (tyramine, dopamine, serotonin, octopamine, tryptamine, etc.). HCAs have been known for a very long time as plant constituents. Caffeoyl-Put, previously named paucine, was found present in seeds of the Leguminosae *Pentachlera macrophylla* in 1894[11], and later in tobacco flowers.[12] Feruloylputrescine was first discovered in *Salsola subaphylla* (Chenopodiaceae) and given the name subaphylline.[13] It and was later identified in the fruits and leaves of *Citrus*,[14] and in species of the Solanaceae and Gramineae.[15] Caffeoyl- , coumaroyl- and feruloyl-Put have been reported in tobacco callus.[16] A caffeoylferuloyl-Spd has been reported in the pollen of *Corylus avellana*.[17] Amides of methyl-Put were found in the leaves of *Kniphofia flavovirens* and identified as *p*-coumaroylmethylputrescine and methoxycinnamoylputrescine.[9] The hydroxy- cinnamic acid amides have now been found throughout the plant kingdom. In a survey of 20 species representing 13 families, HCAs were found to be the main phenolic constituents of the reproductive organs and seeds of all the plants which were analyzed;[15] HCAs were also found present in the inflorescences of several species of Araceae.[18] It is thus generally accepted that HCAs are universal constituents of the reproductive tissues of higher plants. Because several aliphatic and aromatic amines can form amide linkages with at least three different hydroxycinnamic acids, the number of possible chemical structures for HCAs can be quite large. As we will discuss below, however, only a limited number of structures are present in a single species, and the distribution between species and within plant organs is far from random.

B. BIOLOGICAL SIGNIFICANCE OF HYDROXYCINNAMIC ACID AMIDES

An overview of the published literature on HCAs indicates that these compounds are associated with two major developmental/physiological processes: (1) flowering and seed/fruit development, and (2) the hypersensitive response to viral or fungal infection. Because the evidence involving HCAs and such processes/responses is largely correlative at this point, we feel that it is premature to offer a unifying model for the role of HCAs in flowering and disease resistance. It may well be that at least two distinct functions for HCAs have evolved in higher plants, and thus we will treat these two subjects independently. The linkage between HCAs and plant reproduction has been extensively reviewed. In the discussion below, we will provide only a summary of the earlier work,[9,15] and a more detailed analysis of recent work involving *in vitro* culture[19-21] and genetic transformation with *Agrobacterium rhizogenes*.[22,23] Hydroxycinnamic amides have also been implicated in tuberization;[10,24] this work is reviewed in Chapter 12.

1. Hydroxycinnamic Acid Amides and Flowering

Conjugated di- and polyamines do not normally occur in leaves or other vegetative tissues.

When a day neutral tobacco plant becomes "ripe to flower", however, several basic HCAs (*p*-coumaroyl-Put, caffeoyl-Put, feruloyl-Put, *p*-coumaroyl-Spd, caffeoyl-Spd, and feruloyl-Spd) appear in the uppermost leaves.[25,26] From the time of ripeness to flower to the emergence of the inflorescence, only the recently initiated leaves accumulate basic HCAs, but the fully expanded leaves of flowering plants contain little or no HCAs. The observations with tobacco have been confirmed in *Pennisetum typhoides* and *Cichorium intybus*.[27,28] Transient increases in conjugated Put, Spd, and Spm in leaves of *Xanthium strumarium* also occur during photoperiodic induction of this short-day plant, although in this case the chemical identity of the conjugated forms has not been established.[29] Once the flowers develop, HCAs accumulate to very high levels and eventually become a significant fraction of the soluble nitrogen pool.[10] Significantly, the male and female flower organs can be separated by their different amide contents. The basic, water-soluble HCAs predominate in stamens, while ovaries contain mainly neutral, water-insoluble forms.[15,16,30,31] The accumulation of specific patterns of HCAs during flowering thus appears to be a general phenomenon. These earlier studies have been complemented with some striking correlations. First, the male flower parts in cytoplasmic male sterile lines of maize (including both Texas and C cytoplasm lines) contain at most only traces of HCAs; in contrast, the HCA patterns of the female parts are unchanged in both male sterile and fertile lines.[30] A lack of HCAs in sterile male and female flower parts of Araceae has also been reported.[18] Second, when appropriate crosses restore normal fertility, this change is accompanied by the reappearance of normal HCA patterns in anthers.[31] Third, a hybrid of tobacco (RMB_7) which lacks the ability to flower shows no detectable levels of HCAs.[32] Fourth, a rapid and substantial accumulation of feruloyl-Put takes place during the initial phase of grain development in maize, followed by an accumulation of the neutral forms later in development. In contrast to the HCA patterns of male and female flower parts, the HCA profiles during the development of grain in male sterile lines are similar to the controls,[31] although the rate of accumulation is decreased. The HCA content falls dramatically during seed germination.

The above results indicate that HCAs could be used as biochemical markers for the reproductive stage, and lead to the suggestion that these compounds may be involved in developmental phenomena such as flowering and tuberisation.[10] Because correlative evidence suggests but does not prove a causal role, studies during the last five years, conducted mainly by Martin-Tanguy et al.,[30-32] have been aimed at furthering these correlations, and at probing HCA and polyamine functions with polyamine inhibitors and genetic changes induced by the transfer of Ri T-DNA. The results obtained in two different systems are reviewed below.

2. Hydroxycinnamic Acid Amides in *In Vitro* Cultured Tobacco Explants

When leaf explants of tobacco are cultured *in vitro* in a medium promoting bud formation, both growth and morphogenesis can be strongly influenced by potassium nutrition.[33] In K^+-deficient medium and in high K^+ medium, growth and bud formation are markedly inhibited. No apparent relationship is found between free Put and growth and bud formation. In contrast, the accumulation of hydroxycinnamoyl Put correlates well with growth and bud formation.[33] In a subsequent study,[19] tobacco leaf explants were grown in several different media: (1) a basal medium, which did not promote cell differentiation or division; (2) basal medium + benzyladenine (BA), conducive to bud formation; and (3) basal medium + 2,4-dichlorophenoxyacetic acid (2,4-D), or 2,4-D + BA, which led to slow and fast growing callus, respectively. The levels of free and HCA-Put remained low in the explants cultured on basal medium. Intracellular levels of both Put and HCA-Put were markedly enhanced when the cultures were induced to proliferate, and higher levels of these metabolites were reached in callus cultures than in bud cultures.[19] Higher levels of HCA-Put were found in nondifferentiated cultures with a slow growth (2,4-D) rate than in those with a fast growth rate (2,4-D + BA).[19] The formation of buds was accompanied by significant changes in Put and HCA-Put. Increasing levels were found during the first 14 days in culture, when cell multiplication was rapid. During the third week of culture the levels declined as the rate of cell division decreased and bud differentiation occurred.

The biosynthetic inhibitors difluoromethylarginine (DFMA) and difluoromethylornithine (DFMO), which inhibit arginine (ADC) and ornithine decarboxylase (ODC), respectively, had striking effects on leaf explants grown on medium supplemented with 2,4-D + BA.[20] Tissues grown on this medium would normally form fast-growing callus, but after addition of DFMO growth was promoted; furthermore, these explants formed buds after two weeks in culture. In contrast, explants treated with DFMA or DFMA + DFMO showed a significant reduction in growth compared with the controls, after the third week in culture.[20] The two latter treatments caused a significant decrease in the titers of free Put, Spd, and Spm. The accumulation of HCA-Put was reduced in all treatments compared to the control explants, but was most marked in DFMA and DFMA + DFMO. Addition of Put or caffeoyl-Put to the DFMO-treated explants did not reverse the bud stimulating effect of DFMO, but caused a significant decrease in fresh and dry weight. Exogenously added Put and caffeoyl-Put were taken up readily (80% by day four), but this was not reflected in the endogenous levels. Previous studies have shown that addition of HCA-Put at higher levels (2.5 to 5.0 x 10^{-5} *M*) to leaf explants can partially inhibit bud formation but promote callus growth.[34] In contrast, addition of HCA-Put had a strikingly different effect on leaf explants of the tobacco hybrid RMB_7. When grown in medium which will promote bud formation (basal + BA) in wild type explants, RMB_7 explants show only limited cell division and complete absence of morphogenesis. No accumulation of free or conjugated Put is detected under these conditions. Upon addition of HCA-Put (5x10^{-5} *M* to 2.5 x 10^{-4} *M*), rapid cell proliferation occurs. It thus appears that RMB_7 cells have altered sensitivity to exogenous HCA-Put. At higher concentrations (10^{-4} *M* to 10^{-3} *M*), exogenous HCAs inhibited cell multiplication without toxic effects in both wild type and RMB_7 explants.[34]

Root formation in tobacco leaf explants can be promoted by addition of 10^{-5} M indoleacetic acid (IAA) to the culture medium.[21] While free Spd and Spm remained low throughout a seven-week culture period, both free and HCA-Put accumulated. Hydroxycinnamoyl-Put appeared after day one of culture. Feruloyl-Put constituted the major part of the HCA-Put pool (80 to 90%). A rapid and substantial accumulation was observed in foliar explants, reaching a maximum level at day 14, and representing 90 to 95% of the total Put pool.[21] Addition of DFMA or DFMO caused total inhibition of root initiation in these leaf explants, correlated with decreases in free and HCA-Put. This inhibition was completely reversed by exogenous supply of Put or feruloyl-Put. These compounds also reversed the inhibition in fresh and dry weight increase caused by DFMA and DFMO. These data suggest that ODC and conjugated putrescines are involved in auxin-induced root formation, and are consistent with other findings showing a close relationship between rhizogenesis, ODC activity, and increase in endogenous Put and Spd levels in leaf explants of *Datura innoxia*,[35] and during root initiation in stem cuttings of mung bean.[36]

In summary, these *in vitro* studies indicate that a change in the morphology or growth of leaf explants is accompanied by changes in free and conjugated Put levels. It is of interest that, while in the whole plant HCAs are mostly associated with flowering structures, excised explants and *in vitro* cultured callus, roots or shoots would also show very substantial changes in these compounds, in particular cinnamoyl-Put. Similar observations have also been made in cell suspension cultures, as discussed below in the section on HCA metabolism. More recently, striking increases in water-soluble and water-soluble HCAs have been shown in the roots of tobacco; these changes are transient, and occur before flowering of the shoot.[22] While these findings may further complicate analysis of HCA function in plants, it is clear that dramatic changes in the metabolism of free and conjugated Put are demonstrable as the cells undergo an alteration in their differentiation state. Plant reproductive organs, however, remain the sites of highest accumulation and complexity of HCA patterns.[10,15,18,30,31]

3. Hydroxycinnamic Acid Amides in *Agrobacterium rhizogenes*-Transformed Tobacco

The function of polyamines and HCAs has recently been approached by following the genetic and physiological changes which occur during plant cell transformation with *Agrobacterium*.[22,23] This soil microorganism can associate with plants and induce morphological responses in the host cells. The bacterial genes responsible for these effects are relatively accessible for two reasons. First, the bacterial genome is less complex than that of the plants. Second, in the case of *Agrobacterium tumefaciens* and *Agrobacterium rhizogenes,* genetic information of developmental importance to the plant is transferred during infection from a bacterial plasmid to the plant cell.[37] Expression of the foreign DNA (transferred DNA or T-DNA) in the host leads to cell division, and in the case of *Agrobacterium rhizogenes* to differentiation. The root inducing (Ri) T-DNA from *Agrobacterium rhizogenes* affects the transformed plant cells such that they divide and generally organize root meristems, but in some cases shoot meristems are produced.[38] Cells derived from transformed roots have an altered sensitivity to auxin.[38,39] Whole plants can be regenerated from roots containing Ri T-DNA, and display a transformed phenotype characterized by wrinkled leaves, shorter internodes, increased branching, delayed flowering, altered geotropic behavior in roots, etc.[38] This phenotype is sexually transmitted as a set of dominant traits that co-segregate with the transformed genotype. Changes in the pattern and degree of expression of genes encoded by the Ri T-DNA affect the degree of phenotypic alteration, in the absence of detectable alterations in the structure. Two levels have been defined, designated T and T′.[38,40] T′ is an exaggerated form of T; in tobacco, internode distance is further shortened; leaf wrinkling is dramatically increased, and flowering is further retarded. Other alterations in reproductive physiology that affect tobacco plants of T phenotype are also exaggerated in T′ plants: flower size, number and seed production are reduced. This phenotype was sexually transmitted as a set of dominant traits that co-segregated with the transformed phenotype.[38] It has been suggested that Ri T-DNA constitutes a source of genes having precise effects on morphology in a wide range of species and should therefore provide basic information concerning the mechanisms governing plant development.[38,40,41]

In a recent study, we followed the changes in polyamine levels in relation to the various phenotypes induced by the Ri T-DNA of *A. rhizogenes*.[22] The alterations in four morphological parameters (time of flowering, height at time of lowering, final height and number of flowers/plant) (Table 1) were consistent with those found previously.[38] Flowering was retarded in plants of T phenotype, and further retarded in T′ plants. When shoot apices and the young, unexpanded leaves were analyzed, it was found that the levels of the Put, Spd, and Spm were significantly lower in the transformed plants, both T and T′ than in the controls. Levels increased after day 40 in plants of all three phenotypes and remained high until the emergence of the floral complex. Relative to the control (N), aliphatic polyamine content was reduced by 40% in the T and by 75% in plants of the T′ phenotype. Thus, the levels of aliphatic di- and polyamines were reduced in direct proportion to the degree of phenotypic change associated with the presence of Ri T-DNA. As shown in earlier work,[25,26] no conjugated amines were detectable in shoot tops of control plants until after day 47.[22] Conjugates became detectable in T and T′ plants at much later dates (day 60 and 75, respectively). From the end of this initial period to the emergence of the floral complex, water soluble conjugates (caffeoyl-Put and caffeoyl-Spd) accumulated exponentially in the upper, recently initiated leaves. The content and distribution of polyamine conjugates in T and T′ plants followed the same pattern observed in the control, but the levels of these amides were reduced, and as found with free polyamines, they were more reduced in T′ than in T plants. However, unlike the free amines, the kinetics of the accumulation of these conjugates were altered in transformed plants, with a delay of 12 days in the T phenotype and 27 days in T′.[22] The levels of free aliphatic amine levels (Put and Spd) in normal flower buds, flowers (just before

TABLE 1
Morphological Parameters in N, T and T′ Plants[22]

Phenotype	Emergence of the floral complex (cm)	Height at emergence of the floral complex (cm)	Final Height (cm) (120 d after sowing)	Number of flowers per plant at 120 d[a]
N	80 (85)	95 (90)	110 (100)	180 (150)
T	90 (95)	80 (82)	100 (90)	100 (95)[b]
T′	110 (110)	28 (25)	30 (30)	25 (20)[c]

Note: Results are expressed as the mean of measurements of 6 plants from a single experiment, with results from a second experiment involving 3 plants given in paranthesis.

[a] 35 leaves developed on all plants at 120 days.
[b] Occasional sterility observed.
[c] Frequent sterility observed.

TABLE 2
Accumulation of Free Amines during Floral Development[22]

	Flower buds	Before anthesis	Anthesis	After fertilization	Green capsules	Seeds
N phenotype						
Aliphatic polyamines	0.57 ± 0.04[a]	0.81 ± 0.04	0.76 ± 0.04	0.58 ± 0.04	0.20 ± 0.04	0.20 ± 0.04
Tyramine	0.70 ± 0.04	5.00 ± 1.00	5.50 ± 0.50	5.00 ± 0.50	2.00 ± 0.20	1.90 ± 0.20
T phenotype						
Aliphatic polyamines	0.37 ± 0.04	0.51 ± 0.04	0.48 ± 0.04	0.40 ± 0.04	0.10 ± 0.04	0.10 ± 0.04
Tyramine	0.65 ± 0.04	5.00 ± 0.50	5.00 ± 0.05	4.75 ± 0.50	2.00 ± 0.20	2.00 ± 0.20
T′ phenotype						
Aliphatic polyamines	0.07 ± 0.04	0.09 ± 0.04	0.09 ± 0.04	0.08 ± 0.04	0.10 ± 0.04	0.10 ± 0.04
Tyramine	0.35 ± 0.04	2.00 ± 0.50	2.50 ± 0.50	2.00 ± 0.50	1.00 ± 0.20	1.00 ± 0.20

[a] Results are presented in μmol/g fr. wt, as means ± the SD of 4 to 5 replicates.

anthesis, at anthesis and after fertilization), green capsules and seeds are shown in Table 2. Polyamine levels increased slightly during anthesis, followed by a marked decrease after fertilization. Polyamine titers then continued to decline during seed ripening. As shown above, polyamine levels in T and T′ plants remained lower than in N plants. Tyramine represented 80 to 90% of the total amine pool in the reproductive organs (Table 3). In plants of T phenotype, tyramine levels were similar to those observed in the controls, but a major decrease occurred in T′ plants. HCAs were found in large quantities in the inflorescence, mainly in the flowers. The accumulation of water soluble and insoluble HCAs was greatly reduced in T and T′ plants (Tables 3 and 4), and as observed for the other parameters the reduction was greater in T′ plants.

These results clearly show an inverse relationship between the level of phenotypic change and the level of polyamines and their conjugates. Notably, the appearance of the conjugates is delayed in direct proportion to the delay observed in flowering. These results strengthen the hypothesis that HCAs are a driving force in the transition from a vegetative to a floral meristem, and suggest that the effects of genes encoded by the Ri TL-DNA on flowering, and perhaps on other characters as well, may be brought about through an inhibition of polyamine biosynthesis. A recent report supports this suggestion.[23] When DFMO was applied to tobacco plants during

TABLE 3
Accumulation of Water Soluble Polyamine Conjugates during Floral Development[22]

	Flower buds	Before anthesis	Anthesis	After fertilization	Green capsules	Seeds
N phenotype						
Caffeoyl putrescine	6.0 ± 0.5[a]	6.00 ± 0.05	8.50 ± 1.1	2.0 ± 0.1	0.0	0.0
Caffeoyl spermidine	6.5 ± 0.5	14.0 ± 1.5	19.5 ± 1.5	6.0 ± 0.7	0.0	0.0
T phenotype						
Caffeoyl putrescine	3.0 ± 0.5	3.0 ± 0.05	5.0 ± 0.4	1.0 ± 0.04	0.0	0.0
Caffeoyl spermidine	4.0 ± 0.5	9.0 ± 1.1	10.0 ± 1.2	3.0 ± 0.5	0.0	0.0
T′ phenotype						
Caffeoyl putrescine	1.0 ± 0.04	1.0 ± 0.04	2.0 ± 0.04	1.0 ± 0.04	0.0	0.0
Caffeoyl spermidine	2.0 ± 0.20	5.0 ± 0.20	6.0 ± 0.20	1.0 ± 0.04	0.0	0.0

[a] Results are presented in μmol/g fr. wt, as means ± the SD of 4 to 5 replicates.

TABLE 4
Accumulation of Water-Insoluble Amine Conjugates during Floral Development[22]

	Flower buds	Before anthesis	Anthesis	After fertilization	Green capsules	Seeds
N phenotype						
di-*p*-Coumaroyl put	0.75 ± 0.04[a]	1.10 ± 0.2	1.70 ± 0.2	0.0	0.0	0.0
di-*p*-Coumaroyl spd	0.85 ± 0.04	1.50 ± 0.2	3.00 ± 0.5	0.0	0.0	0.0
p-Coumaroyl tyr	0.45 ± 0.04	0.70 ± 0.04	1.20 ± 0.2	0.0	0.0	0.0
Feruloyl tyr	0.0	0.0	0.0	1.60 ± 0.2	4.5 ± 0.2	10.5 ± 1.5
T phenotype						
di-*p*-Coumaroyl put	0.45 ± 0.04	0.60 ± 0.04	0.90 ± 0.0	0.0	0.0	0.0
di-*p*-Coumaroyl spd	0.50 ± 0.04	0.80 ± 0.04	1.10 ± 0.02	0.0	0.0	0.0
p-Coumaroyl tyr	0.30 ± 0.04	0.50 ± 0.04	0.80 ± 0.04	0.0	0.0	0.0
Feruloyl tyr	0.0	0.0	0.0	0.75 ± 0.04	2.7 ± 0.2	9.50 ± 1.2
T′ phenotype						
di-*p*-Coumaroyl put	0.20 ± 0.04	0.25 ± 0.04	0.30 ± 0.04	0.0	0.0	0.0
di-*p*-Coumaroyl spd	0.30 ± 0.04	0.30 ± 0.04	0.45 ± 0.04	0.0	0.0	0.0
p-Coumaroyl tyr	0.10 ± 0.04	0.30 ± 0.04	0.40 ± 0.04	0.0	0.0	0.0
Feruloyl tyr	0.0	0.0	0.0	0.50 ± 0.04	1.8 ± 0.1	5.0 ± 1.2

[a] Results are presented in mmol/g fr. wt, as means ± the SD of 4 to 5 replicates.

floral induction, it induced a phenotype which resembled that obtained by transformation with Ri T-DNA.[22] Treatment of transformed plants showing the T phenotype with DFMO accentuated the phenotypic abnormalities. Treatment with DFMA had no effect; other inhibitors of polyamine biosynthesis such as cyclohexylamine (CHA) and methylglyoxal bis (guanylhydrazone) (MGBG) produced reproductive abnormalities without clearly mimicking the transformed phenotype. Based on these findings, our working hypothesis is that the Ri TL-DNA encodes genes which repress polyamine formation by altering the activity of enzymes encoded by the plant genome (e.g., ornithine decarboxylase and/or arginine decarboxylase).

4. Hydroxycinnamic Acid Amides and Disease Responses

Changes in HCAs also occur during plant-pathogen interactions, especially during the hypersensitive reaction towards viruses or fungi. In tobacco plants, inoculation of leaves with tobacco mosaic virus (TMV) can either produce a systemic infection, as in *Nicotiana tabacum* var. Samsun, or necrotic local lesions, as in *N. tabacum* Xanthi n.c. The latter response is called a hypersensitive reaction, in which the virus remains largely confined to the necrotic lesions.[42,43] In tobacco Xanthi n.c., the behavior of TMV depends on the temperature at which the plants are grown. Below 29°C, TMV infection induces a typical hypersensitive reaction, but at 32°C the infection becomes systemic.[42-44] When the infection occurs at the lower temperature, large amounts of feruloyl-Put, diferuloyl-Put, and feruloyltyramine are formed during the hypersensitive reaction, appearing at the time of lesion formation and accumulating especially in the living cells surrounding the necrotic spot (hypersensitive zone).[45,46] When *N. sylvestris* is infected with TMV strain Aucuba, towards which this plant is hypersensitive, diferuloyl-Put and feruloyltyramine are formed during the response.[45] An increase in temperature, which leads to inhibition of flowering and systemic TMV infection, also suppresses the accumulation of HCAs.[45] Similar correlations are obtained with the *Nicotiana* hybrid RMB_7, obtained from the seventh backcross on *N. rustica* × *N. tabacum* var. Maryland Mammoth.[47] This hybrid, which has lost its capacity to flower, does not synthesize any detectable HCAs at any stage during development. When infected with TMV, it initially produces necrotic lesions as in a typical hypersensitive response, but eventually the virus spreads systemically and the plant dies.[32] No HCAs are produced during infection of RMB_7 plants. The application of coumaroyl-, dicoumaroyl–, and caffeoyl-Put (0.25 to 0.75 mg/ml) to leaf discs of tobacco caused a 90% reduction in the number of local lesions induced by TMV.[46] Because the rate of viral multiplication is inversely correlated with the formation of amides and appears to be inhibited in their presence, these results suggest a role for HCAs in virus resistance. As discussed above, HCAs are present in reproductive parts; it is well known that plant viruses are rarely present in meristems, sex organs, and seeds. Hydroxycinnamic acid amides may thus confer virus resistance to the plant reproductive tissues. Little is known about the biosynthesis of HCAs during the hypersensitive response (see also section below). The activity of ODC increases 20-fold in tobacco leaves following TMV infection at 20°C, but there is little or no change at 32°C, when the infection becomes systemic.[48] Ornithine decarboxylase activity parallels the activity of phenylalanine ammonia lyase (PAL) and tyrosine decarboxylase (TAL) during the hypersensitive reaction.[49] A transferase activity catalyzing the linkage of feruloyl-CoA thioester to tyramine (TFT) has been detected in crude extracts of tobacco leaves ,[50] and increases five- to eight-fold following TMV infection. However, in contrast with ODC, PAL, and TAL, TFT activity reaches its maximum much later during the infection process. It has also been reported,[51] using ^{14}C-tyramine labeling, that HCAs synthesized during the hypersensitive response bind to the cell wall. Because HCAs are good substrates for peroxidase *in vitro*,[52] it has been suggested that they can form insoluble polymers *in vivo* and thus play a role in host-pathogen interactions.[10]

Hydroxycinnamic acid amides also appear to be involved in the hypersensitive response to fungal infection. For example, one of the earliest reactions of potato tuber inoculated with avirulent isolates of *Phytophthora infestans* is the accumulation of *p*-coumaric and ferulic acid amides with tyramine or octopamine.[53] A water-soluble elicitor from *Phytophthora parasitica* can induce resistance at the cut ends of carnation cuttings, or biosynthesis of antifungal phytoalexins.[54] In the former case, a set of highly fluorescent compounds appears, some of which have been identified as HCAs. The resistance of young barley seedlings to infection by *Helminthosporium sativum* has been attributed to the occurrence of dimers of 4-coumaroyl-agmatine in the shoots (see Chapter 1).[9] These compounds, known as hordatines, inhibit the germination of a wide range of fungal spores at low concentrations (1 to 10 μM).[55,56] Finally, HCAs can form complex structures such as macrocyclic alkaloids, for which a function as chemical defenses against pathogens has also been proposed.[9]

C. METABOLISM OF HYDROXYCINNAMIC ACID AMIDES

The results discussed in sections II.B.1 to 3 clearly show that HCAs undergo dramatic changes during plant growth and development. The transient increases and later disappearance of HCAs during seed germination,[9,31] in roots before the plant flowers[22] and in leaf callus before organogenesis,[19-21] suggests that at least under some conditions HCAs can be actively metabolized. We still know very little about the enzymology and regulation of HCA biosynthesis and turnover. The information currently available has been obtained mostly from studies in plant cell suspensions and explants undergoing organogenesis (see Section II.B.2), both of which show relatively simple patterns of HCAs, predominantly cinnamoyl-Put.

The early insights into HCA metabolism in cell cultures come from the work of Palmer, Berlin, and Widholm in the 1970's, on selection of cell lines resistant to amino acid analogs.[57,58] In an attempt to increase the content of phenylalanine (Phe) in carrot and tobacco, cell cultures were selected for resistance to *p*-fluorophenylalanine (PFP).[57] While the PFP-resistant carrot cell line showed the expected increases in the level of free Phe, the PFP-resistant tobacco cells (TX4 line) showed almost normal levels of Phe, but a remarkable increase in phenolics. This was paralleled by 10- to 20-fold rise in PAL activity in the TX4 cells compared to the control ("wild type") line TX1.[58] It was suggested that the basis for PFP-resistance in the TX4 cell was their ability to detoxify PFP by converting it to *p*-fluorocinnamic acid. Further characterization of the TX4 cells showed that the major phenolic compounds were caffeoyl-, feruloyl-, and coumaroyl-Put.[59] Remarkably, TX4 cells showed similar batch growth kinetics to TX1 cells, but were able to maintain very high levels of cinnamoyl-Put throughout the culture period, representing up to 5 to 10% of the cell dry weight. The accumulation of HCAs in TX4 cells was correlated with increased activities of PAL, *trans*-cinnamate-4-hydroxylase, and *p*-coumarate ligase, all involved in the hydroxycinnamic acid pathway.[59] PFP-resistant cells also showed increased activities of ADC and ODC, which presumably contributed to the Put precursor pool. Thus, although the selection pressure should have affected the levels of Phe, the resulting PFP-resistant tobacco cells were also able to coordinately divert a large portion of another amino acid pool (Arg, Orn) into the cinnamoyl-Put end product. Several other independent cell lines selected for resistance to a similar analog, *DL*-metafluorophenylalanine, also accumulated two- to five-fold higher levels of cinnamoyl-Put than the wild type cells.[60]

The conjugation of agmatine and tyramine to cinnamoyl moieties is catalyzed by an enzyme which utilizes HCA-CoA thioesters as the acyl donors,[50,55] and it was thus assumed that HCA-Put may also be formed via CoA thioesters. Because feeding experiments with ^{14}C-labeled Phe and Put suggested that the conjugating enzyme activity might be present at higher levels in TX4 than in TX1 cells,[61,62] the former line was used as the source to search for the enzyme involved in cinnamoyl-Put formation. A hydroxycinnamoyl-CoA:putrescine hydroxycinnamoyl-transferase (PHT) was in fact detected in TX4 cells.[63] The enzyme accepted only the aliphatic diamines Put, cadaverine (Cad), and diaminopropane (Dap) at a ratio of 100:33:8.; the best acyl donors were caffeoyl- and feruloyl-CoA. PHT activity was similar in high and low HCA-producing cells (TX4 and TX1), in contrast with other enzymes (ADC, ODC, PAL, etc.) which are enhanced only in TX4 cells. This result suggests that PHT is not a rate-limiting enzyme for HCA-Put formation. Independently, Negrel et al. have detected a PHT from tobacco callus.[64] The enzyme was purified 1000-fold (19% yield) from tobacco cell suspensions, had a pH optimum of 10.0, and apparent Km values for caffeoyl-CoA and Put of 3 and 10 μM, respectively. The molecular weight estimated by gel filtration was 50 KD. In the same system, an enzyme activity which catalyzes the formation of HCA-Spd conjugates has been recently reported.[65]

In addition to cell lines resistant to Phe analogs, HCA-Put can also accumulate in tobacco cells grown in phosphate-limited medium[66] and upon changes in the nitrogen regime. Plant cell cultures usually grow in media containing inorganic nitrogen sources (nitrate, ammonium, or a combination of both). It is known, however, that at least tobacco cell lines are capable of growth in a variety of organic nitrogen sources.[67] The tobacco XD line, for example, can grow on urea, γ-aminobutyric acid (GABA) or arginine (Arg) as sole nitrogen sources. Although growth on urea may require a period of adaptation, these cells can be transferred into the other organic nitrogen sources mentioned above, and grow at the same rate as the control cells (grown on KNO_3) from the first passage. We have shown that in this system, transfer of the cells to a medium containing an organic nitrogen source, or to one containing ammonium, results in dramatic effects on the patterns of free and conjugated Put.[67] While cells grown on KNO_3 contained low levels of cinnamoyl-Put, those transferred to GABA or urea (14U cell line) accumulated HCA-Put to a level of 5% dry weight. Because all these cell lines are grown in 3 mM nitrogen (organic or inorganic), such a dramatic diversion of nitrogen into a secondary metabolite while maintaining similar growth rates in all cases suggests that these cell suspensions have a metabolic plasticity deserving further study. We have also found that while the XD cell line does not normally grow on Put as the sole nitrogen source, it can be selected for growth on this diamine. A Put-utilizing (PUT) line was thus established and characterized.[67] We found that PUT cells efficiently converted ^{14}C-Put to ^{14}C-GABA; in contrast, XD cells grown in any of the other nitrogen sources were only able to convert Put into polyamines of HCA-Put. Both PUT cells grown on Put and XD cells grown on GABA were killed by 1 to 10 nM gabaculine, an inhibitor of GABA transaminase, while XD cells grown on KNO_3 or urea were about 1000 times less sensitive to the inhibitor. Our results suggest that PUT cells express a pathway which metabolizes Put into GABA. As discussed in Chapter 3, it is well known that Gramineae and Leguminosae species can oxidize Put to GABA via a diamine oxidase and a NAD-dependent γ-aminobutyraldehyde dehydrogenase. However, to our surprise we could not detect any of these enzyme activities in the PUT cells, even though they are capable of converting ^{14}C-Put to ^{14}C-GABA. Several lines of evidence now support the hypothesis that PUT cells are able to metabolize Put to GABA via a cinnamoyl intermediate. First, PUT cells are much more sensitive to aminooxyphenylpropionic acid (AOPP), an inhibitor of PAL, than the XD cells.[68] Because AOPP effectively blocks the synthesis of cinnamic acids and, therefore, of HCA-Put, these results suggest that PUT cells need HCA-Put for an essential process. Second, when ^{14}C-Put is fed to XD cells grown in urea or GABA, the label can be chased into HCA-Put. In contrast, a similar pulse-chase experiment utilizing PUT cells shows an apparent turnover of HCA-Put into GABA.[68] Pulse-chase of detached tobacco flowers with ^{14}C-Put gives similar results to those with PUT cells. Third, a compound identified as caffeoyl-GABA has been isolated from PUT cells, but is absent from XD cells.[69] Because this compound would be a presumptive intermediate in a Put catabolic pathway involving cinnamoyl conjugates, this finding is also consistent with our hypothesis. Caffeoyl-GABA has also been found in tobacco flowers, but is absent from the vegetative organs.[69] Thus, PUT cells appear to behave biochemically like reproductive cells, both synthesizing and turning over the HCA-Put, whereas XD cells have the ability to accumulate HCA but do not seem to metabolize these compounds. In summary, cell cultures have proved useful in the characterization of HCA physiology and biochemistry, and should continue to provide useful insights into the mechanisms regulating the synthesis and turnover of conjugated polyamines.

III. POLYAMINE-DERIVED ALKALOIDS

A. OCCURRENCE AND DISTRIBUTION

Of all the polyamine-derived secondary metabolites found in higher plants, the hydroxycinnamic acid amides discussed above are the most widespread. However, these compounds represent only a small fraction of the secondary metabolites which can be formed using Put, Spd, and Spm as precursors. Several hundred alkaloids derived from di- and polyamines have been reported, distributed among about 56 genera in 30 plant and animal families;[70] the majority of these compounds are found in plant species. The Put-derived alkaloids include the pyrrolidine and tropane alkaloids of the Solanaceae, the pyrrolizidine alkaloids of the Asteraceae, Boraginaceae, and Leguminosae, the *Erythroxylon coca* alkaloids (hygrine and cuscohygrine), and the phenanthro-indolizidines (tylophorine).[70] Spermidine, homoSpd, and Spm can conjugate with fatty acids or cinnamoyl derivatives to form complex structures called macrocyclic alkaloids.[9] Spermidine-derived alkaloids proper include the oncinotines and inandentines of Apocynaceae, the palustrines of *Equisetum,* the maytenins of Celastraceae, and the *Cannabis* alkaloids. Solamines are homoSpd-type alkaloids, while the Spm derivatives include the pithecolobines, verbascenine, and homaline. Cadaverine is a precursor for the piperidine alkaloids of *Nicotiana* and the quinolizidine alkaloids of *Lupinus.* Some plants may contain alkaloids derived from all three aliphatic amines (Put, Spd, and Spm), as in the case of *Ananas comosus*; others such as *Lunaria biennis* (Cruciferae) contain many different alkaloids. Because of their pharmacological interest, intensive investigations have been performed on the alkaloids of *Maytenus*, *Equisetum*, and *Aphelandra*.[70] For the most part, however, we still know very little about the polyamine-derived alkaloid pathways, and even less about how they are integrated with the rest of polyamine metabolism (biosynthesis, turnover, and conjugate formation). We will briefly review a few selected alkaloid pathways.

B. NICOTINE AND TROPANE ALKALOIDS

Most of our knowledge on the synthesis and regulation of Put-derived alkaloids comes from studies of the pyrrolidine (nicotine) and tropane alkaloids of the Solanaceae. Nicotine is the major biologically active component of tobacco leaves and is a feeding deterrent in many animals, as well as a natural insecticide.[71] In fact, it was commercially used as an insecticide for many years until its application was discontinued due to its high toxicity to humans. Although it is usually found in *Nicotiana* and other Solanaceae, nicotine-type alkaloids have also been reported in other plant families and even in lower plants such as the primitive lycopods and horsetails.[72] The tropane alkaloids, present in the nightshade Solanaceae (*Atropa, Brugmansia, Datura, Duboisia,* and *Hyoscyamus*), include the medically important anticholinergic compounds hyoscyamine and scopolamine. In addition to the Solanaceae, these alkaloids are also found in the Convolvulaceae, Cruciferae, and Rhizophoraceae.[73]

The nicotine and tropane alkaloid pathways share several fascinating biological features which unfortunately have been largely neglected, although they have been known for many decades. Both alkaloids are found in large amounts in leaves of Solanaceae, and for many years it was assumed that they were synthesized in these organs. In a series of classic but almost forgotten experiments, Ray Dawson showed that this assumption was invalid.[74] By grafting a nicotine-producing Solanaceae (tobacco) with a non-nicotine producing species of the same family (tomato), he was able to show that the pattern of nicotine alkaloids in the shoots and leaves was dependent on the source of the rootstock. Tobacco shoots grafted onto tomato rootstocks showed little or no nicotine in the leaves. In contrast, tomato shoots grafted over tobacco roots were able to accumulate large amounts of the alkaloid. Based on these results, Dawson suggested that in the tobacco plant the root is the source of most, if not all, the alkaloid pool.[74] At that time, the pioneering work of Phillip White[75] had resulted in the first reproducible plant tissue culture system, the growth in isolation of tomato roots for potentially unlimited periods of time. Dawson applied this system to tobacco, and conclusively showed that isolated root cultures of this plant were capable of synthesizing nicotine.[76] It was later shown that tropane alkaloids are also synthesized in the roots;[72] in both cases, the alkaloids are transported to the shoot via the xylem, were they accumulate in the leaves and are further metabolized. The root-associated expression of these alkaloids is remarkably stable and specific. When nicotine-producing and tropane alkaloid-producing Solanaceae are reciprocally grafted, the pattern of alkaloids in the leaves is still determined by the root component of the graft. This is interesting considering that the nicotine and tropane alkaloids share Arg/Orn as the common precursors for the alkaloid ring.[72,73]

Root cultures were a useful experimental system during the early days of plant tissue culture, until they were displaced in the 1960s when fast growing callus and cell suspension became available. This was in retrospect a poor choice, as far as secondary metabolism is concerned, because it is now well known that in many cases disorganized plant cell cultures produce very low levels of secondary metabolites. This has been particularly valid for nicotine and tropane alkaloids. In recent years, root cultures have made a comeback as an experimental system, largely due to the use of plant cells genetically transformed with *Agrobacterium rhizogenes* (see section II.B.3).[38] The so-called "hairy roots" obtained by transformation with the Ri T-DNA can grow at much faster rates than normal roots, while still being capable of expressing the root-specific metabolic pathways.[67,77-79] Using this approach,[67] we were first able to show that "hairy root" cultures of *Nicotiana tabacum* and *Hyoscyamus muticus* were able to produce nicotine and hyoscyamine, respectively.[67] The original *Hyoscyamus* root clones have been growing and producing tropane alklaloids at the same rate for over eight years.[80] Since this report, the production of Solanaceae alkaloids and other root-specific natural products has been shown by many groups in numerous plant species.[77-79] The availability of normal and transformed root cultures has made possible recent progress in our understanding of the Put-derived alkaloid pathways.

The pyrrolidine ring of the nicotine alkaloids derives ultimately from Arg and/or Orn. For many years it had been assumed that Put and methylPut, which are the immediate precursors of the pyrrolidine ring, were derived mainly from ODC. Several recent findings have questioned the validity of this assumption.[81] While both ADC and ODC activities are present in tobacco callus, it is only the former which closely parallels the changes in alkaloid levels during culture. In addition, ^{14}C-Arg is a much more efficient precursor of nicotine than ^{14}C-Orn. Consistent with the above observations, DFMA is a more effective inhibitor of nicotine and nornicotine accumulation than DFMO. Ornithine decarboxylase profiles show little change under conditions which lead to accumulation of nicotine in tobacco callus,[82] which further supports the above findings. When tobacco roots are fed with 2-^{14}C-Orn, the label is incorporated symmetrically into the carbons adjacent to the N-methyl group of the pyrrolidine ring.[83] These results suggest that Put derived via ADC is the direct, symmetric precursor of the nitrogen ring. Spermidine can be an indirect precursor of nicotine, via its prior oxidation to Put.[84] The currently accepted scheme for the incorporation of putrescine into the pyrrolidine ring of nicotine involves its methylation, oxidation of this derivative to 4-methylaminobutyraldehyde, and cyclization of this product.[72,83] Except for the early methylation and oxidation steps, still little is known about the enzymology of the nicotine pathway. The availability of the "hairy root" system and engineered Ri plasmids have made possible the first attempts in altering this alkaloid pathway by a genetic engineering approach. Hamill et al.[85] recently transformed *Nicotiana rustica* cells with an *A. rhizogenes* strain carrying a gene for yeast ODC in the Ri plasmid, under the control of the constitutive promoter CaMV 35S. The expression of the yeast ODC gene in the transformed root cultures was correlated with a significantly higher ODC activity in a number of independently transformed root clones, compared to the control "hairy roots" (transformed with the same vector, but containing the chloramphenicol acetyl transferase gene instead of the yeast ODC gene under control of the same promoter). The levels of free Put and nicotine were consistently higher in the root clones expressing the yeast ODC gene.[85]

Both the nicotine and tropane alkaloid pathways share the same amino acid precursors for the synthesis of the pyrrolidine and tropane rings, respectively.[72,83] These rings are then joined with a moiety of nicotinic acid (for nicotine) or tropic acid (for hyoscyamine and scopolamine), which is in turn derived from Phe. The enzymology of the final steps in both pathways is poorly understood. In "hairy roots" of *Datura stramonium*, inhibition of ADC activity by DFMA resulted in a dramatic decrease in free and cinnamoyl-Put, as well as in the Put-derived intermediates of hyoscyamine synthesis.[86] In contrast, inhibition of ODC by DFMO resulted in an elevation of ADC activity, but had little effect on the levels of Put and tropane alkaloids. These results are similar to those in tobacco callus,[83] in which Put derived via ADC is associated with alkaloid formation, and also suggest that the two alternative routes for Put formation are closely regulated. Until quite recently, the biosynthetic origin of the tropane ring was thought to be very different than that of the pyrrolidine ring. Using the same approach discussed above for tobacco roots,[83] Leete showed many years ago that when 2-^{14}C-Orn was fed to *Datura stramonium*, the label was found only in the C-1 bridgehead of the tropane ring.[87] This was in sharp contrast with the symmetrical incorporation found in the pyrrolidine ring of nicotine, and was used by Leete to argue for the existence of an asymmetric intermediate. It was proposed that instead of free Put being derived from ODC or ADC, Orn was first methylated at the δ-amino group. δ-*N*-methylOrn was then decarboxylated to methylPut and subsequently oxidized and cyclized.[83] This scheme has been recently challenged by Hashimoto et al. in a series of elegant papers.[88-89] First, using normal root cultures of *Hyoscyamus albus*, these researchers were not able to detect any of the enzymatic activities in the scheme proposed by Leete, namely, δ-*N*-Orn methyltransferase or δ-*N*-methyl-Orn decarboxylase.[88] In contrast, a very sharp peak in the activity of Put-*N*-methyl transferase (PMT) was observed, which preceded the accumulation of hyoscyamine; ADC, ODC, and arginase activities in turn preceded the peak in PMT. When 1,4,^{14}C-Put was fed to root cultures, about 6% of the label was incorporated into *N*-methyl-Put within a few hours, followed by incorporation into hyoscyamine.[89] Incubation with ^{14}C-Arg or -Orn resulted in labeling of Put and *N*-methyl-Put; very little radioactivity was recovered in δ-*N*-Orn. These results thus indicate that in *H. albus* root cultures the tropane ring of hyoscyamine is synthesized from both Orn and Arg, and strongly argue for the synthesis of *N*-methyl-Put and the tropane ring via a symmetric intermediate. Chemical degradation studies of hyoscyamine derived from labeling with ^{14}C-Orn, done between Leete and Yamada's groups indicate that in fact the label from this precursor is incorporated symmetrically.[89] In *Duboisia* root cultures, which make both nicotine and tropane alkaloids,[90] the tropane and pyrrolidine rings both appear to be made via a symmetrical intermediate. It still remains possible, however, that in Solanaceae such as *Datura* an asymmetric pathway is present. Because high activities of PMT are found in *Datura* roots,[91] this issue must be revisited in this species using the approaches discussed above.

Other aspects of the enzymology of tropane alkaloids have also been studied in *Hyoscyamus* root cultures. A diamine oxidase (DAO) has been purified and characterized in *H. niger* roots.[92] Although the enzyme does oxidize free Put and Cad, the specificity constants for these substrates are 11 and 1%, respectively, of those for the putative endogenous substrate, *N*-methyl-Put. It was thus proposed that this root-specific DAO is involved in the synthesis of tropane alkaloids. The other major tropane alkaloid produced in Solanaceae, scopolamine, is the 6,7-epoxide of [*SI*]-hyoscyamine. The first step in this epoxidation is the hydroxylation of hyoscyamine to 6β–hydroxyhyoscyamine. This reaction is catalyzed by hyoscyamine-6β-hydroxylase (H6H), a 2-oxoglutarate-dependent dioxygenase which has been purified from root cultures of *H. niger*.[93] A cDNA coding for H6H has been cloned, and Western blot analysis has confirmed the presence of the enzyme in roots, and shown it to be absent from all other plant organs.[94] Using immunogold staining, it has also been shown that the enzyme is localized in the pericycle, a single cell layer located immediately outside the root vascular cylinder.[94] A 6β-hydroxyhyoscyamine epoxidase has been reported from *H. niger* roots.[95] The enzyme activity is considerably lower than that for H6H; the activities for these two enzymes could not be separated during partial purification, raising the possibility that the epoxidase and hydroxylase functions are carried in the same polypeptide. The significance of this epoxidase for scopolamine synthesis is not yet clear.

C. PYRROLIZIDINE AND QUINOLIZIDINE ALKALOIDS

The pyrrolizidine alkaloids have been reported in several plant families (Asteraceae, Boraginaceae, Leguminosae), and are thought to function as chemical plant defenses.[71,96] Some insects can accumulate pyrrolizidine alkaloids and use them to repel potential predators, while others can modify them into sex pheromones.[97] The biosynthesis of pyrrolizidine alkaloids has been studied primarily in *Crotalaria* and *Senecio*.[96,98] Results from radiolabeling experiments suggest that Orn, Arg, Put, Spd, homoSpd, and Spm can be biosynthetic precursors of the necine base, presumably via a symmetrical C_4 intermediate such as Put.[73] Hartmann et al.[96] showed a close relationship between the biosynthesis of polyamines and senecionine *N*-oxide, the major pyrrolizidine alkaloid produced in root cultures of *S. vulgaris*. Treatment with DFMA resulted in a complete inhibition of the incorporation of labeled Orn and Arg into polyamines (Spd and Spm) and alkaloids, but DFMO had no effect. Combined with the finding that Agm was as efficiently incorporated into senecionine *N*-oxide as Put, these results suggest, as also shown above for nicotine and tropane alkaloids,[81,86] that the pathway flows completely through Arg and Agm, while ODC makes little if any contribution. Detached shoots of *S. vulgaris* were unable to incorporate label from ^{14}C-Arg, ^{14}C-Put, or ^{14}C-Spd into pyrrolizidine alkaloids, supporting the hypothesis that roots are the site of synthesis.[96,99] However, these same precursors were taken up very efficiently by roots and incorporated into senecionine *N*-oxide (3 to 6%). Considerable amounts of this labeled alkaloid were translocated into the shoot and accumulated in the inflorescences. Because the translocation of senecionine *N*-oxide was blocked when the stem was "steam girdled", it was suggested that the root to shoot transport of the alkaloid occurs mainly via the phloem.[99]

The quinolizidine alkaloids are represented by about 70 structures and occur commonly in the Leguminosae, especially in the genus *Lupinus*, but have also been reported in the Berberidaceae, Chenopodiaceae, Papaveraceae, Ranunculaceae, and Solanaceae.[97] The biosynthesis of these alklaloids occurs mainly in the aerial plant organs.[100-104] Biosynthetic studies using an enzyme preparation from a suspension culture of *L. polyphyllus* showed the incorporation of three Cad moieties into a tetracyclic quinolizidine alkaloid, using pyruvate as a substrate for four required transaminations.[100,103] This enzyme as well as lysine decarboxylase (LDC), which catalyzes the first biosynthetic step, were subsequently localized in the chloroplast stroma.[101] Activity of LDC was positively correlated with quinolizidine alklaloid formation in cultures of *Heimia salicifolia*.[102] It has been suggested that, in addition to functions in nitrogen storage and transport,[103] quinolizidine alkaloids can also regulate protein synthesis.[104]

IV. CONCLUDING REMARKS

The metabolism and function(s) of polyamine-derived secondary metabolites is still poorly understood. Recent progress has been made on the physiology and enzymology of hydroxycinnamoyl conjugates, and we can expect that within the next five years we will have an adequate metabolic scheme for HCA synthesis and turnover. This knowledge will in turn help us address the significance of these compounds for growth and development. Although very intriguing at present, the evidence for involvement of HCAs in morphogenesis remains largely correlative. We suggest that careful attention should be given in future studies to the contribution of HCAs to the soluble nitrogen pool, especially in situations when the apparent turnover is very high (e.g., germinating seeds). The availability of purified enzymes and cloned genes for this pathway should be a powerful complement to the tissue culture systems and biosynthetic inhibitors. The metabolism of polyamine-derived alkaloids remains another exciting area of research. We have barely scratched the tip of the metabolic iceberg, and many pathways remain untouched. Because several polyamine-derived alkaloids show promise as pharmacological agents, such studies should be a priority and hold a large biotechnological potential. The pathways for nicotine and tropane alkaloids are just becoming open for molecular inquiries and manipulation, and we foresee exciting developments in the near future. The metabolism of Put in roots also presents us with some fascinating biochemical puzzles. It is known, for example, that tobacco roots can accumulate cinnamoyl-Put during certain developmental stages.[22] Since Put is the precursor of Spd and Spm, which are tightly linked with all rapidly dividing cells including root meristems, and in roots it is also the precursor for nicotine, it should be of interest to uncover the metabolic controls which partition Put pools into "primary" and "secondary" pathways. This allocation of precursor can get even more complicated in certain Solanaceae which synthesize both nicotine and tropane alkaloids in the roots.[90] The implications for the whole plant are no less challenging. It has recently been found[105] that tobacco plants respond to herbivore damage by accumulating substantial amounts of nicotine. This begs the question of how the upper parts signal the underground organs to increase the production of nicotine, and of the metabolic and ecological consequences of tying up such a large fraction of the soluble nitrogen pool into a single compound. Will we be able to modify such responses by genetic engineering? We believe it is reasonable to conclude that, far from being boring or "secondary", the study of polyamine-derived natural products holds at least as many surprises and promises as that of the major intermediary pathways.

ACKNOWLEDGMENTS

We would like to thank Ms. Paula Sgrignoli for editorial help. J. M.-T. is grateful to Dr. C. Martin for suggestions, and to Drs. Pierre Mamont and Peter McCann (Merrel Dow Research Institute, Strasbourg, France, and to Marion Merrel Dow, Cincinnati, OH) for the generous supplies of DFMO and DFMA, respectively. The work discussed by H. E. F. was funded in part by the National Science Foundation (EET-8713078).

REFERENCES

1. **Flores, H.E.,** Studies on the physiology and biochemistry of polyamines in higher plants, Ph.D. thesis, Yale University, New Haven, 1983.
2. **Seiler, N.,** Acetylation and interconversion of the polyamines, in *The Physiology of Polyamines*, Vol. I, Bachrach, U. and Heimer, Y. M., Eds., CRC Press, Boca Raton, FL, 1989, 159.
3. **Morgan, D.M.L.,** Polyamine oxidases and oxidized polyamines, in *The Physiology of Polyamines*, Vol. I, Bachrach, U. and Heimer, Y. M., Eds., CRC Press, Boca Raton, FL, 1989, 203.
4. **Flores, H.E., Protacio, C.M., and Signs, M.W.,** Primary and secondary metabolism of polyamines in plants, *Rec. Adv. Phytochem.*, 23, 329, 1989.
5. **Tiburcio, A.F., Kaur-Sawhney, R., and Galston, A.W.,** Polyamine metabolism, in *The Biochemistry of Plants*, Vol. 16, Miflin, B. J., Ed., Academic Press, Orlando, FL, 1990, 283.
6. **Folk, J. E.,** Transglutaminases, *Annu. Rev. Biochem.*, 49, 517, 1980.
7. **Serafini-Fracassini, D., Del Duca, S., and D'Orazi, D.,** First evidence for polyamine conjugation mediated by an enzymic activity in plants, *Plant Physiol.*, 87, 757, 1988.
8. **Margosiak, S.A., Dharma, A., Bruce-Carver, M.R., Gonzales, A. P., Louie, D., and Kuehn, G.D.,** Identification of the large subunit of ribulose 1,5-bisphosphate carboxylase/oxygenase as a substrate for transglutaminase in *Medicago sativa* L. (alfalfa), *Plant Physiol.*, 92, 88, 1990.
9. **Smith, T.A., Negrel, J., and Bird, C.R.,** The cinnamic acid amides of the di- and polyamines, in *Advances in Polyamine Research*, Vol. 4, Bachrach, U., Kaye, A., and Chayen, R., Eds., Raven Press, New York, 1983, 347.
10. **Martin-Tanguy, J.,** The occurrence and possible function of hydroxycinnamoyl acid amides in plants, *Plant Growth Regul.*, 3, 381, 1985.
11. **Mbadiwe, E.I.,** Caffeoylputrescine from *Pentachlera macrophylla*, *Phytochemistry*, 12, 2546, 1973.

12. **Buta, J.G. and Izac, R.R.**, Solanaceae: caffeoylputrescine in *Nicotiana tabacum. Phytochemistry*, 11, 1188, 1972.
13. **Ryabinin, A.A. and Il'ina, E.M.**, The alkaloid of *Salsola subaphylla. Drob. Dokl., Akad. Nauk. USSR.*, 67, 513, 1949.
14. **Wheaton, T.A. and Stewart, I.**, Feruloylputrescine: isolation and identification from citrus leaves and fruits, *Nature*, 206, 620, 1965.
15. **Martin-Tanguy, J., Cabanne, F., Perdrizet, E., and Martin, C.**, The distribution of hydroxycinnamic acid amides in flowering plants, *Phytochemistry*, 17, 1927, 1978.
16. **Mizusaki, S., Tanabe, Y., and Noguchi, M.**, *p*-Coumaroylputrescine, caffeoylputrescine and feruloylputrescine from callus tissue culture of *Nicotiana tabacum*, *Phytochemistry*, 10, 1347, 1971.
17. **Meurer, B., Wray, V., Grotjahn, L., Wiermann, R., and Strack, D.**, Hydroxycinnamic acid spermidine amides from pollen of *Corylus avellana*, *Phytochemistry*, 25, 433, 1986.
18. **Ponchet, M., Martin-Tanguy, J., Marais, A., and Martin, C.**, Hydroxycinnamoyl acid amides and aromatic amines in the inflorescence of some Araceae species, *Phytochemistry*, 21, 2865, 1982.
19. **Martin-Tanguy, J., Martin, C., Paynot, M., and Rossin, N.**, Effect of hormone treatment on growth bud formation and free amine and hydroxycinnamoylputrescine levels in leaf explant of *Nicotiana tabacum* cultivated *in vitro*, *Plant Physiol.*, 88, 600, 1988.
20. **Burtin, D., Martin-Tanguy, J., Paynot, M., and Rossin, N.**, Effects of the suicide inhibitors of arginine and ornithine decarboxylase activities on organogenesis,growth, free polyamine and hydroxycinnamoylputrescine levels in leaf explants of *Nicotiana* Xanthi n.c. cultivated *in vitro* in a medium producing callus formation, *Plant Physiol.*, 89, 104, 1989.
21. **Burtin, D., Martin-Tanguy, J., Paynot, M., Carre, M., and Rossin, N.**, Polyamines, hydroxycinnamoylputrescines and root formation in leaf explants of tobacco cultivated *in vitro*. Effects of the suicide inhibitors of putrescine synthesis, *Plant Physiol.*, in press, 1991.
22. **Martin-Tanguy, J., Tepfer, D., Paynot, M., Burtin, D., Heister, L., and Martin, C.**, Inverse relationship between polyamine levels and the degree of phenotypic alteration induced by Ri TL-DNA from *Agrobacterium rhizogenes*, *Plant Physiol.*, 92, 912, 1990.
23. **Burtin, D., Martin-Tanguy, and D. Tepfer**, α-DL-Difluoromethylornithine, a specific, irreversible inhibitor of putrescine biosynthesis, induces a phenotype in tobacco similar to that ascribed to the root-inducing, left-hand transferred DNA of *Agrobacterium rhizogenes. Plant Physiol.*, 95, 461, 1991.
24. **Martin-Tanguy, J., Negrel, J., Paynot, M., and Martin, C.**, Hydroxycinnamic acid amides, hypersensitivity, flowering and sexual organogenesis in plants, in *Plant Molecular Biology*, von Wettstein, D. and Chua, N.H., Eds., Plenum Press, New York, 1987, 253.
25. **Cabanne, F., Martin-Tanguy, J., and Martin, C.**, Phenolamines associes a l'induction florale et a l'etat reproducteur du Nicotiana tabacum var. Xanthi, *Physiol. Veg.*, 15, 445, 1977.
26. **Cabanne, F., Dalebroux, M.A., Martin-Tanguy, J., and Martin, C.**, Hydroxycinnamic acid amides and ripening to flower of *Nicotiana tabacum* var. Xanthi n.c., *Physiol. Plant.*, 53, 399, 1981.
27. **Belliard, J., Pernes, J., and Sandmeier, M.**, Les differentes phases du developpement chez le Mil (*Pennisetum typhoides* Stapf et Hubbard) et la recherche de marquers, *Physiol. Veg.*, 17, 387, 1979.
28. **Martin-Tanguy, J., Margara, J., and Martin, C.**, Phenolamides et induction florale de *Cichorium intybus* dans differentes conditions de culture en seret et *in vitro*, *Physiol. Plant.*, 61, 259, 1984.
29. **Galston, A.W., Kaur-Sawhney, R., Tiburcio, A.F., Hamasaki, N., Oshima, T., and Furuya, M.**, The control of morphogenesis by polyamines, in *Polyamines and Ethylene: Biochemistry, Physiology, and Interactions*, Flores, H. E., Arteca, R.N., and Shannon, J.C., Eds., American Society of Plant Physiologists, Rockville, MD, 1990, 224.
30. **Martin-Tanguy, J., Deshayes, A., Perdrizet, E., and Martin, C.**, Hydroxycinnamic acid amides (HCA) in *Zea mays:* distribution and changes with cytoplasmic male sterility, *FEBS Lett* , 108, 176, 1979.
31. **Martin-Tanguy, J., Perdrizet, E., Prevost, J., and Martin, C.**, Hydroxycinnamic acid amides in fertile and cytoplasmic male sterile lines of maize, *Phytochemistry*, 21, 1939, 1982.
32. **Martin, C. and Martin-Tanguy, J.**, Polyamines conjugees et limitation de l'expansion virale chez les vegetaux, *C. R. Seances Acad. Sci. (Paris)*, 292, 249, 1981.
33. **Klinguer, S., Martin-Tanguy, J., and Martin, C.**, K-nutrition, growth bud formation and amine hydroxycinnamic acid amide contents in leaf explants of *Nicotiana tabacum* cultivated *in vitro*, *Plant Physiol.*, 82, 561, 1986.
34. **Martin, C., Kunesch, G., Martin-Tanguy, J., Negrel, J., Paynot, M., and Carre, M.**, Effect of cinnamoyl putrescines *in vitro* cell multiplication and differentiation of tobacco explants, *Plant Cell Rep.*, 4, 158, 1985.
35. **Chriqui, D., D'Orazi, D., and Bagni, N.**, Ornithine and arginine decarboxylases and polyamine involvement during *in vivo* differentiation and *in vitro* dedifferentiation of *Datura innoxia* leaf explants, *Physiol. Plant.*, 68, 589, 1986.
36. **Jarvis, B.C., Shannon, P.R.M., and Yasmin, S.**, Involvement of polyamines in root development in stem cuttings of mung bean, *Plant Cell Physiol.*, 24, 267, 1989.

37. **Morris, R.,** Genes specifying auxin and cytokinin biosynthesis in phytopathogens, *Ann Rev. Plant Physiol.*, 37, 509, 1986.
38. **Tepfer, D.,** Transformation of several species of higher plants by *Agrobacterium rhizogenes:* sexual transmission of the transformed genotype and phenotype, *Cell*, 47, 959, 1984.
39. **Spano, L., Mariotti, D., Cardarelli, M., Branca, C., and Costantino, P.,** Morphogenesis and auxin sensitivity of transgenic tobacco with different complements of Ri T-DNA, *Plant Physiol.*, 87, 479, 1988.
40. **Durand-Tardif, M.R.B., Slightom, J., and Tepfer, D.,** Structure and expression of Ri T-DNA from *Agrobacterium rhizogenes* in *Nicotiana tabacum*: organ and phenotypic specificity, *J. Mol. Biol.*, 186, 557, 1985.
41. **Ooms, G., Karp, A., Burrell, M., Twell, D., and Roberts, J.,** Genetic modification of potato development using Ri T-DNA, *Theor. Appl. Genet.*, 70, 440, 1985.
42. **Martin, C. and Gallet, M.,** Hypersensibilite aux virus, temperature et induction florale chez les vegetaux, *C. R. Acad. Sci.*, 262, 997, 1966.
43. **Martin, C. and Gallet, M.,** Contribution a l'etude de l'action de la temperature sur la reaction d'hypersensibilite de certain hotes a l'egard du virus de la mosaique du Tabac, *C. R. Acad. Sci.*, 262, 646, 1966.
44. **Tanguy, J. and Martin, C.,** Phenolic compounds and hypersensitivity reaction in *Nicotiana tabacum* infected with tobacco mosaic virus, *Phytochemistry*, 11, 19, 1972.
45. **Martin-Tanguy, J., Martin, C., and Gallet, M.,** Presence de composes aromatiques lies a la putrescine dans divers *Nicotiana* viroses, *C. R. Acad. Sci.*, 276, 1433, 1973.
46. **Martin-Tanguy, J., Martin, C., Gallet, M., and Vernoy, R.,** Sur de puissants inhibiteurs naturels de multiplication du virus de la mosaique de tabac, *C. R. Acad. Sci.*, 282, 2231, 1976.
47. **Smith, H.H.,** Plant genetic tumors, *Prog. Exp. Tumor Res.*, 15, 138, 1972.
48. **Negrel, J., Vallee, J.C., and Martin, C.,** Ornithine decarboxylase activity and the hypersensitive reaction to tobacco mosaic virus in *Nicotiana tabacum*, *Phytochemistry*, 23, 2747, 1984.
49. **Paynot, M., Martin, C., and Giraud, M.,** Activite de la phenylalanine-ammoniac lyase et hypersensibilite au virus de la mosaique du Tabac de *Nicotiana tabacum* var. Xanthi n.c., *C. R. Acad. Sci.*, 273, 537, 1971.
50. **Negrel, J. and Martin, C.,** The biosynthesis of feruloyltyramine in *Nicotiana tabacum*, *Phytochemistry*, 23, 2797, 1984.
51. **Negrel, J.,** Aspects du metabolisme de la putrescine et de la tyramine au cours de la reaction hypersensible au virus de la Mosaique de Tabac chez *Nicotiana tabacum*, Ph.D. thesis, Paris, 1984.
52. **Negrel, J. and Smith, T.A.,** Oxidation of *p*-coumaroylagmatine in barley seedling extracts in the presence of hydrogen peroxide or thiols, *Phytochemistry*, 23, 739, 1984.
53. **Clarke, D.D.,** The accumulation of cinnamic acid amides in the cell walls of potato tissue as an early response to fungal attack, in *Active Defense Mechanisms in Plants*, Wood, R. K. S., Ed., Plenum Press, New York, 1982.
54. **Ponchet, M., Martin-Tanguy, J., Andreoli, C., and Martin, C.,** Apparition de substances de type phenolamide lors de l'interaction *Dianthus caryophyllus* L. var. Scania - *Phytophthora parasitica* Dastur, *Agronomie*, 2, 37, 1982.
55. **Bird, C.R. and Smith, T.A.,** The biosynthesis of coumaroylagmatine in barley seedlings, *Phytochemistry*, 20, 2345, 1981.
56. **Smith, T.A. and Best, G.R.,** Distribution of the hordatines in barley, *Phytochemistry*, 17, 1093, 1978.
57. **Palmer, J.E. and Widholm, J.M.,** Characterization of carrot and tobacco cell cultures resistant to *p*-fluorophrenylalanine, *Plant Physiol.*, 56, 233, 1975.
58. **Berlin, J. and Widholm, J.M.,** Correlation between phenylalanine ammonia lyase activity and phenolic biosynthesis in *p*-fluorophenylalanine-sensitive and -resistant cell lines of tobacco, *Z. Pflanzelphysiol.*, 97, 317, 1977.
59. **Berlin, J., Knobloch, K.H., Hofle, G., and Witte, L.,** Biochemical characterization of two tobacco cell lines with different levels of cinnamoyl putrescines, *J. Nat. Prod.*, 45, 83, 1982.
60. **Berlin, J., Kukosche, G., and Knobloch, K.H.,** Selection of tobacco cell lines with high yields of cinnamoyl putrescines, *Planta Med.*, 42, 173, 1981.
61. **Berlin, J.,** Formation of putrescine and cinnamoyl putrescine in tobaco cell cultures, *Phytochemistry*, 20, 53, 1981.
62. **Berlin, J. and Witte, L.,** Metabolism of phenylalanine and cinnamic acid in tobacco cell lines with high and low yields of cinnamoyl putrescine, *J. Nat. Prod.*, 45, 88, 1982.
63. **Meurer-Grimes, B., Berlin, J., and Strack, D.,** Hydroxycinnamoyl-CoA:putrescine hydroxycinnamoyltransferase in tobacco cell cultures with high and low levels of caffeoylputrescine, *Plant Physiol.*, 89, 488, 1989.
64. **Negrel, J.,** The biosynthesis of cinnamoylputrescines in callus tissue cultures of *Nicotiana tabacum* , *Phytochemistry*, 28, 477, 1989.
65. **Negrel, J., Javelle, F., and Paynot, M.,** Separation of putrescine and spermidine hydroxycinnamoyl transferases extracted from tobacco callus cultures, *Phytochemistry*, in press, 1991.

66. **Knobloch, K.H. and Berlin, J.,** Phosphate mediated regulation of cinnamoyl putrescine biosynthesis in cell suspensions of *Nicotiana tabacum*, *Planta Medica*, 42, 1647, 1981.
67. **Flores, H.E. and Filner, P.,** Metabolic relationships of putrescine, GABA and alkaloids in cell and root cultures of Solanaceae, in *Primary and Secondary Metabolism of Plant Cell Cultures*, Neumann, K.H., Barz, W., and Reinhard, E., Eds., Springer-Verlag, Berlin, 1985, 174.
68. **Flores, H.E.,** unpublished data.
69. **Balint, R., Cooper, G., Staebel, M., and Filner, P.** *N*-Caffeoyl-4-amino *n*-butyric acid, a new flower-specific metabolite in cultured tobacco, *J. Biol. Chem.*, 262, 11026, 1987.
70. **Guggisberg, A. and Hesse, M.,** Putrescine, spermidine, spermine, and related polyamine alkaloids, in *The Alkaloids, Chemistry and Pharmacology*, Vol. XXII, Brossi, A., Ed., Academic Press, New York, 1983, 85.
71. **Jacobson, M. and Crosby, D.G.,** *Naturally Occurring Insecticides*, Marcel Dekker, New York, 1971, 573.
72. **Waller, G.R. and Novacki, E.K.,** *Alkaloid Biology and Metabolism in Plants*, Plenum Press, New York, 1977, 293.
73. **Liebistch, H.W. and Schutte, H.R.,** Alkaloids derived from ornithine, in *Biochemistry of Alkaloids*, Mothes, K., Schute, H.R., and Luckner, Eds., Detuscher Verlag de Wissenschaften, Berlin, 1985, 106.
74. **Dawson, R.F.,** The localization of the nicotine biosynthetic mechanism in the tobacco plant, *Science*, 94, 396, 1941.
75. **White, P.R.,** Potentially unlimited growth of excised tomato root tips in a liquid medium, *Plant Physiol.*, 9, 585, 1934.
76. **Dawson, R.F.,** Nicotine synthesis in excised tobacco roots, *Amer. J. Bot.*, 29, 813, 1942.
77. **Flores, H.E., Hoy, M.W., and Pickard, J.J.,** Secondary metabolites from root cultures, *Trends Biotechnol.*, 5, 64, 1987.
78. **Hamill, J.D., Parr, A.J., Rhodes, M.J.C., Robins, R.J., and Walton, N.J.,** New routes to plant secondary products, *Bio/Technology*, 5, 800, 1987
79. **Signs, M.W. and Flores H.E.,** The biosynthetic potential of roots, *BioEssays*, 12, 7, 1990.
80. **Flores, H.E.,** unpublished data, 1991.
81. **Tiburcio, A.F. and Galston, A.W.,** Arginine decarboxylase as the source of putrescine for tobacco alklaloids, *Phytochemistry*, 25, 107, 1986.
82. **Feth, F., Wagner, R., and Wagner, K.G.,** Regulation in tobacco callus of enzyme activities of the nicotine pathway, *Planta*, 168, 402, 19486.
83. **Mothes, K., Schutte, H.R., and Luckner, M.,** *Biochemistry of Alkaloids*, Deutscher Verlag der Wissenschaften, Berlin, 19485, 406.
84. **Leete, E.,** Spermidine: an indirect precursor of the pyrrolidine rings of nicotine and nornicotine in *Nicotiana glutinosa*, *Phytochemistry*, 24, 957, 1985.
85. **Hamill, J.D., Robins, R.J., Parr, A.J., Evans, D.M., Furze, J.M., and Rhodes, M.J.C.,** Over-expressing a yeast ornithine decarboxylase gene in transgenic roots of *Nicotiana rustica* can lead to enhanced nicotine accumulation, *Plant Mol. Biol.*, 15, 27–38, 1990.
86. **Robins, R.J., Parr, A.J., and Walton, N.J.,** Studies on the biosynthesis of tropane alkaloids in *Datura stramonium* L. transformed root cultures. 2. On the relative contributions of L-arginine and L-ornithine to the formation of the tropane ring, *Planta*, 183, 196, 1991.
87. **Leete, E.,** The stereospecific incorporation of ornithine into the tropine moiety of hyoscyamine, *J. Am. Chem. Soc.*, 84, 55, 1962.
88. **Hashimoto, T., Yukimune, Y., and Yamada, Y.,** Putrescine and putrescine *N*-methyltransferase in the biosynthesis of tropane alkaloids in cultured roots of *Hyoscyamus albus*. I. Biochemical studies, *Planta*, 178, 123, 1989.
89. **Hashimoto, T., Yukimune, Y., and Yamada, Y.,** Putrescine and putrescine-*N*-methyltransferase in the biosynthesis of tropane alkaloids in cultured roots of *Hyoscyamus albus*. II. Incorporation of labeled precursors, *Planta*, 178, 131, 1989.
90. **Leete, E., Endo, T., and Yamada, Y.,** Biosynthesis of nicotine and scopolamine in a root culture of *Duboisia leichhardtii*, *Phytochemistry*, 29, 1847, 1990.
91. **Feth, F., Arfmann, H.-A., Wray, V., and Wagner, K.G.,** Determination of putrescine *N*-methyltransferase by high performance liquid chromatography, *Phytochemistry*, 24, 921, 1985.
92. **Hashimoto, T., Mitani, A., and Yamada, Y.,** Diamine oxidase from cultured roots of *Hyoscyamus niger*. Its function in tropane alkaloid biosynthesis, *Plant Physiol.*, 93, 216, 1990.
93. **Yamada, Y., Okabe, S., and Hashimoto, T.,** Homogeneous hyoscyamine 6-β-hydroxylase from cultured roots of *Hyoscyamus niger*, *Proc. Japan Acad. Ser. B*, 66, 73, 1990.
94. **Hashimoto, T., Matsuda, J. Okabe, S., Amano, Y., Yun, D.J., Hayashi, A., and Yamada, Y.,** Molecular cloning, and tissue- and cell-specific expression of hyoscyamine 6β-hydroxylase, in *Progress in Plant Cellular and Molecular Biology*, Nijkamp, H. J. J., Van der Plas, L. H. W., and van Artrijk, J., Eds., Kluwer Academic, Netherlands, 1990, 775.

95. **Hashimoto, T., Kohno, J., and Yamada, Y.**, 6β-hydroxyhyoscyamine epoxidase from cultured roots of *Hyoscyamus niger*, *Phytochemistry*, 28, 1082, 1989.
96. **Hartmann, T., Sanders, H., Adolph, R., and Toppel, G.**, Metabolic links between the biosynthesis of pyrrolizidine alkaloids and polyamines in root cultures of *Senecio vulgaris*, *Planta*, 175, 82, 1988.
97. **Luckner, M.**, *Secondary Metabolism in Microorganisms, Plants, and Animals*, 3rd ed., Springer-Verlag, Berlin, 1990, 563.
98. **Birecka, H., Birecki, M., Cohen, E.J., Bitonti, A.J., and McCann, P.P.**, Ornithine decarboxylase, polyamines, and pyrrolizidine alkaloids in *Senecio* and *Crotalaria*, *Plant Physiol.*, 86, 224, 1988.
99. **Hartmann, T., Ehmke, A., Eilert, U., von Borstel, K., and Theuring, C.**, Sites of synthesis, translocation and accumulation of pyrrolizidine alkaloid *N*-oxides in *Senecio vulgaris* L., *Planta*, 177, 98, 1989.
100. **Wink, M. and Hartmann, T.**, Site of enzymatic synthesis of quinolizidine alkaloids and their accumulation in *Lupinus polyphyllus*, *Z. Pflanzenphysiol.*, 102, 337, 1981.
101. **Wink, M. and Hartmann, T.**, Localization of the enzymes of quinolizidine alkaloid biosynthesis in leaf chloroplasts of *Lupinus polyphyllus*, *Plant Physiol.*, 70, 74, 1942.
102. **Pelosi, L.A., Rother, A., and Edwards, J.M.**, Lysine decarboxylase activity and alkaloid production in *Heimia salicifolia* cultures, *Phytochemistry*, 25, 2315, 1986.
103. **Wink, M.**, Metabolism of quinolizidine alkaloids in plants and cell suspension cultures: Induction and degradation, in *Primary and Secondary Metabolism of Plant Cell Cultures*, Neumann, K.H., Barz, W., and Reinhard, E., Eds., Springer-Verlag, Berlin, 1985, 107.
104. **Twardowski, T., Pulikowska, J., and Wiewiorowski, M.**, Inhibitory effect of selected quinolizidine alkaloids and their derivatives and analogues on the Phe-tRNA binding to ribosomes, *Bull. Acad. Pol Sci.*, 29, 129, 1982.
105. **Baldwin, I.T.**, Damage-induced alkaloids in tobacco: pot-bound plants are not inducible, *J., Chem. Ecol.*, 14, 1113, 1988.

Chapter 5

METABOLIC AND PHYSIOLOGICAL RELATIONSHIPS BETWEEN THE POLYAMINE AND ETHYLENE BIOSYNTHETIC PATHWAYS

M. M. Kushad and E. B. Dumbroff

TABLE OF CONTENTS

I. INTRODUCTION

Although PAs and ethylene are biosynthetically related, their physiological functions are distinct and at times antagonistic, particularly during events that characterize leaf senescence, fruit ripening, and perhaps during the response of plant tissue to chemical and environmental stresses. The dependence of the PA and ethylene biosynthetic pathways on a common precursor, *S*-adenosylmethionine (AdoMet), suggests that shifting demands by each pathway for a limited pool of that precursor could provide an effective means of regulating the biosynthesis of their end products. Inhibition of enzymes by PAs or ethylene in the alternate or competing pathway may also provide a mechanism for the control of physiological processes in which the balance between PAs and ethylene assumes a determinative role. These possibilities have stimulated several studies of the relationships between PAs and ethylene yielding, at times, conflicting results and conclusions. In the current review, we have examined the principal steps in PA and ethylene biosynthesis and considered the evidence for functional links between the two pathways and their consequent effects on specific physiological events.

II. POLYAMINE AND ETHYLENE BIOSYNTHESIS

The biochemical steps leading to the formation of PAs and ethylene have been examined in a number of plant species. In this section we provide a general overview of their biosynthesis and describe some of the enzymatic steps and metabolic intermediates that may regulate activities within and between the two pathways.

A. ADOMET — THE COMMON PRECURSOR

Both PAs and ethylene are synthesized from methionine. Methionine, however, is not the immediate precursor; it must first be converted into AdoMet, a sulfonium compound, before it shunts into the PA or ethylene pathway (Figure 1).

In addition to its involvement in PA and ethylene biosynthesis, AdoMet acts as the primary donor of methyl groups during a large number of biological transmethylation reactions.[1,2] These, and other studies described below, suggest that the amount of AdoMet committed to PA and ethylene biosyntheses is quite limited and constitutes only a small fraction of the AdoMet pool involved in the transmethylation reactions.

The use of AdoMet by plant tissues for the synthesis of PAs or ethylene is largely a function of tissue type, stage of development, and growing conditions. During periods of active cell division, PA levels and rates of PA synthesis are usually much higher than in nondividing tissues or during tissue senescence. For example, PAs and the activities of their biosynthetic enzymes increased during somatic embryogenesis in carrot cell cultures, in developing leaves of *Phaseolus vulgaris*, in the meristematic tissue of potato sprouts, and during ovary development of tomato and avocado fruits.[3-7] In contrast, ethylene production in meristematic and in nonsenescent tissues is usually limited but increases dramatically when tissues age and begin to senesce.[8] Concomitant increases in the synthesis of ethylene and PAs, principally Put, are observed during some physiological events and are discussed in section III C.

B. ADOMET METABOLISM DURING POLYAMINE BIOSYNTHESIS

During periods of active PA biosynthesis, a portion of the AdoMet pool in plants is directed specifically toward the formation of decarboxylated AdoMet (dAdoMet) in a reaction catalyzed by AdoMet decarboxylase[9] (Figure 1). AdoMet decarboxylase in corn seedlings has been assigned a molecular weight of 25,000, as estimated by gel filtration on Sephadex G-200,[10] whereas in the leaves and isolated protoplasts of chinese cabbage, the enzyme was reported to have a molecular weight of approximately 35,000, as determined after a 1500-fold purification using affinity chromatography and Sephacryl S-300 filtration.[11]

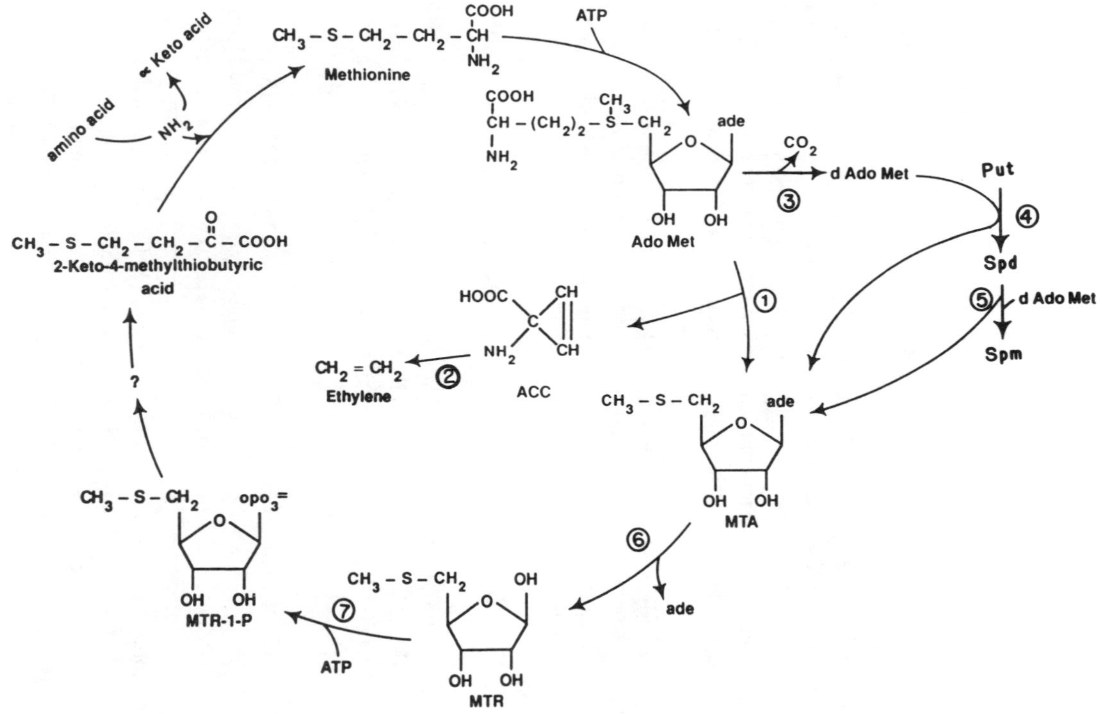

FIGURE 1. Ethylene and polyamine biosynthetic pathways. The circled numbers represent the enzymes involved in catalyzing the reactions as follows: ①, ACC synthase; ②, EFE; ③, AdoMet decarboxylase; ④, Spd synthase; ⑤, Spm synthase; ⑥, MTA nucleosidase; ⑦, MTR kinase. Refer to the text for the full names of the abbreviated intermediates.

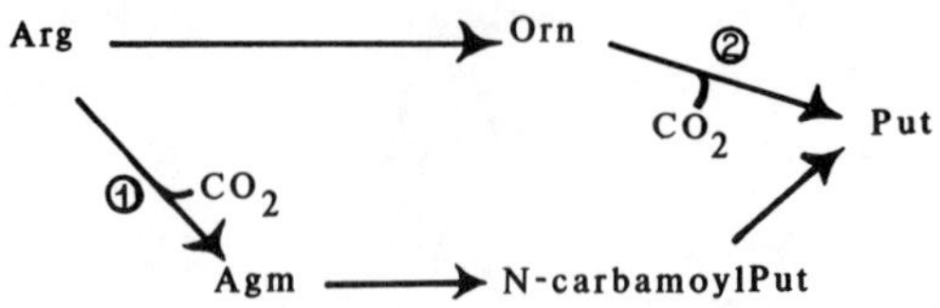

FIGURE 2. Putrescine biosynthesis. The circled numbers represent the enzymes involved in catalyzing the reactions as follows: ①, Arg decarboxylase; ②, Orn decarboxylase. Refer to the text for the full names of the abbreviated intermediates.

The activity of AdoMet decarboxylase is not stimulated by Put nor by divalent cations, but it is competitively inhibited by dAdoMet.[11] The enzyme has a pH optimum between 7.6 to 8.6, a K_m for AdoMet of 5 μM in etiolated corn seedlings[10] and 38 μM in Chinese cabbage,[11] and its activity is strongly inhibited by methylglyoxal-*bis*(guanylhydrazone) (MGBG). Yamanoha et al.[11] concluded that AdoMet decarboxylase is probably the rate-limiting enzyme for the formation of PAs in chinese cabbage.

During PA synthesis (Figure 1), the dAdoMet pool donates an aminopropyl group to Put to form Spd and a second aminopropyl group to Spd to form Spm in consecutive reactions catalyzed respectively by Spd and Spm synthases. The activities of Spd and Spm synthases have been measured in several plant systems, but only Spd synthase has been isolated and purified from plant tissue. Molecular weights of the enzyme ranged from 45,000 to 81,000 and pH optima from 7.2 to 8.8.[11,12]

Most of the Put in plants arises from arginine via one of two alternative decarboxylation pathways (Figure 2). One pathway involves an initial decarboxylation of arginine (Arg) by Arg decarboxylase to form agmatine (Agm), which then converts to *N*-carbamoylputrescine (carbamoylPut), and hence to Put.[13,14] In the alternative pathway, arginine looses a urea molecule to form ornithine (Orn), which is then decarboxylated by Orn decarboxylase to form Put.[15] A minor pathway of Put synthesis involves decarboxylation of citrulline to *N*-carbamoylPut.[16,17]

The transfer of aminopropyl groups from dAdoMet to form Spd and Spm also yields 5′-methylthioadenosine (MTA) as a co-product. MTA is then rapidly recycled for regeneration of methionine and AdoMet (see Section II, D), and as a result of this salvage pathway, MTA does not accumulate to any detectable level in plant tissue (Kushad, unpublished data).

C. ADOMET METABOLISM DURING ETHYLENE BIOSYNTHESIS

Following the discovery by Lieberman's group[18,19] that ethylene was generated from carbons 3 and 4 of methionine, Adams and Yang[20] identified AdoMet as an intermediate in the pathway between methionine and ethylene. They also observed that when climacteric apple tissue was fed radioactive methionine, MTA and 5-methylthioribose (MTR) were formed in parallel with ethylene. However, when younger tissue that was not capable of producing ethylene was fed radioactive methionine, or when the methionine to ethylene conversion was chemically inhibited, no MTA or MTR was formed. The authors[20] concluded that MTA and MTR are products of methionine conversion to ethylene and that both products are shunted into a salvage pathway for regeneration of methionine (Figure 1).

Subsequent examination of the methionine to ethylene conversion by two independent laboratories revealed that 1-aminocyclopropane-1-carboxylic acid (ACC) is an intermediate between AdoMet and ethylene.[21,22] The AdoMet to ACC conversion is catalyzed by ACC synthase,[23] a pyridoxal phosphate enzyme,[21] in what appears to be the rate limiting step in ethylene formation.[24,25] ACC synthase is developmentally regulated and is an inducible enzyme that has been detected in many plant tissues. The partially purified enzyme was reported to have a molecular mass in the range of 45,000 to 67,000 Da, either as a monomer[26,27] or a dimer.[28] Mehta et al.[27] suggested that the discrepancy in the molecular weights reported from different laboratories might have resulted from identification of different isoforms of the enzyme or from differences in the purification procedures.

ACC synthase has an estimated K_m for AdoMet of 9 μM in ripe tomato[26] and 55 μ*M* in mung bean.[28] The complementary DNA sequence for ACC synthase was recently isolated from zucchini[29] and tomato fruits,[30] and these studies indicate that accumulation of the ACC synthase message is coordinated with fruit ripening and stress.

In summary, there is a general agreement that methionine and AdoMet are the primary sources of the aminopropyl moieties of PAs and of ethylene in plants and that their synthesis proceeds through the scheme illustrated in Figure 1. The variability in the K_m values for ACC synthase[26,28] and AdoMet decarboxylase[10,11] suggests that AdoMet metabolism during ethylene and PA biosynthesis proceeds with different catalytic efficiencies. Therefore, it is difficult to conclude from these studies whether AdoMet metabolism favors PA or ethylene biosynthesis.

D. REGENERATION OF METHIONINE AND ADOMET

Although the biosynthesis of PAs and ethylene in many plant tissues can continue for extended periods, the methionine and AdoMet pools are not sufficient to sustain their synthesis for more than a few hours.[31,32] Baur et al.[33] estimated that the size of the methionine pool during the climacteric peak in Fuerte avocado fruit could sustain ethylene production for about 3 h, and yet evolution of ethylene by avocado is known to continue over a period of several days. Concentrations of Spd and Spm in plants are also many-fold higher than the existing AdoMet pools,[32] but the sulfur moieties of AdoMet are not sequestered during PA or ethylene synthesis,[34] thus facilitating their recycling and the rapid regeneration of methionine.

Shapiro[35] used methionine auxotrophs from *Aerobacter aerogenes* to first demonstrate the existence of a mechanism that could utilize MTA for the regeneration of methionine. Subsequent work demonstrated the transfer of [^{35}S] and [$^{14}CH_3$] from MTA into methionine and thereby confirmed the existence of an MTA salvage pathway for the regeneration of methionine.[36,37]

MTA is degraded through nucleosidic cleavage by MTA nucleosidase to MTR[38] (Figure 1). The methylthiol group of MTA remains intact and is effectively recycled to reform the methylthiol group of methionine via MTR.[20] It was subsequently shown that the ribose moiety of MTA and MTR, which derives from ATP, gives rise to the four-carbon backbone (2-aminobutyrate) of methionine and to ethylene.[39,40] Whether the aminopropyl moieties of Spd and Spm are also derived through the same mechanism has not been established.

Examination of the metabolic steps between MTR and methionine in fruit tissue revealed an ATP-dependent phosphorylation of MTR to MTR-1-phosphate (MTR-1-P) by MTR kinase, which is then metabolized to 2-keto-4-methylthiobutyric acid before its conversion to methionine (Figure 1).[41,42] Trackman and Abeles,[43] and more recently Furfine and Abeles,[44] identified 5-methylthioribulose-1-phosphate and 1-phospho-2,3-diketo-5-methylthiopentane as intermediates between MTR-1-P and 2-keto-4-methylthiobutyric acid.

Since ethylene and perhaps the aminopropyl moieties of PAs are derived from the ribose portion of MTA and MTR, as described above and shown in Figure 1, abundant methionine and AdoMet pools are apparently not required for the formation of either PAs or ethylene. After the first turn of the cycle, PAs and ethylene are synthesized from the continual input of ATP, which donates an adenosine group to methionine to form a new molecule of AdoMet. According to this scheme, a balanced reaction equation for ethylene biosynthesis was proposed as follows:[42]

$$2ATP \rightarrow ADP + \text{adenine} + 4pi + 2HCOOH + CO_2 + C_2H_4$$

III. INTERACTIONS BETWEEN POLYAMINES AND ETHYLENE

A. MODULATION OF POLYAMINE BIOSYNTHESIS BY ETHYLENE

The potential for competition between the ethylene and polyamine biosynthetic pathways for AdoMet (Figure 1) has been clearly established,[45,46] and it has raised the possibility that an effective mechanism exists for modulating physiological events in which polyamines and ethylene are both involved. Moreover, PAs and ethylene have each been shown to suppress the other's synthesis, thus providing an additional opportunity for interaction between the two pathways. Nevertheless, the physiological significance of these relationships remains unclear, as is evident by conflicting reports in the literature, perhaps due to innate differences among species and between tissues within the same plant during its course of development.

1. Effects of Inhibiting Ethylene Biosynthesis

The preferential utilization of AdoMet by the ethylene or PA biosynthetic pathways can often be controlled, in excised tissue, by judicious application of specific inhibitors of certain enzymes in the two pathways. Early work with rapidly growing apple suspension cells treated with aminoethoxyvinylglycine (AVG) or aminooxyacetic acid (AOA), which are potent inhibitors of the AdoMet to ACC step and subsequently of ethylene synthesis, did not reveal any increase in the incorporation of 3,4-[^{14}C]methionine into Spd or Spm and little change in AdoMet concentration[47]. In contrast, several more recent studies have shown that inhibiting the conversion of AdoMet to ethylene can strongly enhance PA synthesis.[45,46,48-50] When discs of senescing orange peel were treated with 2.6 mM AVG, the incorporation of radiolabelled methionine into Spd increased nearly 5-fold over controls, although Spm concentrations were not affected.[45] When a combination of AVG and Put was added to the discs, the incorporation values increased nearly ten-fold. In shoot cultures of *Prunus avium,* suppression of ethylene biosynthesis by AVG increased the incorporation of both radiolabelled methionine and Put into Spd.[50] AVG also caused a marked rise in Spd and Spm concentrations and a large drop in ACC levels in protoplasts isolated from chinese cabbage leaves.[48] In cut carnation flowers, treatment with AOA completely blocked ethylene formation, increased the flow of [^{3}H]methionine into Spd and nearly tripled the concentration of Spm in the petals.[46,51] Treatment of barley callus with AOA induced large increases in the activity of AdoMet decarboxylase, in the AdoMet pool, and in PA synthesis, all of which are concurrent with a large decline in the activity of ACC synthase.[49]

2. Inhibition by Ethylene

Ethylene has been shown to be an effective inhibitor of AdoMet and Arg decarboxylases, both of which are key enzymes in the PA biosynthetic pathway via their respective roles in providing aminopropyl groups and Put for the synthesis of Spd and Spm (Figure 1).[52-54] Bakanashvili et al.[55] concluded that, in tomato fruit, suppression of tissue PA levels with ethylene has resulted from a direct inhibition of certain enzymes within the PA biosynthetic pathway and was not merely a side effect on the ripening process. Treatment of nonripening tomato mutants with ethylene did not induce ripening, but it did reduce PA contents and the activity of Orn decarboxylase, the major enzyme for Put synthesis in tomato, by amounts similar to those noted in fruits that produce ethylene and ripen normally.

The effects of high ethylene on PA synthesis are not always negative, but whether ethylene stimulates or inhibits PA accumulation depends on ethylene's function in a tissue and on the physiological stage of development. In young tissues, in which ethylene stimulates growth or is involved in some early phase of development, PA levels often rise. However, under conditions of stress or senescence, processes in which ethylene may inhibit growth and stimulate degradative events, enzymes involved in PA biosynthesis are commonly held in check and

AdoMet can be used preferentially by the ethylene pathway. Examples of these interrelationships between ethylene and the PAs are discussed in Section III C.

B. MODULATION OF ETHYLENE BIOSYNTHESIS BY POLYAMINES

1. Inhibition by Polyamines

Exogenous PAs have been shown to inhibit ethylene production by up to 89% in senescing petals of *Tradescantia* (Table 1) and in tissue slices and isolated protoplasts from apple fruit.[56,57] They also effectively suppressed ethylene biosynthesis induced by exogenous IAA and ACC in etiolated soybean seedlings, and the inhibition was not relieved by the addition of methionine, suggesting that the PAs inhibited the metabolism of ACC and not of AdoMet.[56] Fuhrer et al.[58] reported similar results with oat leaves showing that PAs inhibited ethylene biosynthesis at the AdoMet to ACC step and even more effectively from ACC to ethylene. However, in wounded mesocarp from winter squash fruit, PAs inhibited ACC synthase activity (Table 1) but they did not affect the activity of ethylene forming enzyme (EFE).[59] The inhibition of ACC synthase by Spm, Spd, and Put is apparently noncompetitive, with K_i values of 1.6, 2.9, and 4.9 mM, respectively.[56] In total, these and other studies suggest that polyamines inhibit ethylene biosynthesis by blocking the conversion of AdoMet to ACC and of ACC to ethylene by directly inhibiting the activities of ACC synthase and EFE[56,59,60] and/or by suppressing *de novo* synthesis of ACC synthase.[60] This conclusion is consistent with the observation that PAs suppressed protein synthesis in suspension cultures of fruit cells from pear.[61] Ben-Arie et al.[62] noted that inhibition of ethylene biosynthesis by Spm was temperature dependent and apparently resulted from stabilization of membrane microviscosity, thereby delaying activation of the ethylene biosynthetic system in the microsomal membranes. A different mechanism was proposed by Drolet et al.,[63] who suggested that inhibition of the ACC to ethylene conversion by PAs could derive, in part, from their free radical scavenging properties, since the ACC to ethylene step may involve a free-radical intermediate.[64] For further information about the free radical scavenging properties of PAs, refer to Chapter 14.

In contrast to the previous reports, Pennazio and Roggero[65,66] observed that Spd and Spm increased ethylene production in tobacco and soybean leaves by stimulating the activities of both ACC synthase and EFE. Downs and Lovell[67] also reported that addition of PAs to cut carnation flowers increased ethylene biosynthesis, reduced the number of days for the flowers to reach peak ethylene formation, and decreased bloom longevity. A plausible explanation for the stimulation of ethylene production by PAs is discussed in Section III C 1.

2. Effects of Inhibiting Polyamine Biosynthesis

The development of synthetic inhibitors that interfere with specific steps in PA biosynthesis has helped identify probable sites of interaction between the PA and ethylene pathways. In cut carnation flowers, MGBG, which is an inhibitor of AdoMet decarboxylase[46,68] and thereby of AdoMet conversion to PAs, caused moderate but variable reductions in PA levels and a marked rise in ethylene synthesis in the fully opened flowers.[46] D-Arg and difluoromethylarginine (DFMA), which are both inhibitors of Arg decarboxylase, caused large reductions in Put synthesis and substantial increases in ethylene production.[46] These results confirm, therefore, that suppression of PA biosynthesis can induce a shift of AdoMet utilization into the ethylene biosynthetic pathway.

Although regulation of PA or ethylene biosynthesis by exogenous inhibitors will shift AdoMet metabolism from one pathway to the other, *in vivo* studies in avocado and tomato suggest that activity in the two pathways is not necessarily limited by competition for available AdoMet.[7,15,62] In tomato fruits that ripen slowly and produce little ethylene compared to fruits from other, more common varieties, Put levels rose sharply during ripening, but Spd and Spm levels did not change appreciably in either normal or slow-ripening fruits.[69,70] Thus, AdoMet metabolism in the slow-ripening cultivars was not committed to the synthesis of high levels of

TABLE 1
Effect of Polyamines on ACC Synthase Activity and Ethylene Biosynthesis

Compound	Concentration mM	ACC Synthase[a]	Ethylene[b]
		% control	
Control	0	100	100
Arginine	1	90	—
	10	49	—
Agmatine	1	84	—
	10	61	—
Cadaverine	1	98	82
	10	46	40
Ornithine	1	80	—
	10	35	—
Putrescine	1	76	94
	10	32	34
Spermidine	1	58	42
	10	10	19
Spermine	1	53	33
	10	8	22

[a] See 59 for experimental details.
[b] See 56 for experimental details.

Spd or Spm even though ethylene synthesis was suppressed and Put was not limiting. The lack of an apparent relationship between PA and ethylene biosynthesis is also observed in avocado fruits in which a significant delay occurs between the cessation of PA synthesis during fruit development and the start of ethylene production during ripening.[7] The MTA recycling pathway that is associated with the regeneration of AdoMet remained highly active during peak ethylene synthesis and ripening, exceeding by several-fold the activity levels recorded during the period of maximum PA synthesis.[7]

Current evidence indicates that only a small fraction of the total amount of methionine or AdoMet present in a tissue is directed into PA or ethylene biosynthesis. In experiments with bacteria[71] and animal tissue,[72] only 5 to 10% of the total methionine supplied to the test systems was channelled into the production of polyamines. Moreover, when intact protoplasts from Chinese cabbage leaves were incubated with [2-^{14}C]methionine, only 1% of the radiolabel was incorporated into Spd and Spm with the remainder being incorporated into protein.[73] Accordingly, inhibition of PA biosynthesis should result in, at most, the incorporation of 1 to 10% of the total methionine and AdoMet pools into the production of ethylene. However, it seems likely that during senescence, when protein synthesis declines, larger amounts of methionine may be directed into AdoMet and thence to the ethylene pathway.

C. MODULATION OF PHYSIOLOGICAL EVENTS

The relationships between PA and ethylene metabolism,described in the previous sections, have been evaluated within the context of several physiological processes and in a number of organs, tissues and plant species.

1. Senescence

A large body of experimental evidence attests to the promotive role of ethylene in the complex physiological phenomenon commonly recognized as senescence.[74] In contrast to ethylene, PAs are reported to be effective antisenescence agents and to retard chlorophyll loss, membrane deterioration, and increases in RNAse and protease activities, all of which help

TABLE 2
Effects of Inhibitors of Ethylene and Polyamine Biosynthesis on Polyamine Levels, Ethylene Production and the Senescence of Carnation Flowers

Inhibitor[a,b]	Polyamines[c]	Ethylene[d]	Senescent flowers after 6 days
DFMA[e]	Put ↓ 76%	↑ 79%	↑ 116%
D-arginine[e]	Put ↓ 85%	↑ 58%	↑ 100%
MGBG[f]	Spd ↓ 19% Spm ↓ 38%	↑ 69%	↑ 180%
AOA[g]	Spm ↑ 178%	n.d.[h]	none

[a] See 46 for experimental details.
[b] Pedicels of partly-opened flowers immersed in 10 mM solutions of inhibitors.
[c] Concentration changes in petals from fresh, fully-opened flowers, no ethylene produced.
[d] Cumulative ethylene production of senescing flowers.
[e] Inhibits Arg decarboxylase.
[f] Inhibits AdoMet decarboxylase.
[g] Inhibits ACC synthase.
[h] n.d. = None detected.

characterize the senescence process.[74,75] These observations raise the possibility that senescence is controlled, in some measure, by interactions between PAs and ethylene, perhaps via shifting demands for a limited pool of AdoMet[45,46] and, as noted in Sections III A and B, via mutual suppression of enzymes in their competing pathways, viz. the inhibition of Arg and AdoMet decarboxylases by ethylene[52-54] and of ACC synthase and EFE by PAs.[15,59,76]

During studies with carnation flowers,[46] inhibitors of enzymes in both pathways were used in conjunction with measurements of PA levels and ethylene production to demonstrate the potential for either pathway to increase its utilization of the existing pool of AdoMet. Inhibition of AdoMet decarboxylase with MGBG or of Arg decarboxylase with DFMA or D-Arg increased ethylene production and promoted the onset of senescence, whereas inhibiting ACC synthase with AOA blocked ethylene production, increased Spm by 178% and caused a marked delay in the start and progress of senescence (Table 2). In untreated control flowers, endogenous concentrations of Spd and Spm did not change as flowers aged, but Put, their immediate biosynthetic precursor, increased by 130% in parallel with a sharp rise in ethylene production. Higher levels of Put combined with a drop in PA oxidase activity during senescence[77] may offset the decreased activity of AdoMet decarboxylase,[51] thereby stabilizing total PA levels which, via their inhibitory effects on ethylene synthesis,[59,76] would help modulate the course of senescence.

Although exogenous Spd and Spm have been reported to stimulate ethylene production in some plant systems,[65-67] those results may reflect a buildup of endogenous levels of PAs to concentrations that inhibit the activity of AdoMet decarboxylase,[78] causing a shunt of AdoMet to the ethylene pathway. That interpretation is supported by an observed rise in ACC levels in response to exogenous Spd or Spm but not to high levels of Put, which cannot exert feedback control on AdoMet decarboxylase (Figure 1) to promote ethylene synthesis.[65,66]

The concurrent rise of ethylene and Put observed in senescing carnation flowers is common to a range of plant species. When potato seed-tubers were stored for 24 months, internal ethylene contents rose by 60% and Put levels increased more than two-fold whereas Spd and Spm fell by an average of 35%.[79] The increase in Put and ethylene plus a moderate decline in Spd and Spm is consistent with a shift in the utilization of AdoMet by the ethylene biosynthetic pathway and with a senescence-related decrease in PA oxidase activity.[77] The view that high levels of PAs help regulate the progress of senescence, by modulating ethylene synthesis and other senescence-related events, is strengthened by observations from pea apices that reductions in PA concentration precede the development of advanced senescence.[80] Furthermore, observations with cultured pear fruit cells indicate that, to be effective, high levels of PAs must be present before the rise in ethylene begins and probably function by suppressing macromolecular synthesis and protein turnover in the ethylene pathway.[61] Relationships between PA levels and ethylene production are not readily apparent in some fruits. In avocado, PAs reach peak concentrations during the early stages of fruit development but then decline by 70% or more at full maturity, well before the fruits ripen.[7] However, the low values attained are still comparable to levels of PAs found in other preclimacteric flowers and fruits.[46,69,70] Thus, PA titer in fully mature, unpicked avocado may remain sufficiently high to exert a regulatory influence on ethylene synthesis, which in avocado does not show its normal climacteric rise until the fruits are picked and ripened in storage.

2. Fruit Ripening

Ethylene is the principal phytohormone involved in the ripening of climacteric fruit, and a substantial body of evidence has shown that inhibiting its synthesis and/or its action will delay the ripening process.[8,74] The potential for competition between the PA and ethylene biosynthetic pathways for AdoMet and the reported antisenescence properties of PAs[75] have prompted investigations of their possible roles in modulating both ethylene synthesis and fruit ripening.

Following harvest of winter pear, the fruit requires a definite period of low temperature treatment before ethylene biosynthesis and ripening will proceed. When fruits of d'Anjou and "Backham's Triumph" cultivars did not fullfil their low temperature requirement, they neither ripened nor synthesized ethylene, but they did maintain high levels of Spm.[81] In contrast, in fruit that completed their low temperature requirement, Spm concentrations fell below detectable levels, copious amounts of ethylene were produced and the fruit ripened. The authors[81] concluded that the inverse correlation between Spm levels and ethylene production may reflect some measure of regulation by Spm of the events that lead to fruit ripening.

Dibble et al.[69] and Saftner and Baldi[70] reached a similar conclusion during work with normal-ripening (Rutgers) and slow-ripening (Alcobaca and Liberty) cultivars of tomato fruit. Spd and Spm levels in all three varieties remained relatively stable as the fruits ripened, but Put levels in Alcobaca and Liberty rose sharply from the breaker stage through ripeness, which the authors suggest may account for the long shelf-life and the highly reduced peaks in climacteric ethylene exhibited by these slow-ripening fruits.

Kramer et al.[82] reported that apple fruit placed in controlled-atmosphere storage retained higher PA levels and ripened more slowly than air-stored fruit. They also noted that exogenous PAs effectively suppressed cell wall softening and the activity of polygalactouronase (PG), the principal enzyme involved in the wall softening process (Figure 3). The PAs were assumed to function via a mechanism similar to that reported for Ca^{2+} [83] and involving formation of cation cross-bridges with pectic acid and other polysaccharides, thus limiting accessibility of the cell wall to degradative enzymes. The efficacy of the PAs in blocking PG activity correlated with the number of amine groups per molecule, with Put being significantly less effective than the tri- or tetraamines. At 10 mM Spm, PG activity was completely suppressed whereas the same concentration of Put or Spd blocked PG action by 20% and 67%, respectively.

3. Environmental Stress

Several of the physiological changes that are characteristic of plant senescence are also common to various types of environmental stress and include losses of chlorophyll, protein and nucleic acids, a decline in membrane integrity, and disruption of cell homeostasis.[75,84] Large increases in ethylene and Put also frequently accompany both processes.[15,74] In contrast, however, Spd and Spm levels are often stable during an environmental challenge,[85,86] but they may rise during extended periods of stress, apparently at the expense of Put.[87-89] Unfortunately,

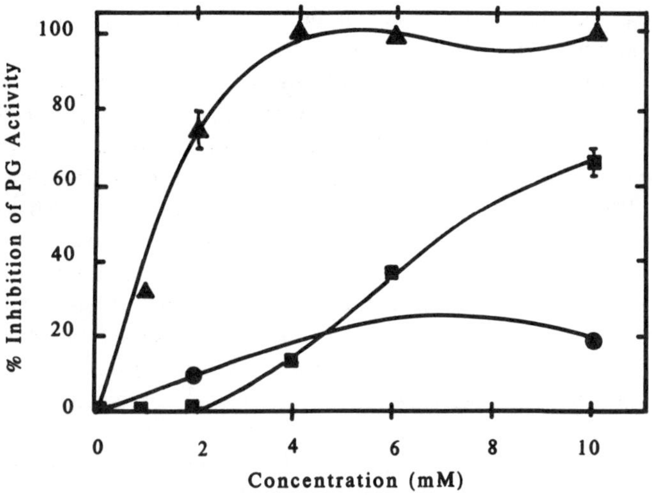

FIGURE 3. Effect of Spm (▲), Spd (■), and Put (●) on the *in vitro* activity of purified polygalacturonase. Vertical bars represent the mean ± SE.[82]

few studies have examined specific relationships among the PAs or between PAs and ethylene during a stress response, and detailed information about these phenomena is limited. In tomato plants grown in nutrient culture, a combination of high levels of NH_4^+ plus low levels of K^+ caused a sharp rise in both ethylene production and Put accumulation by the shoots.[85] When ethylene action was suppressed by addition of Ag^+ to the nutrient medium, the symptoms of NH_4^+ toxicity were almost completely relieved, but Put accumulation continued as a function of K^+ deficiency and of the low pH values induced by NH_4^+. The absence of a close relationship between ethylene synthesis and PA accumulation during a stress event was also noted by Wang[87] when cucumber seedlings were chilled at 5°C for 2 days. Put and Spm showed little change during chilling, but Spd levels doubled. Levels of ACC and ethylene also rose sharply but only after the chilled seedlings had been warmed for 2 h at 20°C. Treatment with paclobutrazol, a triazole-type commercial plant growth regulator, reduced both chilling injury and the accumulation of Spd. However, suppression of ethylene synthesis with AOA did not reduce chilling injury nor did it affect Spd levels, suggesting that increased production of ACC and ethylene was a result of and not the cause of chilling injury. These observations also suggest that potential competition for AdoMet by the ethylene and PA biosynthetic pathways did not limit the activity of either pathway or influence the stress response by the seedlings.

4. Seed Germination

Although increased rates of PA biosynthesis have been recorded during the germination of seeds from several plant species,[90-93] little information is available concerning the functional relationships between PAs and ethylene during the germinative process. Ethylene has been shown to break dormancy in some seeds,[94,95] and PAs are thought to play significant roles in DNA replication, structure and function, and ultimately in cell division and growth.[15,95]

During studies with strongly dormant sugar maple seeds, low levels of Put, Spd, and Spm in the radicles did not change during the full course of a prolonged after-ripening period, but large increases in the three PAs were noted several days after radical emergence and coincident with the period of rapid cell division.[96] Ethylene production rose sharply at the time of radicle emergence and continued its sustained rise concurrent with the subsequent increase in PA titers. The observed changes in ethylene and PAs were clearly more closely associated with the germinative process than with the emergence of seeds from dormancy. The initial rise in ethylene was coincident with the rapid mobilization of food reserves that accompanies radicle emergence,[97] and it continued for several days into the period of rapid cell division, accompanied by the rapid synthesis of Put, Spd, and Spm.

The previous observations are indicative of a regulatory role for ethylene, and perhaps PAs,[98] in the utilization of seed storage compounds, and they also reflect the probable requirement of PAs in the process of cell division.[75] When inhibitors of PA synthesis were applied to newly emerging sugar maple seeds, levels of Put, Spd, and Spm and subsequent mitotic activity were sharply reduced, but cell elongation, as evidenced by growth of the hypocotyls, was not affected.[99] These results suggest that in young, metabolically active tissues of germinating seeds, rapid and concurrent increases in PAs and ethylene are not limited by any competitive demands for AdoMet. Moreover, in rapidly growing tissues, ethylene has been shown to stimulate the activities of both AdoMet and Arg decarboxylase and may thereby contribute to a subsequent rise in PA titer.[100] In germinating chick peas, maximum ethylene production occurred when growth of the embryonic axis was due entirely to cell elongation, i.e., before mitotic activity began, but low concentrations of applied Spm (0.1 mM) increased both fresh weights and ethylene yield. *In toto*, results from this limited number of studies do suggest that a sequential relationship between ethylene and the PAs coordinates processes that control the emergence and establishment of germinating seeds.

IV. CONCLUSIONS AND FUTURE RESEARCH

The temporal relationships that have been observed between PAs and ethylene at different stages of plant development suggest mutual involvement of these metabolites in specific physiological events that are governed by coordinated changes in PA and ethylene metabolism. Three biochemical mechanisms have been invoked to explain the biosynthetic relationships between the two pathways, viz. (1) competitive demand for a limited pool of a common intermediate, (2) feedback inhibition, and (3) inhibition of enzyme systems in one pathway by products of the competing pathway.[11,15,46] However, the degree of linkage between the PA and ethylene pathways remains a moot point, and the significance of these mechanisms in untreated tissues during the course of normal plant development, i.e., in the absence of exogenous inhibitors, PAs or ethylene, has not been demonstrated. Moreover, we still await unequivocal evidence that concurrent changes, whether parallel or divergent, in PA and ethylene levels represent a functional interaction that controls a specific physiological process. The requirement for PAs and ethylene in primary plant metabolism is not in question; rather it is the nature and extent of their relationships at both the physiological and biosynthetic levels that remain unresolved.

Although the results from several studies with different plant species provide convincing evidence that interactions between PAs and ethylene do modulate fruit ripening, senescence and ethylene production,[15,46,49,70,75,79] other studies have not revealed close functional relationships.[75] For example, work on carnation flowers by Roberts et al.[46] established that inhibition of the PA or ethylene pathways induced large increases in the amount of end product in the competing pathway, accompanied by a commensurate increase or delay in the onset of flower senescence. These observations appear to confirm the importance of interactions between PAs and ethylene in modulating senescence, but data from the same study cast doubt on the certainty of that conclusion. In untreated tissue in which neither pathway was artificially suppressed, Spd and Spm levels remained stable and at high levels throughout the period of rapid ethylene synthesis, showing no signs of decline until the flowers were moribund and ethylene production had ceased. The authors proposed a cogent explanation of their results, one that is consistent with published evidence and that adheres to the hypothesis that PAs and ethylene do interact to modulate senescence, but doubt about the hypothesis still persists.

Current uncertainties concerning the proposed relationships between PAs and ethylene could be resolved by probing two or three experimental plant systems in detail and routinely monitoring changes in ethylene and PA levels, the activities of key enzymes in both pathways, and metabolic fluxes of radiolabelled metabolites following judicious application of inhibitors of enzyme systems. The development and use of appropriate mutants and transgenic plants would also provide insight into biochemical mechanisms and the apparent relationships between PAs and ethylene. Mutants might be selected that show delayed senescence or unusually high or low levels of PAs. In addition, transformants could be developed that contain antisense genes for ACC synthase, Spd synthase or AdoMet decarboxylase linked to promoters that are temporally expressed and/or readily inducible by hormonal or environmental cues.

REFERENCES

1. **Cantoni, G.L.,** Biological methylation: selected aspects, *Annu. Rev. Biochem.*, 44, 435, 1975.
2. **Heby, O.,** Role of polyamines in the control of cell proliferation and differentiation, *Differentia*, 19, 1, 1981.
3. **Feirer, R.P., Mignon, G., and Litvay, J.D.,** Arginine decarboxylase and polyamines required for embryogenesis in wild carrot, *Science*, 223, 1433, 1984.
4. **Palavan, N. and Galston, A.W.,** Polyamine biosynthesis and titer during various developmental stages of *Phaseolus vulgaris*, *Plant Physiol.*, 55, 438, 1982.
5. **Kaur-Sawhney, R., Shih, L.M., and Galston, A.W.,** Relation of polyamine biosynthesis to the initiation of sprouting in potato tuber, *Plant Physiol.*, 69, 411, 1982.
6. **Heimer, Y.M., Mizrahi Y., and Bachrach, U.,** Ornithine decarboxylase activity in rapidly proliferating plant cells, *FEBS Lett.*, 104, 146, 1979.
7. **Kushad, M.M., Yelenosky, G., and Knight, R.,** Interrelationship of polyamine and ethylene biosynthesis during avocado fruit development and ripening, *Plant Physiol.*, 87, 463, 1988.
8. **Abeles, F.B.,** *Ethylene in Plant Biology*, Academic Press, New York, U.S.A., 1973.
9. **Smith, T.A,** Polyamines, *Annu. Rev. Plant Physiol.*, 36, 117, 1985.
10. **Suzuki, Y. and Hirasawa, E.,** *S*-Adenosylmethionine decarboxylase of corn seedlings, *Plant Physiol.*, 66, 1091, 1980.
11. **Yamanoha, B. and Cohen, S.S.,** *S*-Adenosylmethionine decarboxylase and spermidine synthase from chinese cabbage, *Plant Physiol.*, 78, 784, 1985.
12. **Hirasawa, E. and Suzuki, Y.,** Biosynthesis of spermidine in maize seedlings, *Phytochemistry*, 22, 103, 1983.
13. **Smith, T. A.,** L-Arginine carboxy-lyase of higher plants and its relation to potassium nutrition, *Phytochemistry*, 2, 241, 1963.
14. **Young, N.D. and Galston, A.W.,** Putrescine and acid stress. Induction of arginine decarboxylase activity and putrescine accumulation by low pH, *Plant Physiol.*, 71, 767, 1983.
15. **Slocum, R.D., Kaur-Sawhney, R., and Galston, A.W.,** The physiology and biochemistry of polyamines in plants, *Arch. Biochem. Biophys.*, 235, 283, 1984.
16. **Crocomo, O.J. and Basso, L.C.,** Accumulation of putrescine and related amino acids in potassium deficient *Sesamum*, *Phytochemistry*, 13, 2659, 1974.
17. **Maretzki, A., Thom, M., and Nickell, L.G.,** Products of arginine catabolism in growing cells of sugar-cane, *Phytochemistry*, 8, 811, 1969.
18. **Lieberman, M., Kunishi, A.T., Mapson, L.W., and Wardale, D.A.,** Ethylene production from methionine, *Biochem. J.*, 97, 449, 1965.
19. **Lieberman, M., Kunishi, A., Mapson, L.W., and Wardale, D.A.,** Stimulation of ethylene production in apple tissue slices by methionine, *Plant Physiol.*, 41, 376, 1966.
20. **Adams, D.O. and Yang, S.F.,** Methionine metabolism in apple tissue: implication of *S*-adenosylmethionine as an intermediate in the conversion of methionine to ethylene, *Plant Physiol.*, 60, 892, 1977.
21. **Adams, D.O. and Yang, S.F.,** Ethylene biosynthesis: identification of 1-aminocyclopropane-1-carboxylic acid as an intermediate in the conversion of methionine to ethylene, *Proc. Natl. Acad. Sci. U.S.A.*, 76, 170, 1979.
22. **Lürssen, K., Naumann, K, and Schröder, R.,** 1-Aminocyclopropane-1-carboxylic acid- an intermediate of the ethylene biosynthesis in higher plants, *Z. Pflanzenphysiol.*, 92, 285, 1979.
23. **Boller, T., Herner, R. C., and Kende, H.,** Assay for and enzymatic formation of an ethylene precursor, 1-aminocyclopropane-1-carboxylic acid, *Planta*, 145, 293, 1979.
24. **Cameron, A.C., Fenton, C.A., Yu, Y., and Adams, D. O.,** Increased production of ethylene by plant tissues treated with 1-aminocyclopropane-1-carboxylic acid, *Hort. Sci.*, 14, 178, 1979.

25. **Yu, Y., Adams, D.O., and Yang, S.F.,** Regulation of auxin-induced ethylene production in mung bean hypocotyls, *Plant Physiol.*, 63, 589, 1979.
26. **Bleeker, A.B., Robinson, G., and Kende, H.,** Studies on the regulation of 1-aminocyclopropane-1-carboxylate synthase in tomato using monoclonal antibodies, *Planta*, 173, 358, 1988.
27. **Mehta, A.M., Jordan, R.L., Anderson, J.D., and Mattoo, A.K.,** Identification of a unique isoform of 1-aminocyclopropane-1-carboxylic acid synthase by monoclonal antibody, *Proc. Natl. Acad. Sci. U.S.A.*, 85, 8810, 1988.
28. **Tsai, D.-S., Arteca, R., Bachman, J.M., and Phillips, A. T.,** Purification and characterization of 1-aminocyclopropane-1-carboxylic acid synthase from etiolated mung bean hypocotyl, *Arch. Biochem. Biophys.*, 264, 632, 1988.
29. **Sato, T. and Theologis, A.,** Cloning the mRNA encoding 1-aminocyclopropane-1-carboxylate synthase, the key enzyme for ethylene biosynthesis in plants, *Proc. Natl. Acad. Sci. U.S.A.*, 86, 6621, 1989.
30. **Van Der Straeten, D., Van Wiemeersch, L., Goodman, H.M., and Van Montagu, M.,** Cloning and sequence of two different cDNAs encoding 1-aminocyclopropane-1-carboxylate synthase in tomato, *Proc. Natl. Acad. Sci. U.S.A.*, 87, 4859, 1990.
31. **Yang, S.F. and Hoffman, N.E.,** Ethylene biosynthesis and its regulation in higher plants, *Annu. Rev. Plant Physiol.*, 35, 155, 1984.
32. **Torget, R., Lapi, L., and Cohen, S.S.,** Synthesis and accumulation of polyamines and *S*-adenosylmethionine in chinese cabbage infected by turnip yellow mosaic virus, *Biochem. Biophys. Res. Commun.*, 87, 1132, 1979.
33. **Baur, A.H., Yang, S.F., and Pratt, H.K.,** Ethylene biosynthesis in fruit tissue, *Plant Physiol.*, 47, 696, 1971.
34. **Baur, A.H. and Yang, S.F.,** Methionine metabolism in apple tissue in relation to ethylene biosynthesis, *Phytochemistry*, 11, 3207, 1972.
35. **Shapiro, S.K.,** Response of *Aerobacter aerogenes* methionine auxotrophs to adenine thiomethyl compounds, *J. Bacteriol.*, 65, 310, 1953.
36. **Schwartz, M. and Shapiro, S.K.,** The mechanism of utilization of thiomethyladenosine in the biosynthesis of methionine, *J. Bacteriol.*, 67, 98, 1954.
37. **Shapiro, S.K., Lohmar, P., and Hertenstein, M.,** Utilization of *S*-adenosylmethionine for the biosynthesis of methionine, *Arch. Biochem. Biophys.*, 100, 74, 1963.
38. **Guranowski, A.B., Chiang, P.K., Cantoni, G.L.,** 5′-Methylthioadenosine nucleosidase. Purification and characterization of the enzyme from *Lupinus luteus* seeds, *Eur. J. Biochem.*, 114, 293, 1981.
39. **Wang, S.Y., Adams, D.O., and Lieberman, M.,** Recycling of 5′-methylthioadenosine-ribose carbon atoms into methionine in tomato tissue in relation to ethylene production, *Plant Physiol.*, 70, 117, 1982.
40. **Yung, K.H., Yang, S.F., and Schlenck, F.,** Methionine synthesis from 5-methylthioribose in apple tissue, *Biochem. Biophys. Res. Commun.*, 104, 771, 1982.
41. **Kushad, M.M., Richardson, D.G., and Ferro, A.J.,** 5-Methylthioribose kinase activity in plants, *Biochem. Biophys. Res. Commun.*, 108, 167, 1982.
42. **Kushad, M.M., Richardson, D.G., and Ferro, A.J.,** Intermediates in the recycling of 5-methylthioribose to methionine in fruits, *Plant Physiol.*, 73, 257, 1983.
43. **Trackman, P C. and Abeles, R.H.,** Methionine synthesis from 5′-*S*-methylthioadenosine. Resolution of enzyme activities and identification of 1-phospho-5-*S*-methylthioribulose, *J. Biol. Chem.*, 258, 6717, 1983.
44. **Furfine, E.S. and Abeles, R.H.,** Intermediates in the conversion of 5′-methylthioadenosine to methionine in *Klebsiella pneumoniae*, *J. Biol. Chem.*, 263, 9598, 1988.
45. **Even-Chen, A., Matoo, A.K., and Goren, R.,** Inhibition of ethylene biosynthesis by aminoethoxyvinylglycine and by polyamines shunts label from 3,4-[^{14}C]methionine into spermidine in aged orange peel discs, *Plant Physiol.*, 69, 385, 1982.
46. **Roberts, D.R., Walker, M.A., Thompson, J.E., and Dumbroff, E.B.,** The effects of inhibitors of polyamine and ethylene biosynthesis on senescence, ethylene production and polyamine levels in cut carnation flowers, *Plant Cell Physiol.*, 25, 315, 1984.
47. **Adams, D.O., Wang, S.Y., and Lieberman, M.,** Effect of ethylene biosynthetic inhibitors on methionine incorporation into spermine and spermidine in apple suspension cells, *Plant Physiol.*, 67, S-50, 1981.
48. **Cohen, S.S., Sindhu, R.K., and Greenberg, M.,** Cellular systems for the study of the biosynthesis of polyamines and ethylene, as well as of virus multiplication, *Plant Growth Regul.*, 3, 227, 1985.
49. **Katoh, Y., Hasegawa, T., Suzuki, T., and Fujii, T.,** Effects of 1-aminocyclopropane-1-carboxylic acid production on the changes in polyamine levels in Hiproly barley callus after auxin withdrawal, *Agric. Biol. Chem.*, 51, 2457, 1987.
50. **Biondi, S., Diaz, T., Iglesias, I., Gamberini, G., and Bagni, N.,** Polyamines and ethylene in relation to adventitious root formation in *Prunus avium* shoot cultures, *Physiol. Plant.*, 78, 474, 1990.
51. **Dumbroff, E.B.,** Polyamines — functions and relationships with ethylene and cytokinins, in *Polyamines and Ethylene: Biosynthesis, Physiology and Interactions*, Flores, H.E., Arteca, R.N., and Shannon, J.C., Eds., Amer. Soc. Plant Physiologists, Bettsville, MD, 1990, 256.

52. **Icekson, I., Goldlust, A., and Apelbaum, A.,** Influence of ethylene on S-adenosylmethionine decarboxylase activity in etiolated pea seedlings, *J. Plant Physiol.*, 119, 335, 1985.
53. **Icekson, I., Bakhanashvili, M., and Apelbaum, A.,** Inhibition by ethylene of polyamine biosynthetic enzymes enhanced lysine decarboxylase activity and cadaverine accumulation in pea seedlings, *Plant Physiol.*, 82, 607, 1986.
54. **Apelbaum, A., Goldlust, A., and Icekson, I.,** Control by ethylene of arginine decarboxylase activity in pea seedlings and its implication for hormonal regulation of plant growth, *Plant Physiol.*, 79, 635, 1985.
55. **Bakanashvili, M., Barkai-Golan, R., Kopeliovitch, E., and Apelbaum, A.,** Polyamine biosynthesis in *Rhizopus*-infected tomato fruits: possible interaction with ethylene, *Physiol. Mol. Plant Path.*, 31, 41, 1987.
56. **Suttle, J.C.,** Effect of polyamines on ethylene production, *Phytochemistry,* 20, 1477, 1981.
57. **Apelbaum, A., Burgoon, A.C., Anderson, J.D., and Lieberman, M.,** Polyamines inhibit biosynthesis of ethylene in higher plant tissue and fruit protoplasts, *Plant Physiol.*, 68, 453, 1981.
58. **Fuhrer, J., Kaur-Sawhney, R., Shih, L.-M., and Galston, A.W.,** Effect of exogenous 1,3-diaminopropane and spermidine on senescence of oat leaves, *Plant Physiol.*, 70, 1597, 1982.
59. **Hyodo, H. and Tanaka, K.,** Inhibition of 1-aminocyclopropane-1-carboxylic acid synthase activity by polyamines, their related compounds and metabolites of *S*-adenosylmethionine, *Plant Cell Physiol.*, 27, 391, 1986.
60. **Apelbaum, A. and Icekson, I,** Applied polyamines inhibit macromolecular synthesis in plant tissue, in *Advances in Polyamine Research,* Vol. 4, Bacharach, U., Kaye, A. and Chayen, R., Eds., Raven Press, New York, 1983, 437.
61. **Ke, D. and Romani, R.J.,** Effect of spermidine on ethylene production and the senescence of suspension-cultured pear fruit cells, *Plant Physiol. Biochem.*, 26, 109, 1988.
62. **Ben-Arie, R., Lurie, S., and Mattoo, A.K.,** Temperature-dependent inhibitory effects of calcium and spermine on ethylene biosynthesis in apple discs correlate with changes in microsomal membrane microviscosity, *Plant Sci. Lett.*, 24, 239, 1982.
63. **Drolet, G., Dumbroff, E.B., Legge, R.L., and Thompson, J.E.,** Radical scavenging properties of polyamines, *Phytochemistry,* 25, 367, 1986.
64. **McRae, D.G., Baker, J.E., and Thompson, J.E.,** Evidence for involvement of the superoxide radical in the conversion of 1-aminocyclopropane-1-carboxylic acid to ethylene by pea microsomal membranes, *Plant Cell Physiol.*, 23, 375, 1982.
65. **Pennazio, S. and Roggero, P.,** Stimulation of ethylene production by exogenous spermidine in detached tobacco leaves in the light, *Biol. Plant.*, 31, 59, 1989.
66. **Pennazio, S. and Roggero, P.,** Exogenous polyamines stimulate ethylene synthesis by soybean leaf tissue, *Ann. Bot.*, 65, 51, 1990.
67. **Downs, C.G. and Lovell, P.H.,** The effect of spermidine and putrescine on the senescence of cut carnations, *Physiol. Plant.*, 66, 679, 1986.
68. **Williams-Ashman, H.G. and Schenone, A.,** Methylglyoxal *bis*(guanylhydrazone) as a potent inhibitor of mammalian and yeast *S*-adenosylmethionine decarboxylase, *Biochem. Biophys. Res. Commun.*, 46, 288, 1972.
69. **Dibble, A.R., Davies, P.J., and Mutschler, M.,** Polyamine content of long-keeping Alcobaca tomato fruit, *Plant Physiol.*, 86, 338, 1988.
70. **Saftner, R.A. and Baldi, B.G.,** Polyamine levels and tomato fruit development: possible interaction with ethylene, *Plant Physiol.*, 92, 547, 1990.
71. **Tabor, H., Tabor, C.W., Cohen, M.S., and Hafner, W.W.,** Streptomycin resistance (*rpsL*) produces an absolute requirement for polyamines for growth of an *Escherichia coli* strain unable to synthesize putrescine and spermidine, *J. Bacteriol.*, 147, 702, 1981.
72. **Dememtriou, A.A., Cohen, M.S., Tabor, C.W., and Tabor, H.,** Identification of pyruvate in *S*-adenosylmethionine decarboxylase from rat liver, *J. Biol. Chem.*, 253, 1684, 1978.
73. **Cohen, S.S., Baline, R., and Sindhu, R.K.,** The synthesis of polyamines from methionine in intact and disrupted leaf protoplasts of virus-infected chinese cabbage, *Plant Physiol.*, 68, 1150, 1981.
74. **Leshem, Y.Y., Halevy, A.H., and Frenkel, C.,** *Processes and Control of Plant Senescence*, Elsevier, Amsterdam, 1986, Chap. 2.
75. **Evans, T.E. and Malmberg, R.L.,** Do polyamines have roles in plant development?, *Annu. Rev. Plant Physiol. Plant Mol. Biol.*, 40, 235, 1989.
76. **Apelbaum, A., Icekson, I., Burgoon, A.C., and Lieberman, M.,** Inhibition by polyamines of macromolecular synthesis and its implication for ethylene production and senescence processes, *Plant Physiol.*, 70, 1221, 1982.
77. **Kaur-Sawhney, R., Flores, H.E., and Galston, A.W.,** Polyamine oxidase in oat leaves: a cell wall-localized enzyme, *Plant Physiol.*, 68, 494, 1981.
78. **Hiatt, A.C., McIndoo, J., and Malmberg, R.L.,** Regulation of polyamine biosynthesis in tobacco. Effects of inhibitors and exogenous polyamines on arginine decarboxylase, ornithine decarboxylase, and *S*-adenosylmethionine decarboxylase, *J. Biol. Chem.*, 261, 1293, 1986.

79. **Mikitzel, L.J, and Knowles, N.R.,** Polyamine metabolism of potato seed-tubers during long-term storage and early sprout development, *Plant Physiol.*, 91, 183, 1989.
80. **Smith, M.A. and Davies, P.J.,** Effect of photoperiod on polyamine metabolism in apical buds of G-2 peas in relation to the induction of apical senescence, *Plant Physiol.*, 79, 400, 1985.
81. **Toumadje, A. and Richardson, D.J.,** Endogenous polyamine concentrations during development, storage and ripening of pear fruits, *Phytochemistry*, 27, 335, 1988.
82. **Kramer, G.F., Wang, C.Y., and Conway, W.S.,** Correlation of reduced softening and increased polyamine levels during low-oxygen storage of "McIntosh" apples, *J. Amer. Soc. Hort. Sci.*, 114, 942, 1989.
83. **Conway, W.S. and Sams, C.E.,** The effects of postharvest infiltration of calcium, magnesium, and strontium on decay, firmness, respiration, and ethylene production in apples, *J. Amer. Soc. Hort. Sci.*, 112, 300, 1987.
84. **Altman, A., Friedman, R., Amir, D., and Levin, N.,** Polyamine effects and metabolism in plants under stress conditions, in *Plant Growth Substances*, Wareing, P.F., Ed., Academic Press, London, 1982, 483.
85. **Corey, K.A. and Barker, A.V.,** Ethylene evolution and polyamine accumulation by tomato subjected to interactive stresses of ammonium toxicity and potassium deficiency, *J. Amer. Soc. Hort. Sci.*, 114, 651, 1989.
86. **McDonald, R.E. and Kushad, M.M.,** Accumulation of putrescine during chilling injury of fruits, *Plant Physiol.*, 82, 324, 1986.
87. **Wang, C.Y.,** Changes of polyamines and ethylene in cucumber seedlings in response to chilling stress, *Physiol. Plant.*, 69, 253, 1987.
88. **Shevyakova, N.I., Kir'yan, I.G., and Strogonov, B.P.,** Enhanced rate of spermidine formation in an NaCl-resistant cell line of *Nicotiana sylvestris, Fiziologiya Rastenii*, 31, 810, 1984.
89. **Tiburcio, A.F., Masde'u, M.A., Dumortier, F.M., and Galston, A.W.,** Polyamine metabolism and osmotic stress. 1. Relation to protoplast viability, *Plant Physiol.*, 82, 369, 1986.
90. **Lin, P.P.C.,** Polyamine anabolism in germinating *Glycine max* (L.) seeds. Dynamics of cadaverine and putrescine formation in the embryonic axis, *Plant Physiol.*, 76, 372, 1984.
91. **Sen, K., Choudhuri, M.M., and Ghosh, B.,** Changes in polyamine contents during development and germination of rice seeds, *Phytochemistry*, 20, 631, 1981.
92. **Anguillesi, M.C., Grilli, I., and Floris, C.,** Rate of synthesis of spermine and spermidine in germinating seeds of *Glycine, Helianthus* and *Triticum, Planta*, 148, 24, 1980.
93. **Villanueva, V.R., Adlakha, R.C., and Cantera-Soler, A.M.,** Changes in polyamine concentration during seed germination, *Phytochemistry*, 17, 1245, 1972.
94. **Ketring, D.L.,** Ethylene and seed germination, in *The Physiology and Biochemistry of Seed Dormancy and Germination*, Khan, A.A., Ed., North-Holland, Amsterdam, 1977, 157.
95. **Ketring, D.L. and Morgan, P.W.,** Physiology of oil seeds. II. Regulation of dormancy in Virginia-type peanut seeds, *Plant Physiol.*, 47, 488, 1971.
96. **Walker, M.A., Roberts, D.R., Waite, J.L., and Dumbroff, E.B.,** Relationships among cytokinins, ethylene and polyamines during the stratification-germination process in seeds of *Acer saccharum, Physiol. Plant.*, 76, 326, 1989.
97. **Shih, C.Y., Dumbroff, E.B., and Peterson, C.A.,** Developmental studies of the stratification-germination process in sugar maple embryos, *Can. J. Bot.*, 63, 903, 1985.
98. **Srivastava, S.K., Kansara, M.S., and Mungre, S.M.,** Effect of polyamines and guanidines on the growth, nitrogen assimilation and reserve mobilization in germinating radish seedlings, *J. Plant Growth Regul.*, 3, 339, 1985.
99. **Walker, M.A., Roberts, D.R., Shih, C.Y., and Dumbroff, E.B.,** A requirement for polyamines during the cell division phase of radicle emergence in seeds of *Acer saccharum, Plant Cell Physiol.*, 26, 967, 1985.
100. **Cohen, E. and Kende, H.,** The effect of submergence, ethylene and gibberellin on polyamines and their biosynthetic enzymes in deepwater-rice internodes. *Planta*, 169, 498, 1986.

Chapter 6

TISSUE AND SUBCELLULAR LOCALIZATION OF POLYAMINES AND ENZYMES OF POLYAMINE METABOLISM

Robert D. Slocum

TABLE OF CONTENTS

I. INTRODUCTION

Polyamines are strongly basic molecules which are positively charged at physiological pH.[1] As such, they have been shown to bind strongly *in vitro* to negatively charged nucleic acids,[2] acidic phospholipids,[3] and many types of proteins, including numerous enzymes whose activities are directly modulated by polyamine binding.[4-8] These ionic interactions are thought to be important in regulating the structure and function of biological macromolecules, as well as their synthesis, *in vivo*.[9,10] Indeed, studies with polyamine deficient mutants suggest that polyamines are essential for normal growth and development in living organisms.[11]

In this context, perhaps, it is surprising that relatively little is known regarding the subcellular distribution of polyamines. In part, this is due to inherent difficulties related to the localization of any small, highly-charged biological molecule, where there is the potential for artifactual associations based on nonspecific binding, leakage from subcellular compartments, etc. Nonetheless, several innovative approaches to this problem, including the characterization of polyamine transport in cells and organelles, have led to a better understanding of the localization of polyamines in various cell types in recent years.

The localization of enzymes involved in polyamine metabolism in plants and other eucaryotic organisms is also poorly understood. Enzyme localization data obtained from subcellular fractionation studies suffer from many of the same shortcomings described for polyamine studies above. However, many enzymes of polyamine metabolism have recently been purified and, as antibodies which specifically recognize these enzymes have become available, immunocytochemical methods have been used to localize them *in situ*. In addition, the use of specific, radiolabeled inhibitors has permitted the autoradiographic localization of some enzymes, while enzyme cytochemistry has been used to examine the tissue and subcellular distribution of others.

Taken together, these localization studies have offered important new insights into polyamine metabolism and the possible biological functions of polyamines themselves. This chapter will provide a brief overview of this work and, wherever possible, the limitations of existing methods and possible new approaches to old problems will be examined.

II. LOCALIZATION OF POLYAMINES

A. SUBCELLULAR FRACTIONATION AND TRANSPORT STUDIES

One approach to studying the subcellular distribution of polyamines involves the fractionation of the cell into various membrane, organellar, and cytosolic fractions. In the case of plants and other walled organisms, it is also necessary to estimate the amounts of polyamines associated with the cell wall fraction.

Goldberg and Perdrizet[12] reported that polyamines in mature plant tissues were associated primarily with the cell wall, where they were bound to polyuronic acids of the pectic fraction and also to lignin. However, these studies were carried out with cell wall-enriched fractions of tissue pellets and the potential for nonspecific binding of intracellular polyamines to wall polymers cannot be ruled out. Studies comparing uptake of radiolabeled polyamines by carrot cells versus protoplasts, suggest that approximately two thirds of the total polaymine is bound to the cell wall,[13] suggesting that the wall exhibits a large binding capacity for polyamines. Li and McClure[14] detected polyamines in buffered 0.2 *M* NaCl washing solutions which removed apoplastic constituents of intact oat leaf tissues without significant leakage of cellular marker enzymes, suggesting that polyamines are indeed normal consituents of the wall. However, they also noted polyamine oxidation products in apoplastic exudates, which is not surprising since plant diamine and polyamine oxidases are also known to be cell wall localized (see

Section III C). This observation suggests that some care must be taken to inhibit amine oxidase activity during the extraction of wall polyamines if quantitative estimates are to be made. More recent studies by Bagni and co-workers, examining *in vitro* interactions between polyamines and cell wall fractions, are reviewed in Chapter 7. Another type of evidence that supports the existence of extracellular polyamines in plant tissues is the finding that xylem sap contains large amounts of Put (0.15 to 9.2 μM concentrations) and smaller amounts of Spd.[15]

Subcellular fractionation studies examining the distribution of intracellular polyamines in plants have been scarce. In tobacco cell cultures, Walker et al.[16] documented only 2 to 3% of the total free putrescine (Put) and spermidine (Spd) in pellet fractions obtained by differential centrifugation; the rest occurring in the supernatant. In this study, however, it was not clear to what extent label in the supernatant fraction represented polyamines from the cytosol vs. those leaking from organelles, since the distribution of various marker enzymes suggested significant disruption of organelles occurred during cell fractionation.

Pistocchi et al.[13] took a somewhat different approach, examining the uptake of ^{14}C-labeled Put/Spd by carrot cell protoplasts or vacuoles isolated by gentle lysis of the protoplasts. Following short-term (5 min) uptake of labeled polyamine, protoplasts or vacuoles were rapidly centrifuged through a silicone oil layer to estimate total uptake and surface membrane binding. In parallel, protoplasts and vacuoles were centrifuged through oil layers separated by an isotonic washing solution containing 100 mM unlabeled Put or Spd, in order to remove surface-bound labeled polyamine. These studies showed that approximately 50% of the ^{14}C-Spd bound to the tonoplast and 92% was associated with the plasma membrane, the rest being taken up by the cell or vacuole. Uptake by the cell appeared to be passive, although uptake into vacuoles was carrier-mediated (pH 7 optimum, $K_m = 62\ \mu M$). In long-term labeling studies, significant quantities of polyamine accumulated within isolated vacuoles, although precise measurements of polyamine uptake proved to be difficult, due to their leakiness. There seems to be little doubt, however, that a large portion of the intracellular polyamine pool in plants is sequestered within the vacuole. In *Neurospora crassa*, about a third of the total cellular Spd was located in the vacuole.[17]

Polyamines have also been found in isolated mitochondria and chloroplasts[18] and an energy-dependent uptake of polyamines by mitochondria isolated from *Helianthus tuberosus* tissue has recently been characterized.[19] Specific information regarding the polyamine contents of nuclei isolated from plant tissues has not been reported, although it seems likely that much of the intracellular polyamine pool would be associated with nuclear chromatin, as has been shown in many other organisms. In animal cells, Mach et al.[20] found that Put, Spd, and spermine (Spm) were normal constituents of nuclei, although the Spm content was about fivefold higher than that of Put or Spd. Datta et al.[5] reported that Spm stimulated the activity of a nuclear kinase in isolated pea nuclei, but provided no data regarding endogenous nuclear polyamine titers.

B. CYTOCHEMICAL AND IMMUNOCYTOCHEMICAL METHODS

In view of problems associated with cell fractionation techniques, some workers have attempted to localize cell and tissue polyamines *in situ*. For example, Hougaard et al.[21] have used antibodies produced against Spd or Spm in immunocytochemical localization studies with animal cells. These studies suggest polyamines are mainly associated with highly condensed chromatin in the cell. They have also used two fluorescence localization methods for polyamines, based upon their reaction with *o*-phthalaldehyde, or formaldehyde/fluorescamine reagents. It is not clear that these reagents preferentially react with polyamines vs. other types of amines but, in general, the distribution of fluorescence in the cell is similar to that seen using immunocytochemistry. As with cell fractionation studies, these methods afford some opportunity for diffusion and redistribution or loss of polyamines from the tissue during fixation and the authors stress the need for well-controlled fixation methods. None of these cytochemical or immunocytochemical methods appear to have been used to localize polyamines in plant tissues.

C. AUTORADIOGRAPHIC LOCALIZATION OF RADIOLABELED POLYAMINES

A third method for the localization of polyamines involves the use of autoradiographic techniques to examine the cellular- and tissue-level distribution of radioactivity after supplying exogenous labeled polyamine. Dion and Herbst[22] used this method to demonstrate the subcellular compartmentation of ^{3}H-spermidine within *Drosophila* salivary gland cells. While similar studies have not been carried out with plants, several investigators have applied radiolabeled polyamines to plant tissues in order to study their uptake and transport (see Chapter 7, Section II). However, a major problem with all of these studies is that the metabolism of the labeled polyamines *in vivo* was not studied and it is not known to what extent localized radioactivity is correlated with the distribution of polyamines, as opposed to their metabolites. In plants and some fungi, cell wall-localized diamine and polyamine oxidases (Section III C) would be expected to oxidize a considerable portion of the applied polyamine and this problem needs to be addressed in future studies of this type.

An attractive alternative to the above studies may be offered by the recent development of Spd and Spm analogs designed for use in photoaffinity labeling studies of polyamine binding to cellular constituents.[23] Assuming that the transport properties of these polyamine analogs are similar to those of Spd and Spm, these compounds may be useful in demonstrating the subcellular compartmentation of polyamines, minimizing potential localization artifacts inherent to other methods.

III. LOCALIZATION OF ENZYMES OF POLYAMINE METABOLISM

A. ARGININE, ORNITHINE AND LYSINE DECARBOXYLASES

Torrigiani et al.[18] reported that arginine decarboxylase (EC 4.1.1.19; ADC) and ornithine decarboxylase (EC 4.1.1.17; ODC) activities were localized in isolated chloroplasts from pine cotyledons and in mitochondria from *Helianthus tuberosus* tuber tissue. However, less than 10% of the total ADC and ODC activities were actually found in the chloroplast fraction, the remaining activity being about equally divided between 200 × g pellet and a 26,000 × g supernatant fractions, perhaps resulting from chloroplast aggregation and breakage during isolation.

In tobacco cells, Walker et al.[16] detected approximately 21% of the total ADC activity in a 16,000 × g pellet fraction (containing chloroplasts and mitochondria); the rest was found in a 100,000 × g pellet and the supernatant. In contrast, they reported that about 16% of the ODC activity was found in a 1,000 × g pellet containing nuclei, while only 4% of the activity was found in the 16,000 × g pellet and 75% was found in the supernatant. These results suggest that ADC activity occurs mostly in the cytosol and, to a lesser extent, in chloroplasts and mitochondria, while ODC activity is predominantly cytosolic, with a small nuclear-localized activity. In barley seeds, Panagiotidis et al.[24] have also reported distinct cytosolic and chromatin-bound ODC activities.

Lysine decarboxylase (EC 4.1.1.18; LDC) activity has also been localized to the stromal fraction of chloroplasts isolated from lupine.[25]

Persson et al.[26] used immunofluorescence microscopy to examine patterns of ODC expression in mouse kidney tissues relative to the induction of ODC activity by testosterone. Similar immunocytochemical methods do not yet appear to have been employed to localize ODC, ADC, or LDC in plant tissues, primarily due to the fact that antibodies which recognize these enzymes have only recently become available.[27,28] These antisera should permit both light microscopic and high-resolution electron microscopic localization of these enzymes in plants.

One innovative approach to the localization of ODC activity in animal tissues is based upon the specific binding of an enzyme-activated, irreversible inhibitor DL-a-difluoromethylornithine (DFMO). Seely and Pegg[29] have shown that ^{3}H-DFMO binds to a single 53,000 Da protein in mouse kidney tissues *in vivo* and that, in tissue extracts, quantitative precipitation of the ^{3}H-DFMO-labeled enzyme parallels loss of ODC activity in the presence of the inhibitor. Based on these observations, several workers have used light microscopic autoradiographic methods to localize ODC in animal tissues.[30,31] This method was also used localize ODC in rapidly-growing tobacco ovary tissues[32] (Figure 1) and in corn root meristems.[33]

In our laboratory, the subcellular distribution of ^{14}C-DFMO-labeled ODC in tomato ovaries has recently been investigated using electron microscopic autoradiography (unpublished results). In these tissues, approximately 45% of the total radioactivity is localized within the vacuole, consistent with the findings of Mengoli et al.[34] for ^{14}C-DFMO uptake by carrot cells. Of the remaining radioactivity in tomato tissues, about 55% of it is found within the nucleus and 45% within the cytoplasm. These observations support data from subcellular fractionation studies (see above), indicating that separate cytosolic and chromatin-associated ODC activities occur in plant cells. Zagon et al.[30] also reported that a small fraction of ^{3}H-DFMO label was localized in the nucleus in mouse kidney proximal tubule cells.

As with other types of studies, some caution must be exercised in interpreting the localization of ODC based on labeled DFMO binding to the enzyme. In all studies, labeled tissues have been fixed with aldehydes to stabilize tissue morphology. Potentially, this could result in the nonspecific fixation of nonODC bound DFMO in the tissue via its amine functions. We attempted to minimize this problem by using formaldehyde instead of glutaraldehyde (a bifunctional cross-linking reagent) as a primary fixative, but some nonspecific localization may have occurred. In our studies, a loss of large amounts of DFMO from the vacuolar compartment would have been expected during fixation of tomato tissues if this inhibitor was unbound. However, actual losses of labeled inhibitor were $< 10\%$ of the total tissue label, suggesting the much of the inhibitor was bound to proteins or other macromolecules. ODC has been reported to have an extremely short half-life ($t_{1/2} = 20$ min)[29] and much of the enzyme protein may have been degraded during the long (6 to 16 h) labeling times used in this study, perhaps resulting in the accumulation of labeled ODC degradation products within the vacuole.

A second potential problem with labeled DFMO localization studies involves the possible metabolism of the inhibitor in tissues. Slocum et al.[35] reported that DL-α-difluoromethylarginine (DFMA), a mechanism-based inhibitor of ADC which is specific for this enzyme *in vitro*, was largely metabolized to DFMO in mouse liver and tobacco ovary tissues containing high arginase activities. This metabolism of DFMA to DFMO resulted in the unexpected inhibition of ODC activity by DFMA in tobacco tissues[36] and a similar phenomenon has been reported in DFMA-treated pea seedlings.[37] Based on these results, it seems prudent that the use of ^{3}H-DFMA to localize ADC activity be restricted to those tissues which did not contain appreciable arginase activities.

^{3}H-DFMO has also been reported to be metabolized to unknown compounds in mouse fibroblasts[38] and in bean tissues,[39] though to a much smaller extent and over longer periods of time. The degree to which the formation of such metabolites might compromise the interpretation of ODC localization data based upon labeled DFMO autoradiography is probably insignificant, but warrants further study.

The inhibition of pea LDC by DL-α-difluoromethyllysine (DFML) has recently been reported.[37] It may be possible to usc radiolabeled DFML to localize LDC, as has been described for ODC, but the possible metabolism of this inhibitor also should be investigated prior to such studies.

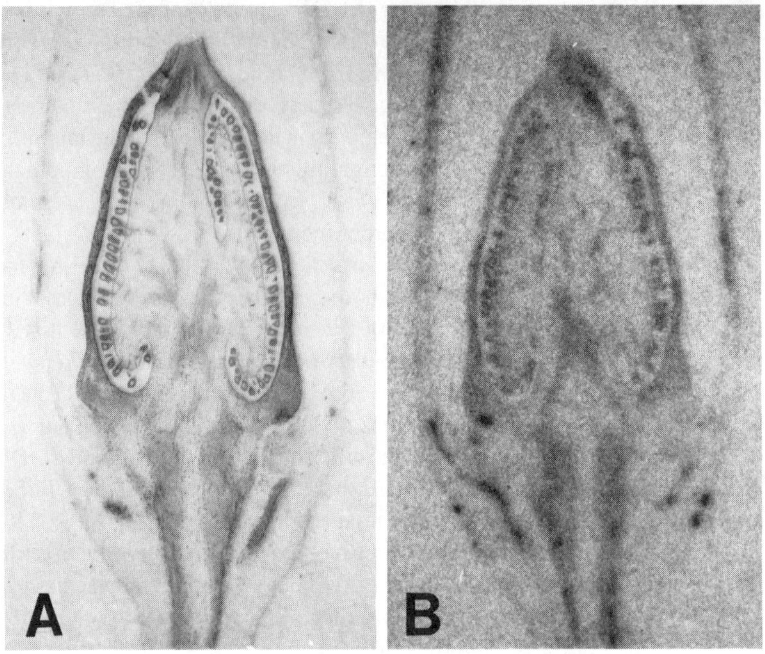

FIGURE 1. Autoradiographic localization of ^{14}C-DFMO in tobacco flowers, 8 days post-anthesis. (A) Median section through flower showing pedicel and ovary tissues. (B) Autoradiograph of same section showing distribution of ^{14}C-DFMO label in flower tissues. Note heavy labeling of ovules.

B. *S*-ADENOSYLMETHIONINE DECARBOXYLASE, SPERMIDINE, AND SPERMINE SYNTHASES, AND OTHER ENZYMES INVOLVED IN POLYAMINE BIOSYNTHESIS

Two other enzymes involved in Put biosynthesis via the ADC pathway are agmatine iminohydrolase (EC 3.4.3.12, AIH) and *N*-carbamamoylputrescine amidohydrolase (EC 3.5.1.-). The subcellular location of AIH has not been reported, while NCPut amindohydrolase from corn and barley appears to be a cytosolic enzyme.[40,41] Similarly, *S*-adenosylmethionine decarboxylase (EC 4.1.1.50, SAMDC) activity has been reported for the cytosolic fractions of pine cotyledons[18] and tobacco cells.[16] Spermidine synthase (EC 2.5.1.16) activity in chinese cabbage was also cytosolic.[42] The subcellular location for spermine synthase (EC 2.5.1.22) has not been reported. All of these data were obtained from cell fractionation studies only.

C. DIAMINE AND POLYAMINE OXIDASES

Diamine oxidase (EC 1.4.3.6, DAO) and polyamine oxidase (EC 1.5.3.3) are enzymes involved in polyamine catabolism in plants and other organisms (see Chapter 3). In the last few years there has been a renewed interest in these enzymes with regard to plant growth regulation. Most evidence now suggests that both DAO and PAO are cell wall-localized enzymes and, since they both produce H_2O_2 as a reaction product, it has been proposed that their activities may influence growth via coupled H_2O_2-dependent, peroxidase-mediated cross-linking of cell wall polymers, such as lignin.[43]

Several types of evidence support a cell wall location for these enzymes. Kaur-Sawhney et al.[44] demonstrated that the PAO activity of oat leaf tissues was completely absent in oat protoplast preparations, although about 80% of the activity could be recovered in the cell wall digest. Similarly, Flores and Filner[45] demonstrated that oat PAO and pea DAO activities both resided primarily in the cell wall digests of protoplast preparations, while the activity of

Δ-pyrroline dehydrogenase, another enzyme involved in the catabolism of Put, was localized within the protoplasts.

Federico and Angelini[46] reported that pea DAO was the major protein in extracellular fluids extracted from pea tissues. Since pea DAO was removed by low ionic strength buffer washes, the authors concluded that this enzyme was loosely bound to the cell wall. Using a similar method, Li and McClure[14] found that up to 99% of oat PAO activity was extracted in buffered 0.2 *M* NaCl intercellular washing solutions. Since the extracellular fluids containing DAO and PAO in these two studies were not contaminated with cytoplasmic marker enzymes, both enzymes were believed to be entirely wall-localized. Conversely, intercellular washing fluid from barley leaf tissue did not contain PAO activity, while nearly all of the activity was found in barley mesophyll protoplasts.[14] On this basis, the authors deduced that the barley PAO was localized within the cell. The reason for this discrepency in closely-related species of gramineae is unclear. There have been other reports of cytosolic DAO activities in pea[47] and *Lathyrus*,[48] based upon cell fractionation studies. However, in the case of the pea DAO, which is loosely bound to the cell wall,[46] it seems likely that the enzyme was extracted from the wall into the cytoplasmic fraction during tissue homogenization.

DAO in lentil has also been localized by immunofluorescence microscopy.[49] In this study, it was claimed by the authors that the lentil enzyme was localized in the cell wall, although this is not clear from their micrographs, which rather suggest a cytoplasmic location for DAO in this plant. It might be possible to more clearly demonstrate the cell wall-localization for lentil DAO using immunogold electron microscopic localization methods in future studies. Since the corn[50] and oat[51] PAOs have recently been purified, antibodies may soon be available to carry out immunohistochemical localization studies for this enzyme as well.

DAO and PAO have also been localized in plant tissues using several different histochemical methods. For example, a soluble starch/KI method, which detects the production of amine oxidase-generated H_2O_2, was used by Kaur-Sawhney et al.[44] to localize PAO in oat leaves. These authors reported that high PAO activities were associated with lignified vascular tissues of the leaf. Scorer et al.[52] used this same method to examine growth-related changes in DAO activity in *Canavalia* seedlings.

A somewhat different method was used by Angelini and Federico[43] to localize DAO activity in *Cicer arietinum* and PAO activity in corn. These authors employed a coupled amine oxidase-peroxidase method in which DAO- or PAO-generated H_2O_2 was utilized by endogenous tissue peroxidases to oxidize various substrates in color-forming reactions. A possible drawback to this indirect method for visualizing amine oxidase activities is that it is assumed that the tissue distributions of DAO or PAO and peroxidase are coincident. There is no reason to expect this to be the case in all tissues and this method might lead to the underestimation of amine oxidase activities where peroxidase activities were low or absent.

In each of the above histochemical or immunohistochemical localization studies, the subcellular localization of DAO and PAO could not be demonstrated unambiguously, due to the limited resolution inherent in such methods. Recently, our laboratory has described a new electron microscopic cytochemical method for the localization of amine oxidases, based on the precipitation of H_2O_2 by $CeCl_3$ (see Figure 2), which permits high-resolution localization of plant DAO and PAO activities at the ultrastructural level.[53] This approach is similar to the soluble starch/KI method in that it involves the direct visualization of H_2O_2 formation in the tissue in the presence of an appropriate diamine or polyamine substrate.

In corn coleoptile, PAO activity was found exclusively in the cell wall (Figure 3), supporting earlier biochemical studies. PAO activity staining was concentrated in the middle lamella region of the wall. In pea epicotyl and root tissues, DAO activity also occurred only in the cell wall. In both plants, amine oxidase activities were greatly elevated in lignified tissues, consistent with their proposed role in lignin formation.[43]

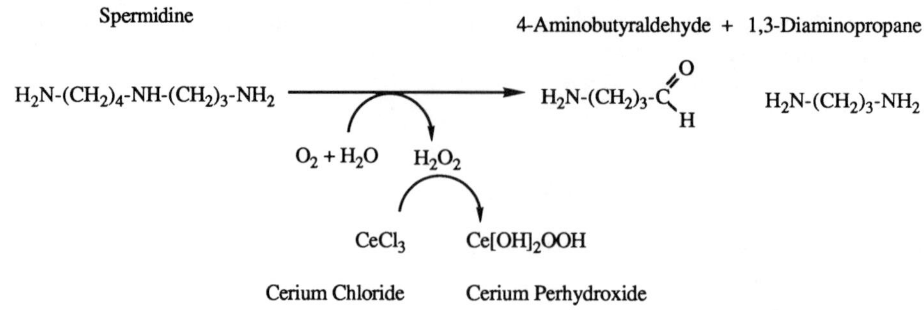

FIGURE 2. Scheme depicting the $CeCl_3$ method used to localize PAO activity in plant tissues. PAO oxidation of putrescine results in the production of H_2O_2, which is precipitated by $CeCl_3$. The cerium perhydroxide precipitate can be visualized by electron microscopy, permitting high resolution localization of PAO activity in the tissue.

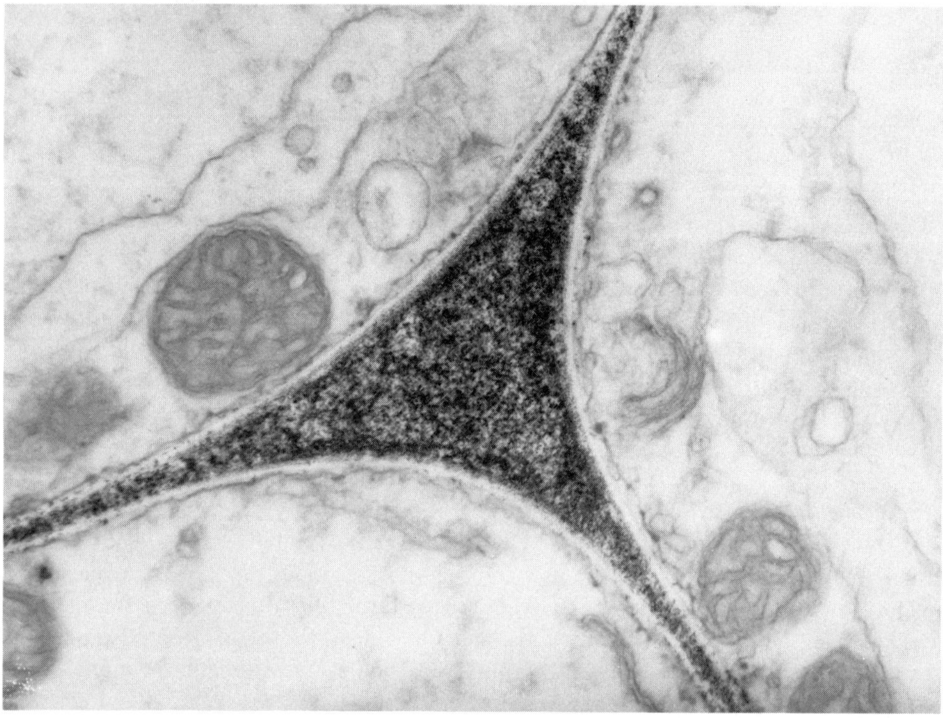

FIGURE 3. Cytochemical localization of PAO activity in corn coleoptile vascular parenchyma tissue using the $CeCl_3$ method. Staining is concentrated in the middle lamella region of the cell wall.[53] (Magnification × 34,300.)

One limitation with the $CeCl_3$ method, or other histochemical methods, is their inability to distinguish one or more amine oxidases with differing diamine or polyamine substrate specificities in the same tissue. This has not been considered to be a problem, since most plants studied exhibit either DAO (e.g., the legumes) or PAO (e.g., the Gramineae) activities, but not both. Recently, however, Cogoni et al.[54] have purified a DAO from barley which has a broad substrate specificty, oxidizing both Put and Spd, while a distinct barley PAO oxidizes only Spd or Spm, but not diamines.[55] In such plants, both DAO and PAO activities could be localized on the basis of Spd oxidation and resultant H_2O_2 production.

IV. CONCLUDING REMARKS

Information about the tissue and subcellular distributions of polyamines and enzymes of polyamine metabolism in plants is still rather limited, as is evident from the preceding discussion. Subcellular fractionation techniques generally seem to be unsuitable for estimating the cellular compartmentalization of polyamines, although indirect approaches, such as the characterization of polyamine transport properties of isolated organelles, have provided much new information. It may also be possible to use nonaqueous fractionation methods to more accurately determine cellular partitioning of polyamines within various organelles. Similarly, alternate methods for tissue fixation and preparation, such as freeze-substitution and/or cryomicrotomy may permit immunocytochemical localization of polyamines with minimal artifacts related to nonspecific fixation and redistribution of polyamines within the tissue.

Our understanding of the localization of enzymes of polyamine metabolism in plants has been advanced through the application of several innovative methods, ranging from cellular fractionation to enzyme cytochemistry. Many of these methods have serious limitations and its seems clear that immunocytochemistry has the greatest potential to provide clear, unambiguous information about the tissue and subcellular localization of these enzymes. Nearly all of the enzymes involved in plant polyamine metabolism have now been purified and biochemically characterized and one anticipates that specific antibodies for these enzymes will be available for localization studies in the near future.

ACKNOWLEDGMENTS

This work was supported by a grant from the National Aeronautics and Space Administration (NAGW-1382).

REFERENCES

1. **Takeda, Y., Samejima, K., Nagano, K., Watanabe, M., Sugeta, H., and Kyogoku, Y.,** Determination of protonation sites in thermospermine and in some other polyamines by ^{15}N and ^{13}C nuclear magnetic resonance spectroscopy, *Eur. J. Biochem.*, 130, 383, 1983.
2. **Feurstein, B.G. and Marton, L.J.,** Specificity and binding in polyamine/nucleic acid interactions, in *The Physiology of Polyamines*, Vol. I, Bachrach, U. and Heimer, Y.M., Eds., CRC Press, Boca Raton, FL, 1989, 109.
3. **Tadolini, B. Cabrini, L., Varani, E., and Sechi, A.M.,** Spermine binding and aggregation of vesicles of different phospholipid composition, *Biol. Amines*, 3, 87, 1985.
4. **Corley, E., Wolosiuk, R.A., and Hertig, C.M.,** Regulation of the activation of chloroplast fructose-1,6-*bis* phosphatase (E.C. 3.1.3.11). Inhibition by spermidine and spermine, *Biochem. Biophys. Res. Commun.*, 115, 707, 1983.
5. **Datta, N., Schell, M.B., and Roux, S.J.,** Spermine stimulation of a nuclear NII kinase from pea plumules and its role in the phosphorylation of a nuclear polypeptide, *Plant Physiol.*, 84, 1397, 1987.
6. **Eichler, W. and Corr, R.,** Polyamine effects on DNA-directed RNA polymerases in the ciliate *Tetrahymena thermophila. In vivo* and *in vitro* experiments suggesting highly specific regulative interactions, *Biol. Chem. Hoppe-Seyler*, 370, 451, 1989.
7. **Fink, J., Jeblick, W., Blaschek, W., and Krauss, H.,** Calcium ions and polyamines activate the plasma membrane-located 1,3-β-glucan synthase, *Planta*, 171, 130, 1987.
8. **Igarashi, K., Kashiwagi, K., Kobayashi, H., Ohnishi, R., Kakegawa, T., Nagasu, A., and Hirose, S.,** Effect of polyamines on mitochondrial F1 ATPase-catalyzed reactions, *J. Biochem.*, 106, 294, 1989.
9. **Barbiroli, B, Mezzetti, G., and Moruzzi, M.S.,** Polyamines and initiation of protein synthesis in eucaryotes, in *The Physiology of Polyamines*, Vol. I, Bachrach, U. and Heimer, Y.M., Eds., CRC Press, Boca Raton, FL, 1989, 125.

10. **Jacob, S.T. and Stetler, D.A.**, Polyamines and RNA synthesis, in *The Physiology of Polyamines*, Vol. I, Bachrach, U. and Heimer, Y.M., Eds., CRC Press, Boca Raton, FL, 1989, 133.
11. **Tabor, C.W. and Tabor, H.**, Polyamines, *Annu. Rev. Biochem.*, 53, 749, 1984.
12. **Goldberg, R. and Perdrizet, E.**, Ratio of free to bound polyamines during maturation in mung-bean hypocotyl cells, *Planta*, 161, 531, 1984.
13. **Pistocchi, R., Keller, F., Bagni, N., and Matile, P.**, Transport and subcellular localization of polyamines in carrot protoplasts and vacuoles, *Plant Physiol.*, 87, 514, 1988.
14. **Li, Z.-C. and McClure, J.W.**, Polyamine oxidase of primary leaves is apoplastic in oats but symplastic in barley, *Phytochemistry*, 28, 2255, 1989.
15. **Friedman, R., Levin, N., and Altman, A.**, Presence and identification of polyamines in xylem and phloem exudates of plants, *Plant Physiol.*, 82, 1154, 1986.
16. **Walker, M.A., Ellis, B.E., Chapple, C.C.S., and Dumbroff, E.B.**, Subcellular localization of amines and activities of their biosynthetic enzymes in *p*-fluorophenylalanine resistant and wild type tobacco cell cultures, *Plant Physiol.*, 85, 78, 1987.
17. **Paulus, T.J., Cramer, C.L., and Davis, R.H.**, Compartmentation of spermidine in *Neurospora crassa*, *J. Biol. Chem.*, 258, 8608, 1983.
18. **Torrigiani, P., Serafini-Fracassini, D., Biondi, S., and Bagni, N.**, Evidence for the subcellular localization of polyamines and their biosynthetic enzymes in plant cells, *J. Plant Physiol.*, 124, 23, 1986.
19. **Pistocchi, R., Antognoni, F., Bagni, N., and Zannoni, D.**, Spermidine uptake by mitochondria of *Helianthus tuberosus*, *Plant Physiol.*, 92, 690, 1990.
20. **Mach, M., Ebert, P., Popp, R., and Ogilvie, A.**, Compartmentation of polyamines in mammalian cells, *Biochem. Biophys. Res. Commun.*, 104, 1327, 1982.
21. **Hougaard, D.M., Bolund, L., Fujiwara, K., and Larsson, L.-I.**, Endogenous polyamines are intimately associated with highly-condensed chromatin *in vivo*. A fluorescence, cytochemical and immunocytochemical study of spermine and spermidine during the cell cycle and in reactivated nuclei, *Eur. J. Cell Biol.*, 44, 151, 1987.
22. **Dion, A. and Herbst, S.**, The localization of spermidine in salivary gland cells of *Drosophila melanogaster* and its effect on H^3-uridine incorporation, *Proc. Natl. Acad. Sci. U.S.A.*, 58, 2367, 1967.
23. **Nagarajan, S. and Ganem, B.**, Chemistry of naturally occurring polyamines 9. Synthesis of spermidine and spermine photoaffinity labeling reagents, *J. Org. Chem.*, 50, 5735, 1986.
24. **Panagiotidis, C.A., Georgatsos, J.G, and Kyriakidis, D.A.**, Superinduction of cytosolic and chromatin-bound ornithine decarboxylase activities of germinating barley seeds by actinomycin D, *FEBS Lett.*, 146, 193, 1982.
25. **Wink, M. and Hartmann, T.**, Localization of enzymes of quiolizidine alkaloid biosynthesis in leaf chloroplasts of *Lupinus polyphyllus*, *Plant Physiol.* 70, 74, 1982.
26. **Persson, L., Rosengren, E., and Sundler, F.**, Localization of ornithine decarboxylase by immunocytochemistry, *Biochem. Biophys. Res. Commun.*, 104, 1196, 1982.
27. **Beier, H., Fecker, L.F., and Berlin, J.**, Lysine decarboxylase from *Hafnia alvei*: purification and molecular data, and preparation of polyclonal antibodies, *Z. Naturforsch.*, 42c, 1307, 1987.
28. **Bell, E. and Malmberg, R.L.**, Analysis of a cDNA encoding arginine decarboxylase from oat reveals similarity to the *Escherichia coli* arginine decarboxylase and evidence of protein processing, *Mol. Gen. Genet.*, 224, 431, 1990.
29. **Seely, J.E. and Pegg, A.E.**, Changes in mouse kidney ornithine decarboxylase activity are brought about by changes in the amount of enzyme protein, as measured by radioimmunoassay, *J. Biol. Chem.*, 258, 2496, 1983.
30. **Zagon, I.S., McLaughlin, P.J., Seely, J.E., Hoeksema, G.W., and Pegg, A.E.**, Autoradiographic localization of ornithine decarboxylase in mouse kidney by use of radiolabeled α-difluoromethylornithine, *Cell Tissue Res.*, 235, 371, 1984.
31. **Morris, G., Nadler, J.V., and Slotkin, T.A.**, Autoradiographic localization of ornithine decarboxylase in cerebellar cortex of the developing rat with tritiated α-difluoromethylornithine, *Neuroscience*, 17, 183, 1986.
32. **Slocum, R.D. and Galston, A.W.**, ^{14}C-DFMO autoradiographic localization of ornithine decarboxylase in rapidly growing plant tissues, *Plant Physiol. Suppl.*, 75, 152, 1984.
33. **Schwartz, M., Altman, A., Cohen, Y., and Arzee, T.**, Localization of ornithine decarboxylase and changes in polyamine content in root meristems of *Zea mays*, *Physiol. Plant*, 67, 485, 1986.
34. **Mengoli, M., Bagni, N., and Biondi, S.**, Effect of α-difluoromethylornithine on carrot cell cultures, *J. Plant Physiol.*, 129, 479, 1987.
35. **Slocum, R.D., Bitonti, A.J., McCann, P.P., and Feirer, R.P.**, DL-α-Difluoromethyl[3,4-^{3}H]arginine metabolism in tobacco and mammalian cells. Inhibition of ornithine decarboxylase activity after arginase-mediated hydrolysis of DL-α-difluoromethylarginine to DL-α-difluoromethylornithine, *Biochem. J.*, 255, 197, 1988.
36. **Slocum, R.D. and Galston, A.W.**, *In vivo* inhibition of polyamine biosynthesis and growth in tobacco ovary tissues, *Plant Cell Physiol.*, 26, 1519, 1985.
37. **Birecka, H., Birecki, M., Bitonti, A.J., and McCann, P.P.**, Inhibition of arginine, ornithine and lysine decarboxylases in pea seedlings, *Phytochemistry*, 30, 99, 1991.

38. **Erwin, B.G. and Pegg, A.E.,** Uptake of α-difluoromethylornithine by mouse fibroblasts, *Biochem. Pharmacol.*, 31, 2820, 1982.
39. **Slocum, R.D., Osmeloski, J.F., and Weinstein, L.H.,** Transport and metabolism of polyamine biosynthesis inhibitors DL-α-difluoromethylornithine (DFMO) and DL-α-difluoromethylarginine (DFMA) in bean plants, *Physiol. Plant.*, submitted.
40. **Smith, T.A.,** *N*-Carbamylputrescine amidohydrolase of higher plants and its relation to potassium nutrition, *Phytochemistry*, 4, 599, 1965.
41. **Yanagisawa, H. and Suzuki, Y.,** Purification and properties of *N*-carbamylputrescine amidohydrolase from maize shoots, *Phytochemistry*, 21, 2201, 1982.
42. **Sindhu, R.K. and Cohen, S.S.,** Subcellular localization of spermidine synthase in the protoplasts of Chinese Cabbage leaves, *Plant Physiol.*, 76, 219, 1984.
43. **Angelini, R. and Federico, R.,** Histochemical evidence of polyamine oxidation and generation of hydrogen peroxide in the cell wall, *J. Plant Physiol.*, 135, 212, 1989.
44. **Kaur-Sawhney, R., Flores, H.E., and Galston, A.W.,** Polyamine oxidase in oat leaves: a cell wall-localized enzyme, *Plant Physiol.*, 68, 494, 1981.
45. **Flores, H.E. and Filner, P.,** Polyamine catabolism in higher plants: characterization of pyrroline dehydrogenase, *Plant Growth Regul.*, 3, 277, 1985.
46. **Federico, R. and Angelini, R.,** Occurrence of diamine oxidase in the apoplast of pea epicotyls, *Planta*, 167, 300, 1986.
47. **Suzuki, Y.,** Intracellular distribution of amine oxidase in pea seedlings, *Naturwissenschaften*, 57, 94, 1970.
48. **Suresh, M.R., Ramakrishna, S., and Adiga, P.R.,** Diamine oxidase of *Lathyrus sativus* seedlings, *Phytochemistry*, 15, 483, 1976.
49. **Federico, R., Angelini, R., Argento-Ceru, M.P., and Manes, F.,** Immunohistochemical demonstration of lentil diamine oxidase, *Cell. Mol. Biol.*, 31, 171, 1985.
50. **Federico, R., Alisi, C., and Forlani, F.,** Properties of the polyamine oxidase from the cell wall of maize seedlings, *Phytochemistry*, 28, 45, 1989.
51. **Federico, R., Alisi, C., Forlani, F., and Angelini, R.,** Purification and characterization of oat polyamine oxidase, *Phytochemistry*, 28, 2045, 1989.
52. **Scorer, K.N., Caamal, R.M., Oropeza, C. and Loyola-Vargas, V.M.,** Amine oxidase activity in *Canavalia ensiformis* during early growth: a histochemical study, *J. Plant Physiol.*, 135, 346, 1989.
53. **Slocum, R.D. and Furey, M.J., III,** Electron-microscopic localization of diamine and polyamine oxidases in pea and maize tissues, *Planta*, 183, in press,.
54. **Cogoni, A., Piras, C., Farci, R., Melis, A., and Floris, G.,** *Hordeum vulgare* seedlings amine oxidase. Purification and properties, *Plant Physiol.*, 93, 818, 1990.
55. **Smith, T.A.,** Purification and properties of the polyamine oxidase of barley plants, *Phytochemistry*, 11, 899, 1972.

Chapter 7

UPTAKE AND TRANSPORT OF POLYAMINES AND INHIBITORS OF POLYAMINE METABOLISM IN PLANTS

Nello Bagni and Rossella Pistocchi

TABLE OF CONTENTS

I. BIOCHEMICAL CHARACTERIZATION OF ACTIVE AND PASSIVE COMPONENTS OF POLYAMINE UPTAKE BY PLANT CELLS AND COMPARISON WITH OTHER ORGANISMS

A. INTRODUCTION

In animal and bacterial systems it has been known for many years that exogenous polyamines are transported and accumulated by cells. In particular, it has been established that transport is enhanced in tissues which have a high demand for polyamines such as prostate,[1] tumors[2] and rapidly proliferating cells[3] as well as by tissues depleted in polyamines.[4]

A large number of animal cells and tissues have been investigated with respect to polyamine transport and they were recently reviewed by Seiler and Dezeure.[5] The general characteristics are the following: (1) polyamine transport is energy- and temperature-dependent, except for that in lungs which occurs via diffusion, and (2) it is saturable, with K_m values in the μM range and the affinity for the carrier increases, in order, from putrescine to spermidine to spermine. Most cells appear to have a single transporter for all three polyamines, but there are also exceptions. The transporter can also mediate the uptake of other polyamines (e.g., cadaverine, and homospermidine) or analogs (methylglyoxal-*bis*-(guanylhydrazone), MGBG paraquat). From preliminary data, it is presumed that *N*-acetylated polyamines are mainly substrates of polyamine export, whereas free putrescine and spermidine may be transported in both directions.[6]

Macromolecular synthesis is required for the adaptation of polyamine transport to an increased polyamine requirement and transport is activated by thiols and inhibited by thiol reagents. Hormones and growth factors (e.g., insulin, serum, and EGF) can enhance the uptake rates of exogenous polyamines, besides increasing their intracellular *de novo* synthesis, and it has been established that uptake and synthesis, although frequently coupled functionally, are regulated by different genes.[5]

Scarce attention has been given to polyamine binding to constituents of the animal cell. This was not so for bacterial systems. In fact, Tabor and Tabor [7] found two different mechanisms of polyamine accumulation in *E. coli*: (1) an energy-independent one, occurring at 0°C and removed by washing with unlabeled polyamine solution, and (2) another which occurred only at 37°C, was not removed by washing, and was inhibited by 2,4-DNP. In *E. coli* it is likely that at least two transport systems are involved; the transport is apparently unidirectional and dependent on the proton motive force.[8]

In plants, studies on polyamine transport were initiated only later, although it represents an important point for consideration of these substances as hormones rather than as second messengers.

We have extensively studied the characteristics and mechanisms of polyamine transport in various plant systems, such as petals of *Saintpaulia ionantha*[9,10] (African violet), and in a more active and rapidly dividing system such as carrot cell cultures.[11,12] Our studies were later extended to carrot protoplasts[13] and to subcellular organelles such as vacuoles[13] and mitochondria.[14] The main features of polyamine transport in such systems, which previously have been examined in detail only in *Neurospora crassa*,[15-18] are reviewed here.

B. GENERAL CHARACTERISTICS OF POLYAMINE UPTAKE

The rate of polyamine uptake turned out to be slow only in the petals of *Saintpaulia ionantha*, where it was first studied and where it was linear for 1 to 2 h.[9,10] In carrot cells [11,12] and protoplasts,[13] however, uptake was very rapid, reaching saturation after 1 to 2 min, and the same occurred in the organelles that were examined (vacuoles[13] and mitochondria[14]). These data are in contrast with those obtained in cowpea protoplasts,[19] where a lag period of 20 h was required before observing a rapid uptake of spermidine, but which are in agreement with those obtained with maize cell lines[20] that took up virtually all extracellular putrescine within 8 min. Long-term

TABLE 1
K_m Values for Polyamine Uptake in Various Plant Systems

Plant system	Polyamine	K_m value in uptake system I	II
Saintpaulia petals[a]	Putrescine	8.6 m*M* (pH 5.5)	
		2.4 m*M* (pH 8)	
	Spermidine	1.2 m*M*	
	Spermine	2.1 m*M*	
Carrot cell cultures	Putrescine	41.9 μ*M*	29.2 m*M*
	Spermidine	27.3 μ*M*	linear
	Spermine	7.7 m*M*	15.9 m*M*
Carrot protoplasts	Spermidine	linear	linear
	Spermine	292.0 μ*M*	linear
	Spermine+Ca^{2+}	122.0 μ*M*	n.d.[b]
Carrot vacuoles	Spermidine	61.8 μ*M*	linear
H. tuberosus mitochondria	Spermidine	89.0 μ*M*	n.d.

[a] Uptake system is not biphasic.
[b] n.d. = not determined.

treatment of carrot cell cultures with 1 to 5 m*M* putrescine concentrations did not affect the time-course or uptake of low concentrations of putrescine, although the cells accumulated high levels of the polyamine (about 11 m*M*) which was presumably localized in the vacuole.[21]

The uptake of all three polyamines in *Saintpaulia* petals[9,10] and in carrot cell cultures[11] was dependent on external pH and two maxima were usually observed: one at pH 4 to 5 and another at pH 8. Similar results were obtained for the uptake of putrescine in the lichen *Evernia prunastri*[22] and for the transport of polyamines in sugar beet seedlings,[23] while in *Neurospora* there was little pH dependence.[17]

In all the systems investigated by us, except *Saintpaulia ionantha*, polyamine uptake displayed a biphasic pattern with saturation kinetics in system I (up to 5 m*M* external concentration) and a saturable or linear component in system II (from 5 to 100 m*M*).[11-13] K_m values are reported in Table 1, where it can be seen that the affinity is lower (K_m in the millimolar range) only for uptake in *Saintpaulia* and for spermine uptake in carrot cell cultures. In the first case, we have a mature and differentiated tissue which is probably not suitable for transport studies. In the second case, one explanation could be that spermine is not an endogenous constituent of the cell line utilized. A particular aspect concerning protoplasts is that the K_m value is affected by the presence of calcium; this is further described below. A multiple component for putrescine uptake was found also in the cellular slime mold *Dictyostelium discoideum* involving adsorptive pinocytosis at low concentrations and fluid-phase pinocytosis at high concentrations, the first system being saturable with a K_m of 9.1 μ*M* and the second nonsaturable.[24] In *Neurospora crassa*, the kinetic components for putrescine and spermidine were resolved into a nonsaturable and saturable component; K_m values were obtained after subtracting the nonsaturable component and were in the order of $10^{-4}M$.[17]

C. BINDING TO CELL CONSTITUENTS

An important problem associated with the study of polyamine uptake by cells is that polyamines have the well-known ability to bind noncovalently with different strengths to many cellular constituents, especially negatively charged ones such as pectic substances, phospholipids, proteins, nucleic acids, etc. This phenomenon can, therefore, alter the results of transport experiments and must be identified in order to be subtracted or eliminated.

TABLE 2
Distribution of Supplied [^{14}C] Polyamines in Fraction of Cultured Carrot Cells Obtained by Differential Centrifugation

Putrescine

Fraction	pmol/10^3 cell clusters	%
Cell walls	0.84	24.2
Particulate	0.24	6.9
Cytoplasmic	2.39	68.9
Total pmol taken up	3.47	

Spermidine

Fraction	pmol/10^3 cell clusters	%
Cell walls	11.0	73.0
Particulate	2.9	19.3
Cytoplasmic	1.2	7.7
Total pmol taken up	15.1	

From our early investigations on *Saintpaulia* petals, it was observed that binding of polyamines to the external surface of the cell was not an extensive phenomenon. In fact, fractionation analysis revealed that, at most, 4.5% of the absorbed putrescine was localized in the cell wall fraction [10] and this was probably due to the low content in pectic substances of epidermal cells. To the contrary, results obtained with carrot cells (Table 2) indicated that the percentage of putrescine found in the cell wall fraction was relatively high (24.2%), an amount that was even larger for spermidine (73%)[11] and spermine (77.5%).[12] Consequently, in all subsequent uptake experiments, the cells were washed extensively with an unlabeled putrescine solution.

A study directed towards a better understanding of these interactions showed that endogenous polyamines were not covalently bound to cell walls isolated from carrot cell cultures; the content was dependent upon the age of the cultures and was affected by the presence of Ca^{2+} during wall extraction.[25] An *in vitro* study on interactions between polyamines and pectic substances showed that the number of galacturonic units per binding site is proportional to the number of positive charges in the polyamine molecules.[26] A similar result was also obtained for the *in vivo* binding of polyamines to cells of the red alga *Porphyridium* and for the *in vitro* binding to its cell wall polysaccharide.[27]

In order to distinguish between spermine uptake and binding by carrot cells, we examined spermine uptake in whole cells and in the cell walls isolated from them. The data were normalized with respect to gram fresh weight and to polygalacturonic acid content. As can be seen in Figure 1, uptake increased with time only in whole cells, while in cell walls it remained constant. The value reached after 10 s, which was the same for both, could represent the external binding.[12]

In the experiments performed with protoplasts or subcellular organelles, the problem of polyamine interactions with the cell wall was eliminated, but the problem of a possible binding to membranes was not. In protoplasts and vacuoles, the extent of this binding during spermidine uptake was determined by washing them with an osmotically adjusted solution of 100 m*M* unlabeled spermidine, which was inserted through two layers of silicon oil through which the protoplasts and vacuoles had to pass during the centrifugation step performed in order to remove them from the incubation medium. The wash-out radioactivity (Figure 2) was higher for

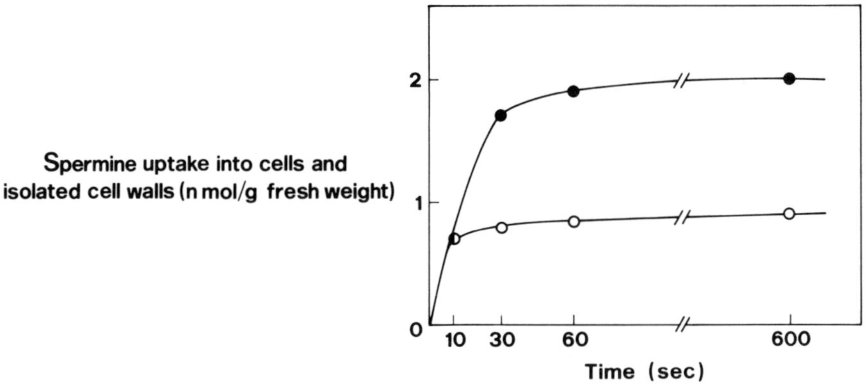

FIGURE 1. Time course of spermine uptake in whole cells (●) and isolated cell walls (○) from carrot cultures.

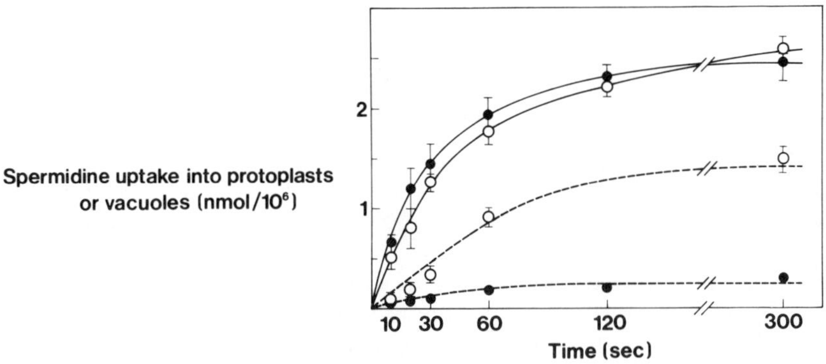

FIGURE 2. Time course of spermidine uptake into carrot protoplasts (●) and vacuoles (○), with (---) or without (—) a washing layer containing 100 mM unlabeled spermidine.

protoplasts (91.5%) than for vacuoles (50%) but it decreased, on a percent basis, with incubation with higher polyamine concentrations so that the ratio between the polyamine absorbed and that externally bound, in this case, was much greater.[13] The washing procedure was therefore performed in all subsequent experiments.

With regard to mitochondria, the extent of binding to membranes was easily assessed by incubation in the absence of the respiratory substrate (succinate) or in the presence of a respiratory chain inhibitor (antimycin A) or uncouplers (FCCP, valinomycin) as is shown in Figure 3. It represented 50% of total radioactivity and this value was subtracted in all experiments.[14]

D. EFFLUX

The absorbed polyamines were not permanently trapped in the cells, but efflux occurred preferentially in the presence of external molecules. In carrot cell cultures, 80% efflux of the absorbed putrescine was obtained in the presence of 2.5 m*M* external putrescine, probably by means of an exchange mechanism; the same exchange occurred when external putrescine was replaced with Ca^{2+}. Spermidine efflux in the presence of 2.5 m*M* external spermidine was also 80% but the exchange with calcium was limited.[11] In contrast to the other two polyamines, spermine efflux did not exceed 20% and, as stated earlier, this was probably due to the absence of spermine as an endogenous constituent of our cell line.[12] When carrot cells adapted to grow in media supplemented with 5 m*M* putrescine were transferred to putrescine-free medium, the natural release of putrescine was very low, suggesting a mechanism of compartmentation and/

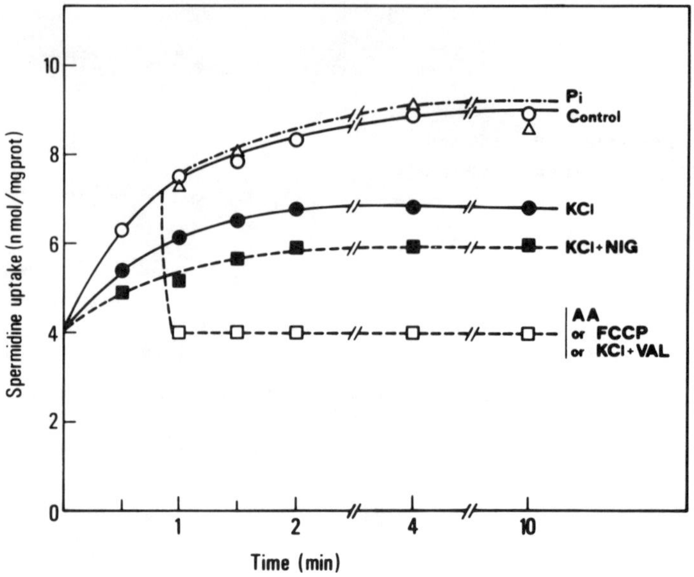

FIGURE 3. Time course of spermidine uptake into *H. tuberosus* mitochondria and effect of different substances (NIG, nigericin; VAL, valinomycin).

or a control of efflux in the adapted cells. Also, the efflux of labelled putrescine absorbed by these cells in short-term experiments was similar to that of untreated cells, suggesting that the putrescine accumulated during the long-term treatment was not easily exchangeable.[21]

Some data were also obtained for spermidine efflux from the vacuole: 58 and 69% of accumulated spermidine was released after 30 and 40 s respectively, in the presence of 100 μ*M* external spermidine.[13]

With regard to other organisms, in *Neurospora*, putrescine and spermidine efflux from preloaded cells was low, but increased significantly in the presence of 5 mM external molecules. The efflux was nearly complete in cells treated with butanol, showing that polyamines are bound in exchangeable fashion to cell constituents.[17] In this respect, plant cells are very different from bacteria, in which polyamine efflux was not observed either in the presence of unlabeled polyamines, calcium, CCCP, or the calcium ionophore A23187.[8]

E. COMPETITION EXPERIMENTS

Competition studies have, at present, been performed only with petals or cell cultures, which cannot exclude interference by the cell wall. Nevertheless, since living plant cells are endowed with cell walls, these results should not be disregarded. In *Saintpaulia ionantha*, the presence of one polyamine did not seem to inhibit the uptake of the others.[10] Table 3 shows the results of putrescine and spermidine uptake by carrot cells in the presence of other polyamines or an analog. It appears that putrescine uptake was more sensitive to the presence of other compounds even when those are present at the same low concentrations. Spermidine uptake was affected by MGBG only when both compounds were present at elevated concentrations, while it was less affected by putrescine.[11] Spermine uptake was not affected by 100-fold higher concentrations of putrescine or spermidine.[12] The uptake of spermidine and spermine appears, therefore, to be more specific than that of putrescine. This could suggest that higher molecular weight polyamines utilize a different transport system.

Monoacetylpolyamines were also employed as analogs having a reduced number of positive charges; the presence of acetyl-putrescine or -spermine did not reduce the uptake of the corresponding labeled nonacetylated amine.[12]

TABLE 3
Effects of Several Antagonists on Putrescine and Spermidine Uptake in Carrot Cell Cultures

Compound (μM)	Concentration (μM)	Antagonist % Inhibition: 1,3 DAP	Cadaverine	Spermidine
Putrescine				
25	25	33.3 ± 1.2	0	0
25	100	40.0 ± 2.3	0	66.7 ± 3.0
25	250	n.d.[a]	33.3 ± 1.6	81.2 ± 3.1
25	500	80.0 ± 3.3	36.4 ± 2.1	89.6 ± 3.5
		Putrescine	**MGBG**	
Spermidine				
25	25	0	0	
25	100	0	0	
25	250	0	0	
25	500	0	0	
500	500	0	26.4 ± 3.1	
500	1000	20.0 ± 2.3	57.2 ± 3.5	

[a] n.d. = not determined.

In *Neurospora*, equimolar concentrations of polyamines were mutually inhibitory, suggesting that they shared components of the same transport system. MGBG, arginine, and ornithine also partially inhibited polyamine uptake.[17]

F. EFFECT OF CATIONS

Our first studies on the effect of cations on polyamine uptake showed that putrescine and spermidine uptake in *Saintpaulia* petals was not affected by the presence of Ca^{2+} or Mg^{2+} at equimolar concentrations while, at higher concentrations, spermidine uptake was inhibited by Ca^{2+} by 53%.[10] In carrot cells, results differed depending on the polyamine. In fact, in the presence of increasing Ca^{2+} concentrations, the uptake of putrescine was slightly stimulated only by the 10 μ*M* concentration and inhibited at higher levels. To the contrary, the uptake of spermidine and spermine was increasingly enhanced in the Ca^{2+} concentration range from 10 μ*M* to 1 m*M* (Table 4).

The effect of La^{3+} was also investigated as it is a competitive inhibitor of Ca^{2+} transport, which exerts its effect without entering the cell. It nullified the stimulatory effect of Ca^{2+} on putrescine uptake and at 100 μ*M* to 1 m*M* concentrations, it mimicked the stimulatory effect of Ca^{2+} on spermidine and spermine uptake respectively.[11]

The effect of Ca^{2+} was investigated in greater detail on spermine uptake and it was found that it did not affect the K_m nor the V_{max} of system I. Other experiments have ruled out the possibility of either a co-transport or an antiport mechanism between the two cationic species so that the effect of Ca^{2+} appears to be unspecific. Furthermore, washing out the externally bound Ca^{2+} with 0.5 m*M* EGTA prior to uptake experiments results in an enhancement instead of the expected inhibition of spermine uptake probably because the washed out Ca^{2+} was replaced by spermine. This result seems to illustrate the presence of interferences with the cell wall.[12]

With regard to protoplasts, the uptake of all three polyamines was enhanced by 1 m*M* Ca^{2+}. In the case of spermine, it was enhanced throughout the spermine concentration range of 10 μ*M* to 1 m*M* (Figure 4) and thus, in contrast to the data obtained with carrot cell cultures, Ca^{2+}

TABLE 4
Effect of Ca^{2+} and La^{3+} on Spermidine and Spermine Uptake in Carrot Cell Cultures

Cation concentrations (m*M*)	Spermidine uptake (pmol/10^3 cell clusters)		
	Ca^{2+}	Ca^{2+} + La^{3+}	La^{3+}
0		5.4	
0.01	6.2	4.7	5.6
0.1	8.1	10.5	13.3
1.0	14.3	1.8	1.4

Cation concentrations (m*M*)	Spermine uptake (pmol/10^3 cell clusters)		
	Ca^{2+}	Ca^{2+} + La^{3+}	La^{3+}
0		5.2	
0.01	6.2	5.2	5.8
0.1	7.6	8.2	7.6
1.0	14.7	12.9	11.0
10.0	11.7	0.9	0.4

affected the V_{max} which increased from 45 to 363 nmol/mg prot/30 s as well as the K_m which decreased from 292 to 122 μ*M* (Table 1). Since the uptake value was very low in the absence of Ca^{2+}, this cation seems to be necessary for polyamine transport in protoplasts.[13]

La^{3+} up to 1 m*M* also stimulated spermine uptake in protoplasts by fourfold. Mg^{2+} (10 μ*M* to 10 m*M*) had no effect on uptake either in cell cultures or protoplasts;[12] in mitochondria (0.2 to 1 m*M*) it reduced the amount of spermidine absorbed, as did K^+ (20 to 100 m*M*), while Ca^{2+} had no effect.[14]

G. COMPARTMENTATION OF POLYAMINES

Data on the compartmentation of exogenous polyamines in cell cultures obtained by fractionation analysis have been presented above (Table 2) and indicate that the cell wall can bind most of the spermidine and spermine supplied. A report from Goldberg and Perdrizet[28] showed that also endogenous polyamines, extracted and separated at low ionic concentration, are present in this compartment in mung-bean hypocotyls, so that the cell wall no doubt represents one compartment which is able to sequester these compounds.

Another main compartment of the plant cell is the vacuole. It is, therefore, important to know if the polyamines present inside the cell are preferentially located in it. The compartmentation of endogenous molecules (Table 5) showed the presence of 42 and 28% of putrescine and spermidine, respectively, in the vacuole. The compartmentation of exogenously supplied polyamines was studied in two different ways; first, protoplasts were incubated with 100 m*M* unlabeled spermidine, then the vacuoles were isolated from them and it appeared that spermidine content had increased sixfold with respect to the endogenous content; second, protoplasts were supplied with a low amount of labeled spermidine. After isolating the vacuoles, 27% of it was found in these organelles. As already reported, leakage of polyamines from the vacuoles can also occur, so this phenomenon could have altered both the endogenous and the exogenous data. At any rate the vacuole appears to be a possible storage site for polyamines.[13]

Other data on polyamine compartmentation in plant cells are those of Walker et al.[29] who

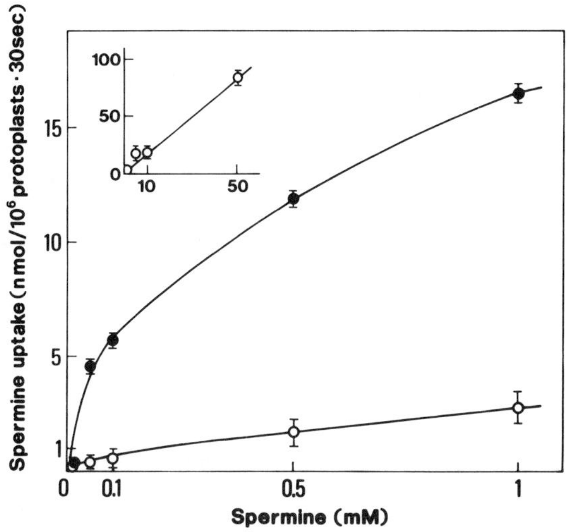

FIGURE 4. Spermine uptake in carrot protoplasts over an external spermine concentration range of 10 μ*M* to 1 m*M* and of 1 to 50 m*M*, in the presence of (●) or absence (○) of 1 m*M* Ca^{2+}.

found nearly all of the free endogenous putrescine and spermidine in the supernatant fraction of two cell lines of *Nicotiana tabacum*, as well as the enzymes involved in their biosynthesis (ADC, ODC, and SAMDC). In *Neurospora*, 28% of the cellular spermidine was vacuolar, and since more than 70% of it was not accessible to enzymatic activities, it was supposed that the remainder of the sequestered polyamines was strongly bound to anionic sites in the cell.[16]

With regard to other subcellular compartments it was shown that polyamines are present inside mitochondria and chloroplasts,[30] and this presence was mostly attributed to the activity of the biosynthetic enzymes also found in these organelles. We demonstrated, at least with mitochondria, that polyamine compartmentation can be attributed also to transport from the cytoplasm. The uptake of spermidine is an energy-dependent process and, in contrast to animal mitochondria, independent from the presence of inorganic phosphate.[14]

Polyamines are also present in nuclei, but Walker et al.[29] did not find substantial amounts of putrescine and spermidine in this fraction, despite the well-known function of polyamines in the stabilization of nucleic acids and in the induction of cell division.

H. EFFECT OF METABOLIC INHIBITORS

The uptake of polyamines supplied at low external concentrations, as in most of our experiments, occurred against a concentration gradient. Therefore metabolic inhibitors were tested in order to determine if the transport observed was energy-dependent. Using *Saintpaulia* petals, 0.01 and 0.1 m*M* 2,4-DNP and CCCP 0.01 and 0.1 m*M* had no effect on putrescine uptake. At 0°C, however, there was a 35% inhibition of spermidine uptake and a 68% inhibition was obtained with 20 m*M* NaSCN.[10]

In carrot cells, results were obscured by the fact that the inhibitors used were dissolved in ethanol or DMSO, which themselves inhibited uptake at some concentrations. Nonetheless, maximum inhibition was obtained on spermidine uptake with 0.1 m*M* CCCP (28%) and on putrescine uptake with 1 m*M* 2,4-DNP (34%) at a low concentration of DMSO, while DCCD, DES, vanadate, and NEM were ineffective both in the presence or absence of calcium.[11] The low percentage of inhibition observed in carrot cells could be due to interactions between the inhibitors and some cell wall constituents. In protoplasts, spermidine and spermine uptake was

TABLE 5
Endogenous Polyamines Content of Carrot Taproot Cells, Protoplasts, and Vacuoles and Subcellular Localization of Exogenously Applied Spermidine (100 m*M* Cold or 6.6 μ*M* Labeled Spermidine to Protoplasts)

Polyamine content in	Putrescine	Spermidine
	nmol/10^6 Cells or Protoplasts or Vacuoles	
Cells	3.57 ± 0.54	10.67 ± 1.44
Protoplasts	3.23 ± 0.15 (100%)	3.74 ± 0.38 (100%)
Vacuoles	1.36 ± 0.23 (42%)	1.06 ± 0.16 (28%)
	Protoplasts Preloaded With 100 m*M* Spermidine	
Protoplasts	2.98 ± 0.38 (100%)	73.70 ± 2.87 (100%)
Vacuoles	1.28 ± 0.33 (43%)	6.04 ± 0.10 (8%)
	Protoplasts Preloaded with 6.6 m*M* Spermidine	
Protoplasts	n.d.[a]	0.41 ± 0.06 (100%)
Vacuoles	n.d.[a]	0.11 ± 0.01 (27%)

[a] n.d. = not determined.

inhibited by 50% by 10 μ*M* FCCP only in the presence of 1 m*M* Ca^{2+} (Table 6), while it was not affected in the presence of 1 m*M* La^{3+} or spermine alone. Thus, it appears that the Ca^{2+}-stimulated uptake is an energy-dependent mechanism and that the function of Ca^{2+} is more specific than that of La^{3+}.[13]

The driving force for polyamine uptake was more clearly identified in mitochondria by using different ionophores, such as valinomycin and nigericin, which abolish K^+ and pH gradients across the membranes, respectively; and FCCP, which causes the suppression of the total electrochemical gradient. The complete inhibition of spermidine uptake was obtained only with valinomycin and FCCP, while nigericin only slightly affected it (Figure 3). The uptake of spermidine appears to be driven by the membrane potential, while the pH gradient seems to play a secondary role.[14]

In *Neurospora*, cyanide and azide (1 m*M*), which deprive the cell of ATP, reduced the saturable and nonsaturable components of polyamine uptake by 75 to 85%; a 75% reduction was obtained also at 4°C.[17]

I. HORMONES

As polyamine biosynthesis is reported to be stimulated by auxin,[31] it is of considerable interest to establish whether or not auxin also has an influence on polyamine uptake as do animal hormones. As yet unpublished results indicate that IAA (10^{-7} to 10^{-4} *M*) enhances spermidine uptake into carrot protoplasts only in the presence of 1 m*M* Ca,$^{2+}$ while in the presence of La^{3+} the effect was similar but with much lower uptake values. This IAA-stimulated uptake was inhibited by the presence of vanadate, an ATPase inhibitor, suggesting that the mechanism is energy-dependent. Preliminary results showed that cytokinin also enhanced polyamine uptake, but this stimulation did not seem to be directly dependent on calcium.[32]

TABLE 6
Effect of 10 μ*M* FCCP on Calcium- and Lanthanum-Stimulated Spermine Uptake in Carrot Protoplasts

Spermine Uptake (nmol/10^6 protoplasts)

Spermine concentrations	Additions		
μ*M*	none	Ca^{2+}	Ca^{2+} + FCCP
10	0.47 ± 0.06	0.83 ± 0.12	0.24 ± 0.04
100	1.20 ± 0.08	5.45 ± 0.73	0.24 ± 0.04
1000	3.18 ± 0.46	25.50 ± 1.13	11.61 ± 0.50
μ*M*	none	La^{3+}	La^{3+} + FCCP
10	0.45 ± 0.05	1.15 ± 0.18	0.94 ± 0.08
100	1.32 ± 0.25	5.04 ± 0.77	4.40 ± 0.80
1000	5.20 ± 0.20	20.24 ± 3.82	19.04 ± 1.36

II. LONG-DISTANCE TRANSPORT OF POLYAMINES IN PLANT TISSUES

Long-distance transport is probably an essential part of hormone function. It is important, therefore, to determine if polyamines are translocated inside plants if one is to attribute to them an hormone-like action rather than that of second messengers.

One of the problems associated with polyamine translocation is the presence of di- and polyamine oxidases in cell walls[33] and, in fact, in at least one case, this presence could have influenced the results of an experiment. Young and Galson[34] found very little transport of putrescine and spermidine to shoots and roots of pea seedlings 4 h after the injection of labeled polyamines into the cotyledons. This treatment could have damaged the plasmalemma and caused an enhancement of diamine oxidase activity, since this enzyme is known to be very active in the Leguminosae.[33]

In other systems, the occurrence of polyamine translocation has been demonstrated. The earliest evidence was obtained with a single corymb of *Malus domestica*[35] where [^{3}H]putrescine was absorbed by leaves and translocated to the fruitlet and vice versa. The absorbed putrescine was metabolized to spermidine and spermine and the translocation occurred via the peduncle and did not appear to be polar. These results could explain the observed effect of polyamines in increasing fruit growth, fruit set, and yield per tree of apple trees.[36]

Further characterization of polyamine translocation was obtained by using seedlings of various species.[37] With tomato and maize seedlings, it was shown that [^{3}H]putrescine was absorbed by roots and translocated to the upper parts (cotyledons and coleoptiles respectively) within 30 min and continued to do so up to 24 h. After 24 h, tomato seedlings had taken up 5% of the putrescine supplied and maize, 14%. During this time, 46% of the absorbed putrescine was metabolized to spermidine and spermine so that the data are referred to polyamines absorbed instead of to putrescine alone.

The translocation was dependent on relative humidity insofar as plants kept under conditions of 100% humidity absorbed less than controls and it was also temperature-dependent since translocation increased from 4 to 30°C.

The occurrence of a basipetal transport was investigated by feeding labeled putrescine to cotyledons or to a primary leaf of tomato seedlings in two different stages of growth. In the first case, most of the polyamine remained in the cotyledons and only 6% was translocated to the hypocotyl and root. In the second case, the amount of putrescine taken up by the leaf was higher, being 13% of that supplied, and 12% of it was exported.

One-year old pine seedlings were also used in experiments designed to establish the path taken by polyamines to reach the upper parts. This was achieved by cutting a 1 cm wide ring in the middle of the hypocotyl in order to exclude the cortical parenchyma and phloem. In these plants, uptake over 72 h was compared with that of plants which had not been ringed. Results indicated that polyamine content in the parenchyma and phloem above the ring was lower than that found below the ring and also lower than that of the corresponding part of control plants. Conversely, the polyamine content in the uppermost parts (epicotyl and needles) was only slightly reduced, showing that long distance transport was not significantly affected by the treatment.[37]

The results reported above indicate that polyamine translocation occurs mainly via the xylem vessels and is dependent on the rate of transpiration. Phloem transport has a minor role in that the basipetal transport was lower than the acropetal one. It has to be pointed out that, in the case of the forementioned study on pea seedlings,[34] only phloem transport was studied and this could also explain why these authors failed to observe polyamine transport.

Evidence for a physiological role of polyamine translocation was given by Friedman et al.[38] who found endogenous polyamines in xylem and phloem exudates of various plant species. High concentrations of putrescine and spermidine were found in xylem exudates of sunflower, mung-bean, and orange stems, while grape vine had much lower concentrations. Spermine was found only in trace amounts in exudates of all species. More putrescine was found in exudates of older than in younger sunflower plants and in plants to which a salt stress was applied. Putrescine and spermidine were also found in the phloem sap of sunflower and mung-bean plants.

A different approach to verify the occurrence of polyamine transport was that of Masse' and coworkers[39] who studied the uptake and translocation of a synthetic homolog of spermine, during the *in vitro* growth of potato and corn plantlets. This synthetic tetramine (*N,N'-bis*(3-aminoproplylamino)ethane) was added to the incubation medium and was later found in roots and shoots. Uptake of the tetramine was concentration-dependent and a degradation product was also found in both roots and shoots of the treated plants. These studies show that polyamines synthesized in one part of the plant can be translocated to other organs of the plant via the xylem and phloem. Changes in concentrations of polyamines found in xylem sap following salt stress[38] further suggest that the physiological status of the plant can influence transport.

Two types of evidence suggest that translocation of polyamines may be physiologically necessary. In *Nicotiana glauca*, the biosynthesis of cadaverine, the precursor of anabasine, the main alkaloid in this plant, takes place mainly in roots.[40] Cadaverine biosynthesis is catalyzed by the enzyme lysine decarboxylase, the activity of which is twofold higher in roots than in stems and leaves. However, cadaverine levels are not elevated in roots. This could be explained by the observation that labeled cadaverine moves quickly from the roots to the leaves, and not vice versa.[40] Alternatively, low Cad levels in the root may result from high rates of incorporation into anabasine. In addition, the activities of polyamine biosynthetic enzymes are high in cotyledons of germinating seeds. However, the presence on polyamines in this organ is not correlated with growth stimulation. Therefore, it is likely that polyamines are exported to other parts of the seedling. It has been shown, in fact, that labeled polyamines injected into sugar beet cotyledons are transported to other parts of the plant, though only in small amounts.[23]

III. UPTAKE AND TRANSPORT OF INHIBITORS OF POLYAMINE METABOLISM

Studies on the uptake of inhibitors of polyamine synthesis are very scarce except when they are used as polyamine analogs in competition studies with polyamines themselves. In general, MGBG, the most common inhibitor utilized for this purpose, was found to competitively inhibit polyamine uptake in different systems, such as *Neurospora crassa*[17] and carrot cell cultures.[11]

In animal cells, MGBG was not transported and metabolized and it was actively concentrated up to 1000-fold across the cell membrane, using an uptake system apparently designed for polyamines. The uptake rate of the drug was critically dependent on the growth rate of the cells; rapidly growing tumor cells (Ehrlich ascites carcinoma) most effectively accumulated MGBG.[41] In addition, intracellular putrescine and spermidine deprivation by DFMO enhanced the uptake of the natural polyamines as well as of MGBG.[42]

In animals, DFMO uptake has been examined in *Trypanosoma brucei brucei*[43] and in mouse fibroblasts.[44] In both these systems, DFMO uptake was nonsaturable, not antagonized by basic amino acids and probably occurred via a passive diffusion mechanism.

In plants, the uptake of polyamine inhibitors was studied utilizing CHA and DFMO with the aim of better understanding their role in growth and morphogenesis. For both compounds, new procedures were developed in order to dansylate them and separate them by TLC. It was found that one molecule of CHA binds one dansyl chloride molecule and, under our experimental conditions, this binding was stable. The quantitative determination of this drug based on fluorescence was linear between 0.5 and 20 nmol.[45]

In the case of DFMO, the dansylation procedure of Smith and Best[46] for lysine determination was utilized with some modifications. Using optimal pH (9.3 to 9.5), temperature (22°C) and time (18 h), only one fluorescent product appeared after separation by TLC. When [5-^{14}C]DFMO was dansylated, an additional nonfluorescent spot (about 15% of total radioactivity) appeared above the dansyl-NH_2.[47]

In excised cotyledons of radiata pine (*Pinus radiata*) cultured *in vitro*, CHA was rapidly absorbed since the drug was detectable within 6 h. The rate of uptake, which was initially high, fell rapidly between 6 and 24 h. CHA uptake closely followed the growth pattern of the cotyledons and occurred against a concentration gradient as early as 6 h in culture. In fact, the internal concentration exceeded the external one by several-fold, reaching a cellular concentration of 5.1 mM after 48 h in a medium supplied with 1 mM CHA. Cotyledons cultured on benzylamine-containing medium accumulated more CHA than those in benzylamine-free medium.[45]

Uptake of DFMO was also studied in a carrot cell line adapted to 5 mM DFMO. The uptake of ^{14}C-DFMO supplied in trace amounts was about 75% lower in the treated cells than in the controls, and occurred against a concentration gradient since the internal concentration of unlabeled DFMO in DFMO-treated cells, was 1000-fold higher than the external ^{14}C-DFMO concentration.[47] The reduction of ^{14}C-DFMO uptake in DFMO-treated cells was similar to the results obtained for MGBG uptake in Chinese hamster ovary and rat myoblast cells which were resistant to MGBG.[48] More highly resistant lines selected from low level MGBG resistant lines were totally deficient in the ability to take up the drug.[49]

A different and interesting approach to the study of DFMO uptake was recently published. It was carried out in order to explain the ability of this inhibitor to block powdery mildew infection in barley seedlings. DFMO uptake by intact roots was saturable and it was translocated to the shoots. Uptake into the roots, as a function of DFMO concentration in the external medium, was biphasic giving K_m values for system I of 1 mM and for system II of 53 mM. DFMO uptake was reduced in darkness and was greatly influenced by transpiration rate.[50] DFMO (or a DFMO metabolite) was also shown to be transported slowly through the phloem from the shoot to roots and from lower to upper leaves. In addition, the inhibitor was taken up from host cells by the fungus, *Erysiphe graminis* f. sp. *hordei*, and by feeding DFMO to roots or to excised leaves at different times after infection, a substantial reduction in the disease symptoms observed was obtained.[50]

IV. CONCLUDING REMARKS

The characterization of the transport system for polyamines in plant cells is complicated by the existence of the cell wall which can sequester and interact with supplied polyamines. Nevertheless, by also studying the subject at the subcellular level, it was shown that polyamines are indeed transported into cells. This transport is bidirectional and, at least for the inward movement, it is energy- and Ca^{2+}-dependent. The absorbed polyamines appeared to be compartmented inside the cell and the transport was under hormonal control.

Studies on long-distance transport are not numerous but they are sufficient to prove the existence of a nonpolar polyamine translocation within plants.

Unfortunately, up to now, very few results have been obtained on the uptake and transport of polyamine inhibitors. It is not possible, therefore, to have a general picture of the mechanism involved and the physiological significance of their translocation needs to be further clarified.

REFERENCES

1. **Clark, R.B. and Fair, W.R.,** The selective *in vivo* incorporation and metabolism of radioactive putrescine in the adult male rat, *J. Nucl. Med.*, 16, 337, 1975.
2. **Volkow, N., Goldman, S.S., Flamm, E.S., Cravioto, H., Wolf, A.P., and Brodie, J.D.,** Labeled putrescine as a probe in brain tumors, *Science*, 221, 673, 1983.
3. **Jänne, J. and Raina, A.,** Further observations on the biosynthesis of polyamines in regenerating rat liver, *Acta Chem. Scand.*, 20, 1174, 1966.
4. **Chaney, J.E., Kobayashi, K., Goto, R., and Digenis, G.A.,** Tumor-selective enhancement of radioactivity uptake in mice treated with α-difluoromethylornithine prior to administration of ^{14}C-putrescine, *Life Sci.*, 32, 1237, 1983.
5. **Seiler, N. and Dezeure, F.,** Polyamine transport in mammalian cells, *Int. J. Biochem.*, 22, 211, 1990.
6. **Seiler, N.,** Potential roles of polyamine interconversion in the mammalian organism, in *Progress in Polyamine Research. Novel Biochemical, Pharmacological and Clinical Aspects*, Zappia, V. and Pegg, A.E., Eds., Plenum Press, New York, 1988, 127.
7. **Tabor, C.W. and Tabor, H.,** Transport systems for 1,4-diaminobutane, spermidine, and spermine in *Escherichia coli*, *J. Biol. Chem.*, 241, 3714, 1966.
8. **Kashiwagi, K., Kobayashi, H., and Higarashi, K.,** Apparently unidirectional polyamine transport by proton motive force in polyamine-deficient *Escherichia coli*, *J. Bacteriol.*, 165, 972, 1986.
9. **Bagni, N. and Pistocchi, R.,** Putrescine uptake in *Saintpaulia* petals, *Plant Physiol.*, 77, 398, 1985.
10. **Pistocchi, R., Bagni, N., and Creus, J.A.,** Polyamine uptake, kinetics and competition among polyamines and between polyamines and inorganic cations, *Plant Physiol.*, 80, 556, 1986.
11. **Pistocchi, R., Bagni, N., and Creus, J.A.,** Polyamine uptake in carrot cell cultures, *Plant Physiol.*, 84, 374, 1987.
12. **Pistocchi, R. and Bagni, N.,** Effect of calcium on spermine uptake in carrot cell cultures and protoplasts, *J. Plant Physiol.*, 136, 728, 1990.
13. **Pistocchi, R., Keller, F., Bagni, N., and Matile, Ph.,** Transport and subcellular localization of polyamines in carrot protoplasts and vacuoles, *Plant Physiol.*, 87, 514, 1988.

14. **Pistocchi, R., Antognoni, F., Bagni, N., and Zannoni, D.,** Spermidine uptake by mitochondria of *Helianthus tuberosus*, *Plant Physiol.*, 92, 690, 1990.
15. **Paulus, T.J. and Davis, R.H.,** Metabolic sequestration of putrescine in *Neurospora crassa*, *Biochem. Biophys. Res. Commun.*, 104, 228, 1982.
16. **Paulus, T.J., Cramer, C.L., and Davis, R.H.,** Compartmentation of spermidine in *Neurospora crassa*, *J. Biol. Chem.*, 258, 8608, 1983.
17. **Davis, R.H. and Ristow, J.L.,** Polyamine transport in *Neurospora crassa*, *Arch. Biochem. Biophys.*, 267, 479, 1988.
18. **Davis, R.H. and Ristow, J.L.,** Uptake, intracellular binding, and excretion of polyamines during growth of *Neurospora crassa*, *Arch. Biochem. Biophys.*, 271, 315, 1989.
19. **Joshi. S., Pleji, C.W.A., Haenni, A.-L., and Bosh, L.,** Age dependence of cowpea protoplasts for uptake of spermidine and infectibility by alfalfa mosaic virus, *Plant Mol. Biol.*, 2, 89, 1983.
20. **Hiatt, A.,** Polyamine synthesis in maize cell lines, *Plant Physiol.*, 90, 1378, 1989.
21. **Mengoli, M., Pistocchi, R., and Bagni, N.,** Effect of long-term treatment of carrot cell cultures with millimolar concentrations of putrescine, *Plant Physiol. Biochem.*, 27, 1, 1989.
22. **Escribano, M.I. and Legaz, M.E.,** Putrescine accumulation does not affect RNA metabolism in the lichen *Evernia prunastri*, *Endocyt. C. Res.*, 2, 239, 1985.
23. **Christ, M., Harr, J., and Felix, H.,** Transport of polyamines in sugar beet seedlings, *Z. Naturforsch.*, 44c, 59, 1989.
24. **Turner, R., North, M.J., and Harwood, J.M.,** Putrescine uptake by the cellular slime mould *Dictyostelium discoideum*, *Biochem. J.*, 180, 119, 1979.
25. **Mariani, P., D'Orazi, D., and Bagni, N.,** Polyamines in primary walls of carrot cells: endogenous content and interactions, *J. Plant Physiol.*, 135, 508, 1989.
26. **D'Orazi, D. and Bagni, N.,** *In vitro* interactions between polyamines and pectic substances, *Biochem. Biophys. Res. Commun.*, 148, 1259, 1987.
27. **Scoccianti, V., Bagni, N., Dubinsky, O., and Arad(Malis), S.,** Interaction between polyamines and cells of the marine unicellular red alga *Porphyridium* sp., *Plant Physiol. Biochem.*, 27, 899, 1989.
28. **Goldberg, R. and Perdrizet, E.,** Ratio of free to bound polyamines during maturation in mung-bean hypocotyl cells, *Planta*, 161, 531, 1984.
29. **Walker, M.A., Ellis, B.E., Chapple, C.C.S., and Dumbroff, E.B.,** Subcellular localization of amines and activities of their biosynthetic enzymes in *p*-fluorophenylalanine resistant and wild-type tobacco cell cultures, *Plant Physiol.*, 85, 89, 1987.
30. **Torrigiani, P., Serafini-Fracassini, D., Biondi, S., and Bagni, N.,** Evidence for subcellular localization of polyamines and their biosynthetic enzymes in plant cells, *J. Plant Physiol.*, 124, 23, 1986.
31. **Bagni, N. and Serafini-Fracassini, D.,** Involvement of polyamines in the mechanism of break of dormancy in *Helianthus tuberosus*, *Bull. Soc. Bot. Fr.*, 132; *Actual. Bot.*, 1, 119, 1985.
32. **Kanchanapoom, M., Antognoni, F., Pistocchi, R., and Bagni, N.,** Effect of hormones on spermidine uptake into carrot protoplasts. I. Auxins, *Physiol. Plant*, submitted.
33. **Smith, T.A. and Barker, J.H.A.,** The di- and polyamine oxidases in plants, in *Progress in Polyamine Research*, Zappia, V. and Pegg, A.E.., Eds., Plenum Press, New York, 1989, 537.
34. **Young, N.D. and Galston, A.W.,** Are polyamines transported in etiolated peas?, *Physiol. Plant*, 73, 912, 1983.
35. **Bagni, N., Baraldi, R., and Costa, G.,** Uptake, translocation and metabolism of aliphatic polyamines in leaves and fruitlets of *Malus domestica* (cv. Ruby spur), *Acta Hortic.*, 149, 173, 1984.
36. **Costa, G. and Bagni, N.,** Effects of polyamines on fruit-set of apple, *Hort. Sci.*, 18, 59, 1983.
37. **Rabiti, A.L., Pistocchi, R., and Bagni, N.,** Putrescine translocation in higher plants, *Physiol. Plant*, 77, 225, 1989.
38. **Friedman, R., Levin, N., and Altman, A.,** Presence and identification of polyamines in xylem and phloem exudates of plants, *Plant Physiol.*, 82, 1154, 1986.
39. **Masse´, J., Laberche, J.-C., and Jeanty, G.,** Translocation of polyamines in plants growing on media containing N,N′-*bis*-(3-aminopropylamino)ethane, *Plant Physiol. Biochem.*, 27, 489, 1989.
40. **Bagni, N., Creus, J.A., and Pistocchi, R.,** Distribution of cadaverine and lysine decarboxylase activity in *Nicotiana glauca* plants, *J. Plant Physiol.*, 125, 9, 1986.
41. **Jänne, J., Alhonen-Hongisto, L., Kapyaho, K., and Seppanen, P.,** Experimental approaches to the systemic and topical use of polyamine antimetabolites as antiproliferative agents, in *Advances in Polyamine Research*, Vol. 4, Bachrach, U., Kaye, A., and Chayen, R., Eds., Raven Press, New York, 1983, 17.
42. **Alhonen-Hongisto, L., Seppanen, P., and Jänne, J.,** Intracellular putrescine and spermidine deprivation induces increased uptake of the natural polyamines and methylglyoxal *bis*(guanylhydrazone), *Biochem. J.*, 192, 941, 1980.
43. **Bitonti, A.J., Bacchi, C.J., McCann, P.P., and Sjoerdsma, A.,** Uptake of α-difluoromethylornithine by ***Tripanosoma brucei***, *Biochem. Pharmacol.*, 35, 351, 1986.

44. **Erwin, B.G. and Pegg, A.E.,** Uptake of α-difluoromethylornithine by mouse fibroblasts, *Biochem. Pharmacol.*, 31, 2820, 1982.
45. **Biondi, S., Bagni, N., and Sansovini, A.,** Dicyclohexylamine uptake and effect on polyamine content in cultured cotyledons of radiata pine, *Plant Physiol.*, 66, 41, 1986.
46. **Smith, T.A. and Best, G.R.,** The estimation of lysine by a simplified dansylation procedure, *J. Sci. Food Agric.*, 32, 378, 1981.
47. **Mengoli, M., Bagni, N., and Biondi, S.,** Effect of α-difluoromethylornithine on carrot cell cultures, *J. Plant Physiol.*, 129, 479, 1987.
48. **Mandel, J.-L. and Flintoff, W.F.,** Isolation of mutant mammalian cells altered in polyamine transport, *J. Cell Physiol.*, 97, 335, 1978.
49. **Heaton, M.A. and Flintoff, W.F.,** Methylglyoxal *bis*(guanylhydrazone)-resistant Chinese hamster ovary cells: genetic evidence that more than a single locus controls uptake, *J. Cell Physiol.*, 136, 133, 1988.
50. **Walters, D.R. and Kingham, G.,** Uptake and translocation of α-difluoromethylornithine, a polyamine biosynthesis inhibitor, by barley seedlings: effects on mildew infection, *New Phytol*, 114, 659, 1990.

Chapter 8

UNCOMMON POLYAMINES IN PLANTS AND OTHER ORGANISMS

Gregory C. Phillips and Glenn D. Kuehn

TABLE OF CONTENTS

I. INTRODUCTION

The common aliphatic polyamines, putrescine, spermidine, and spermine, are ubiquitously distributed in living cells. Their occurrence, biosyntheses, and numerous aspects of their respective biochemical behaviors *in vitro* have been reviewed with regularity.[1-5] Since 1971, several uncommon polyamines, which differ from the familiar common polyamines in the number of methylenic moieties between the amine functions, have been detected in a broad spectrum of biological systems including plants,[6-9] algae,[10-13] bacteria,[14-18] and animals.[19-21] A listing of these polyamines is presented in Table 1. The common polyamines are essential for normal growth of all organisms,[22,23] but their precise molecular functions *in vivo* are still unclear. However, the association of the uncommon polyamines with the capacity of some biological systems to grow or function under extreme environments, has provided opportunities for new investigations into their potential functions. The purpose of this review is twofold: first, it is to summarize those investigations that have been conducted on the discoveries of the uncommon polyamines in plants, microorganisms, or animals; and second, to emphasize the impact that these studies have had to date on broadening the knowledge of uncommon polyamines in higher plants.

II. CADAVERINE AND DERIVATIVES

Cadaverine (1,5-diaminopentane), the five-carboned diamine homologue of putrescine, has been known to occur in many biological tissues for over a century. A description of the early discovery of cadaverine in 1885 was provided by Cohen.[24] The occurrence of this diamine in plants, however, appears to be sporadic and far more uncommon than the classical representative, putrescine. It has been found in the plant genera *Pisum, Glycine, Lathyrus, Trifolium, Lupinus,* and *Vicia,* all in the family Leguminosae; in *Nicotiana* and *Lycopersicon* (Solanaceae); in *Hordeum* and *Oryza* (Gramineae); in *Zantedeschia* (Araceae); and in *Sedum* (Crassulaceae). Specific plants from these genera include some cereals[25,26] such as barley[25] and rice,[27] arum lily,[28] lupine,[25,29] maple pea seedlings,[25] soybean,[25] clover,[25] *Sedum*,[30] tobacco,[31] and tomato.[31]

The biosynthetic origin of cadaverine is from the amino acid lysine. Decarboxylation of lysine by lysine decarboxylase provides a direct route to cadaverine.[32] The biosynthesis of cadaverine appears in some systems to be a compensatory route to production of essential cellular diamines when putrescine biosynthesis via decarboxylation of ornithine is limiting. Advantage was taken of this property in the initial discovery in bacteria of the uncommon derivative of cadaverine, *N*-3-aminopropyl-1,5-diaminopentane, or aminopropylcadaverine, as noted below.[33]

To date, no derivatives of cadaverine have been reported from plant sources. However, there is ample precedence from reports on their occurrence in bacteria to suggest that they might also exist in plants. The uncommon triamine aminopropylcadaverine, also called *asym*-homospermidine,[34] was initially discovered in *Escherichia coli* K12 mutants that were deficient in agmatine ureohydrolase and which were grown in a medium containing arginine and lysine.[33] Under these growth conditions, arginine in the medium prevented the biosynthesis of putrescine from glutamate. Compensatory synthesis of cadaverine as a diamine substitute for putrescine is promoted by the decarboxylation of lysine. Interestingly, *E. coli* cultures from which amino-propylcadaverine was isolated were also infected with phage. Cadaverine and aminopropylcadaverine have both recently been identified in another bacterium, *Paracoccus denitrificans*[35] (formerly called *Micrococcus denitrificans*). Moreover, when this organism was cultured in a growth medium supplemented with 1 m*M* cadaverine, aminopropylcadave-rine and a newly identified tetraamine, aminopentylnorspermidine ($H_2N(CH_2)_5NH(CH_2)_3NH(CH_2)_3NH_2$), accumulated in the bacterium.[35] Although *Agrobacterium tumefaciens* and *Halococcus acetoinfaciens* have not been found to synthesize cadaverine, both could produce aminopropyl derivatives of cadaverine when cultured in media

TABLE 1
Structures of Naturally Occurring Aliphatic Diamines and Polyamines

Name	Structure
	Diamines
1,3-Diaminopropane	$H_2N(CH_2)_3NH_2$[a]
Putrescine	$H_2N(CH_2)_4NH_2$
Cadaverine	$H_2N(CH_2)_5NH_2$[a]
	Triamines
Norspermidine (caldine)	$H_2N(CH_2)_3NH(CH_2)_3NH_2$[a]
Spermidine	$H_2N(CH_2)_3NH(CH_2)_4NH_2$
Aminopropylcadaverine	$H_2N(CH_2)_3NH(CH_2)_5NH_2$
Homospermidine	$H_2N(CH_2)_4NH(CH_2)_4NH_2$[a]
	Tetraamines
Norspermine (thermine)	$H_2N(CH_2)_3NH(CH_2)_3NH(CH_2)_3NH_2$[a]
Spermine	$H_2N(CH_2)_3NH(CH_2)_4NH(CH_2)_3NH_2$
Thermospermine	$H_2N(CH_2)_3NH(CH_2)_3NH(CH_2)_4NH_2$
Canavalmine	$H_2N(CH_2)_4NH(CH_2)_3NH(CH_2)_4NH_2$[a]
Homospermine	$H_2N(CH_2)_3NH(CH_2)_4NH(CH_2)_4NH_2$
	Pentaamines
Caldopentamine	$H_2N(CH_2)_3NH(CH_2)_3NH(CH_2)_3NH(CH_2)_3NH_2$[b]
Homocaldopentamine	$H_2N(CH_2)_3NH(CH_2)_3NH(CH_2)_3NH(CH_2)_4NH_2$[b]
	Hexaamines
Caldohexamine	$H_2N(CH_2)_3NH(CH_2)_3NH(CH_2)_3NH(CH_2)_3NH(CH_2)_3NH_2$
Homocaldohexamine	$H_2N(CH_2)_3NH(CH_2)_3NH(CH_2)_3NH(CH_2)_3NH(CH_2)_4NH_2$

[a] These uncommon polyamines have been confirmed to occur in higher plants.
[b] There is presumptive evidence for the occurrence of one or both of these uncommon polyamines in higher plants.

supplemented with cadaverine.[36] Both of these bacteria also were found to contain a new tetraamine, *N,N′-bis*(3-aminopropyl)cadaverine ($H_2N(CH_2)_3NH(CH_2)_5NH(CH_2)_3NH_2$), after growth in cadaverine-supplemented media.[37,38]

The capacity of some bacteria to synthesize a variety of uncommon triamines and tetraamines, such as derivatives of cadaverine and others discussed below, suggests the existence of a biosynthetic system with broad substrate specificity for aminopropylation of diamines and triamines. In most bacterial species, the common polyamines spermidine and spermine are synthesized by transfer of an aminopropyl moiety from decarboxylated *S*-adenosylmethionine (d-SAM) to putrescine. Aminopropyltransferase enzymes of generally narrow substrate specificities catalyze these transfers to putrescine or spermidine to form the common polyamines, spermidine or spermine, respectively.[32] However, an alternative pathway for addition of an aminopropyl moiety to a primary amine occurs in certain bacteria as shown by Tait.[39] Nonphotosynthetic *Paracoccus denitrificans,* purple photosynthetic *Rhodobacter sphaeroides* (formerly *Rhodopseudomonas sphaeroides*), a nonphotosynthetic relative *Agrobacterium tumefaciens,*[40] and some *Vibrio* species[41] have all been found to demonstrate this alternative pathway. In these bacteria, L-aspartic β-semialdehyde, and not d-SAM, is the donor of the aminopropyl group to spermidine. Putrescine and L-aspartic β-semialdehyde condense to form a Schiff-base complex which is reduced to form the basic amino acid carboxyspermidine. Carboxyspermidine yields spermidine upon decarboxylation (Scheme 1). Extracts of these organisms convert putrescine to spermidine, and spermidine to spermine and thermospermine when incubated with L-aspartic β-semialdehyde, NADPH, and pyridoxal phosphate.[40] The enzymes which mediate these conversions have not been thoroughly characterized from any source. Aminopropylation of cadaverine to yield highly uncommon triamines and tetraamines by reactions of Scheme 1 is an attractive possibility.

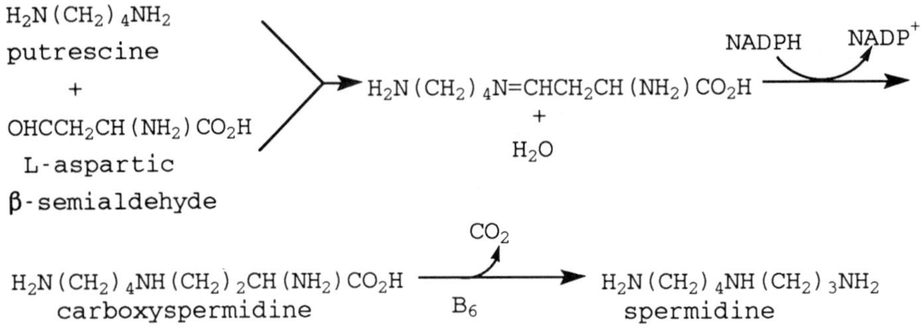

[1]

The reactions of scheme [1] have also been found to be expressed in a higher plant. In fact, seedlings of the plant *Lathyrus sativus* contain two pathways for aminopropylation.[42] Spermidine in *Lathyrus* was synthesized from the pathway using d-SAM as an aminopropyl group donor and the pathway using L-aspartic β-semialdehyde as a similar donor. The latter pathway has not been reported in any other plant, but its occurrence is an excellent candidate for investigation where broad specificity of substrate usage seems to occur in uncommon polyamine biosynthesis.

Due to the narrow substrate specificities of aminopropyltransferases, it could be expected that the wide variety of uncommon polyamine syntheses that have been found in plants and bacteria may be mediated by the Schiff-base complex formations with L-aspartic β-semialdehyde and 4-aminobutyraldehyde (see Scheme 2 below). A presumptive test for involvement of these intermediates could be made by using the reagents cyclohexylamine and 5′-methylthioadenosine, which are inhibitors of aminopropyltransferases. Uncommon triamine and tetraamine synthesis from diamine precursors, such as 1,3-diaminopropane, putrescine or cadaverine, should not be affected by these inhibitors if pathways utilizing Schiff-base complex formations are involved.

[2]

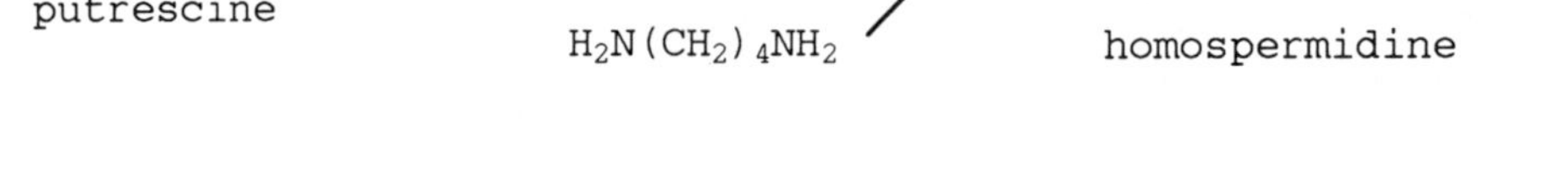

III. HOMOSPERMIDINE AND HOMOSPERMINE

The first uncommon triaminic polyamine identified in a higher plant was homospermidine, or *sym*-homospermidine (1,9-diamino-5-azanonane), found in *Santalum album* L. or sandalwood, where it comprised 0.5 to 1.5% of the mass of dried leaves.[6] Subsequently, this triamine was identified in other higher plants including root nodules of legumes,[7] water hyacinth,[43] *Heliotropium*,[44] grass pea,[45] and sword bean.[8] It was also found in lower photosynthetic systems including mosses,[46] ferns,[46] lichens,[47] and unicellular and multicellular eukaryotic algae.[10-13] A fungal acellular slime mold[48] also contained this polyamine. Homospermidine seems to occur widely among different bacteria.[7,18,49] For example, it was found in the thermophilic eubacterium *Thermus*,[17] in chemolithotrophs such as *Thiobacillus*,[50] in photosynthetic eubacteria such as *Rhodobacter*,[18] *Rhodospirillum*, and *Chromatium*,[51] in the N_2-fixing eubacterium *Rhizobium*,[7] in N_2-fixing cyanobacteria,[52] and in some methanogenic archaebacteria.[34] Homospermidine was the major polyamine found in twelve strains of eight species of the genus *Flavobacterium*.[53]

Interestingly, homospermidine also was confirmed in an amphibian, the Japanese newt,[21] and in arthropods[54] such as the cricket, cockroach, fly, and midge. Finally, it was even shown to occur in a mammalian tissue, hamster epididymis.[55]

The biosynthetic pathway for homospermidine in higher plants has not been fully elucidated. Following the discovery of the polyamine by Kuttan et al.,[6] radiolabeling studies were reported which indicated that all of its carbon and nitrogen atoms were derived from putrescine.[56] A likely enzymic sequence to account for homospermidine biosynthesis was demonstrated in extracts of the bacterium, *Rhodopseudomonas viridis*,[18] an organism that contains about three times more homospermidine than putrescine and lacks spermidine. In this photosynthetic bacterium, a single enzyme of molecular weight ca. 73,000 catalyzes a three-step conversion of two molecules of putrescine to form homospermidine (Scheme 2). Initially, this enzyme catalyzes the oxidation of one molecule of putrescine by NAD^+ to form enzyme-bound 4-aminobutyraldehyde and NADH. The aldehyde intermediate forms a Schiff base with a second molecule of putrescine, which is then reduced by the bound NADH, to yield homospermidine. Homospermidine synthase, as this enzyme is termed, has been found in extracts from six different species of purple photosynthetic bacteria.[40] A similar enzyme activity has been identified and partially purified from grass pea seedlings.[45]

The uncommon tetraamine, homospermine (1,13-diamino-4,9-diazatridecane), has been detected in the thermophilic bacterium *Thermus thermophilus*.[57] Although its enzymic derivation was not studied, a comparison of its structure with that of homospermidine (Table 1) would suggest that homospermine is formed by an aminopropyl transfer to homospermidine. The only other report on the occurrence of homospermine to date has been in a mammalian tissue, hamster epididymis.[55]

IV. CANAVALMINE

Canavalmine (1,13-diamino-5,9-diazatridecane = *N,N′-bis*(4-aminobutyl)1,3-diaminopropane) was initially discovered as one of the major constituents of polyamines in the seed of the higher plant, *Canavalia gladiata*, commonly known as the sword bean.[8] The capacity to synthesize canavalmine appears to be very restricted in nature. Reports of its occurrence in higher plants have been limited to the sword bean, and it has not been reported in any bacteria.[40] Moreover, although *C. gladiata* is a leguminous plant, a search for canavalmine in the seeds of fourteen different species of other leguminous plants failed to detect the tetraamine.[8] This finding suggested that the capacity to biosynthesize canavalmine may have originated in the sword bean and that it was not a product of symbiotic association with nitrogen-fixing bacteria. A search for canavalmine among select mosses, ferns, and lichens did not detect any canavalmine.[58]

In an important report by Matsuzaki et al.,[55] canavalmine and other uncommon polyamines were found in epididymis tissue of the Syrian hamster. This finding was of special interest since it was the first report showing that tetraamines other than spermine are present in mammalian tissues.

The biosynthetic reactions which produce canavalmine have not been elucidated from any source. Clearly, the presence of two aminobutyl terminal groups in the structure of canavalmine is indicative of the existence of enzyme system(s) with the capacity to transfer an aminobutyl moiety to spermidine. If this is, in fact, the manner by which canavalmine is synthesized, then the enzymes(s) involved in the transfer of the aminobutyl moiety must mediate this addition on the aminopropyl terminus of spermidine. This specificity is unlike the required specificity for transfer of an aminopropyl moiety to the aminobutyl terminus of spermidine as catalyzed by spermine synthase (spermidine aminopropyltransferase).[59] It is possible that uncommon polyamines which differ from common polyamines because they bear additional aminobutyl groups, may be formed by a reaction sequence analogous to that described in Scheme 2. For example, the aldehyde derived from oxidized putrescine could be coupled with free spermidine to form canavalmine.

V. THERMOSPERMINE

Thermospermine (1,12-diamino-4,8-diazadodecane) is a structural isomer of the common polyamine spermine. The free amine forms of both compounds have the chemical formula, $C_{10}H_{26}N_4$. Structural and chromatographic similarities between thermospermine and the common tetraamine spermine obscured the identity of thermospermine, until Oshima[16] successfully resolved the two polyamines and elucidated the structure of purified thermospermine from an extreme thermophilic bacterium, *Thermus thermophilus* HB8 (ATCC 27634). In general, no separation methods based on ion exchange have been reported which can successfully resolve spermine and thermospermine. However, the two have been separated as underivatized amines by thin-layer chromatography,[60] by gas chromatography after prior derivatization,[55] or by combined enzymatic-gas chromatographic methods.[37,38,50]

Few studies have been conducted with the specific purpose of evaluating the occurrence of thermospermine in higher plants.[58] Consequently, knowledge of the natural distribution of thermospermine is far from complete. At this time, there have been no formal reports of its existence in higher plants. However, it has been reported in a relatively small number of bacteria including the extreme thermophile *Thermus thermophilus*,[61] the halophile *Halococcus acetoinfaciens*,[38] the chemolithotroph *Paracoccus denitrificans*,[35] and numerous species of agrobacteria[36] including *Agrobacterium tumefaciens, A. rhizogenes, A. radiobacter*, and *A. rubi*. Oshima[61] first reported the unusual occurrence of thermospermine in an insect, the cockroach (*Periplaneta*). Subsequently, it was found to be widely distributed in various organs of other insects such as crickets (*Teleogryllus* and *Gryllus*), the fly (*Drosophila*), and the midge (*Orthocladius*).[54]

VI. THE "TRIMETHYLENIC" POLYAMINES

A. 1,3-DIAMINOPROPANE

The first reported observation of the natural occurrence of 1,3-diaminopropane was made by Weaver and Herbst[62] who found the diamine in several bacteria, yeast and molds, as well as in liver tissue from cow, rat, and guinea pig, and in human semen. These authors also identified the only enzymatic reaction which is known to produce 1,3-diaminopropane in bacteria[62,63] and which was later confirmed to be present in plants (see below, this section). This reaction is mediated by an oxygen-dependent polyamine oxidase that cleaves spermidine or spermine between the secondary nitrogen atom and the methylenic carbon comprising the aminobutyl group. The oxidation of spermidine by this polyamine oxidase initially yields 1,3-diaminopropane, hydrogen peroxide, and an aminal product, 4-aminobutyraldehyde. The aldehyde undergoes spontaneous intramolecular Schiff-base formation to yield pyrroline[64] as shown in Scheme 3. When spermine is oxidized by this enzyme, the products 1,3-diaminopropane and hydrogen

[3]

$H_2N(CH_2)_3NH(CH_2)_4NH_2$ + O_2 ⟶ $H_2N(CH_2)_3NH_2$ + H_2O_2 +

spermidine 1,3-diaminopropane

1-pyrroline

peroxide are again formed, but the structure of the initial aminal product obtained under physiological conditions has been controversial. The product arises from the Schiff-base mediated cyclization of a diamine aldehyde. Initially the product was reported to be 1-(3-aminopropyl)-2-pyrroline.[65] A re-evaluation of the properties of the reaction product, however, led to the conclusion that in aprotic organic solvents the oxidation product was 1,5-diazabicyclo[4.3.0]nonane (or, octahydropyrrolo[1,2-*a*]),[66,67] shown in Scheme 4. Most recently,

[4]

$H_2N(CH_2)_3NH(CH_2)_4NH(CH_2)_3NH_2$ + O_2 ⟶ $H_2N(CH_2)_3NH_2$ + H_2O_2 +

spermine 1,3-diaminopropane

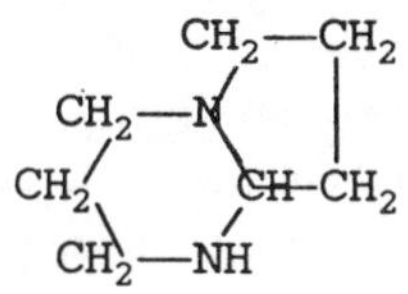

1,5-diazabicyclo[4.3.0]nonane

the physiological form of 1,5-diazabicyclo[4.3.0]nonane that exists in protic media was found to consist of two rapidly interconverting conjugate acids of the secondary and tertiary amines of the bicyclic nonane molecule,[68,69] shown in Scheme 5. Thus, 32 years after the initial recognition by Weaver and Herbst[62] of the problematic identity of this reaction product from polyamine oxidase, the true identity appears to have been firmly established.

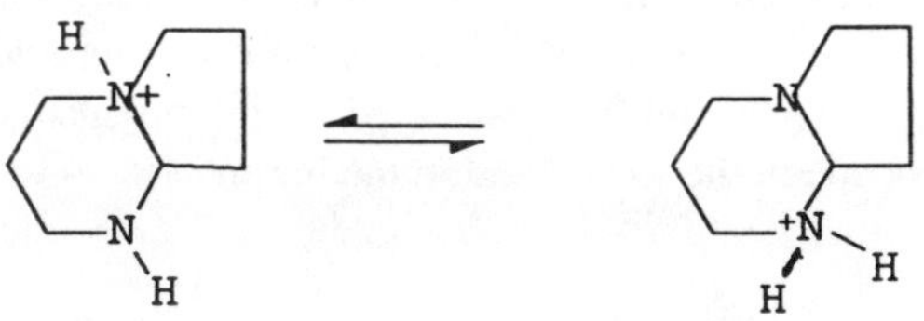

[5]

Although Weaver and Herbst[62] first discovered 1,3-diaminopropane in natural microbial sources in 1958, ten years passed before the diamine was formally reported to exist in a plant source. In an extensive study of the occurrence of spermidine and spermine in the seeds of 54 different plant species, Bagni[70] reported the occurrence of 1,3-diaminopropane in extracts of seeds from peas. Subsequently, Smith[64] discovered the prevalent occurrence of a polyamine oxidase, which catalyzed the reactions shown in Schemes 3 and 4 above, among cereal plants in the Gramineae family. These included oats, maize, wheat, barley, and rye. This study was preceded by the finding of 1,3-diaminopropane in extracts of the monocotyledonous species barley, wheat, and maize, as well as in extracts of leaves from dicotyledonous species including radish, pea, black currant, tobacco, strawberry, apple, pyracantha, tomato, marrow, spinach, and willow.[71]

1,3-Diaminopropane has now been reported from scores of different higher plants, both monocotyledonous and dicotyledonous. As with the discovery and detection of all uncommon polyamines, improvements in techniques for separating polyamines from one another and the development of sensitive methods for quantitating their amounts and determining their structures, have led to the realization that 1,3-diaminopropane is far more prevalent in higher plants than early work suggested. This presents an interesting enigma in plant biochemistry: Although 1,3-diaminopropane has been demonstrated in many plant sources, the enzyme polyamine oxidase has not. Since the only enzymatic reaction known to produce 1,3-diaminopropane is that catalyzed by polyamine oxidase (Schemes 3 and 4), the presence of the diamine reaction product in a broad spectrum of plants, mammals, and microorganisms suggests the expression of a similar oxidase enzyme in these biological systems. However, the detection of a polyamine oxidase with the specificities depicted in Schemes 3 and 4 has been restricted to very few sources.[72] The enzyme was first described in extracts of barley leaves,[64] where it was recognized to provide the only known specific pathway of spermidine and spermine degradation in higher plants. The enzyme was subsequently found to be strongly expressed among many monocotyledonous species of the plant family Gramineae[73] from which it was characterized in oats,[74] barley,[75] corn,[76] and millet.[77] However, the enzyme is not restricted to the Gramineae for it has also been purified from water hyacinth (*Eichhornia crassipes* Solms),[78] which does not belong to the Gramineae but which is also monocotyledonous. Neither is the enzyme restricted to monocots, since its occurrence has recently been demonstrated in dicotyledonous alfalfa (*Medicago sativa* L.), where it is inducible by the environmental stress of drought.[79] Is it possible that the functioning of polyamine oxidase in many plants requires special conditions that elicit either its genetic or enzymatic expression?

The occurrence of 1,3-diaminopropane in microorganisms was reported with increasing regularity after its initial discovery by Weaver and Herbst[62] in *Neisseria perflava* and 20 years later in the extreme acidothermophile, *Caldariella acidophila*, as reported by DeRosa et al.[80] Examples of other bacteria which were shown to contain this diamine were *Mycobacterium smegmatis*,[81] *Serratia marcescens*,[82] *Clostridium thermohydrosulfuricum*,[83] photosynthetic bacteria *Rhodobacter sphaeroides* and *Rhodospirillum rubrum*,[51] and the methanogenic bacteria *Methanococcus vannielii* and *Methanococcus humgatei*.[34] The diamine also was found in the red alga, *Cyanidium caldarium*[84] and in many eukaryotic algae[13] including *Porphyra tenera, Ecklonia cava, Sargassum thunbergii, Euglena gracilis, Trachelomonas* sp., *Chlorella ellipsoidea, Scenedesmus* sp., and others. Its occurrence even extends to insects, where it was reported in dominant amounts in the midge, cricket, cockroach, and fly.[54] As noted earlier, bovine tissues were found also to contain 1,3-diaminopropane in the liver[62] and in the brain.[85]

B. NORSPERMIDINE (CALDINE)

The triamine norspermidine (1,7-diamino-4-azaheptane), also called *sym*-norspermidine and trivially named caldine, differs from other triaminic polyamines in that the primary and secondary nitrogen atoms are all separated by three methylenic carbon atoms. In 1962, Johnson and Markham[86] first reported the isolation of norspermidine from turnip yellow mosaic virus and several other purified plant viruses, in which it appeared to be associated with the viral RNA.[87] The polyamine was reported to be absent from healthy, uninfected plants.[86] Association of norspermidine with turnip yellow mosaic virus RNA was found to stabilize the RNA from heat denaturation.[87] However, a later study failed to confirm the presence of norspermidine in turnip yellow mosaic virus.[88] The discovery of this amine in bacteria was first reported in the extreme thermophile, *Caldariella acidophila*.[15] The diamine 1,3-diaminopropane is considered to be a potential precursor for the biosynthesis of norspermidine, since this conversion requires only an aminopropyl group addition to one of the terminal primary amines of the diamine.[15] Although 1,3-diaminopropane was found in a broad spectrum of plants, as noted in the previous section, the occurrence of norspermidine was not confirmed in a higher plant until recently.[9] In alfalfa, norspermidine was found to accumulate in genetic lines tolerant to drought stress and in cotton it was associated with genetic lines tolerant to heat stress,[79] upon exposure to the respective stress.

The occurrence of norspermidine in biological systems was not fully appreciated until it was discovered with other larger trimethylenic polyamines in extreme thermophilic eubacteria[14] and in extreme thermoacidophilic archaebacteria.[15] In eubacteria, norspermidine was first detected in *Thermus thermophilus, T. aquaticus*,[61,89] and *T. flavus*[49] by Oshima et al. In archaebacteria, it was first found in *Sulfolobus acidocaldarius*[61,89] and *Caldariella acidophila*.[15] The coexistence of norspermidine, norspermine, caldopentamine, and caldohexamine in a single organism,[61,89] immediately suggested that these novel polyamines might be derived from a common biosynthetic pathway involving sequential addition of aminopropyl moieties to the primary precursor, 1,3-diaminopropane.[80] Decarboxylated *S*-adenosylmethionine (d-SAM) was suggested to be the source of the aminopropyl groups and the product 5′-methylthioadenosine (MTA) for the reaction sequence shown in Scheme 6. Indeed, Paulin et al.[83] were able to demonstrate an aminopropyltransferase activity in extracts of the anaerobic acidothermophile, *Clostridium thermohydrosulfuricum*, which had the capacity to convert 1,3-diaminopropane to norspermidine. Similarly, Cacciapuoti et al.[90] reported the purification to homogeneity of a single aminopropyltransferase enzyme from *Sulfolobus solfataricus* which catalyzed transfer of the aminopropyl moiety from decarboxylated *S*-adenosylmethionine to 1,3-diaminopropane, putrescine, norspermidine and spermidine, to form norspermidine, spermidine, norspermine, and spermine, respectively. This enzyme has a molecular weight of ca. 112,000 and is composed of three apparently identical 35 kDa subunits.[90] The purification of this aminopropyltransferase validated the earlier proposal of DeRosa et al.[80] that the entire molecular skeleton of the symmetrical polyamines, norspermidine, and norspermine, could derive from decarboxylated *S*-adenosylmethionine. This enzyme is peculiar in its broad specificity of primary amine

[6]

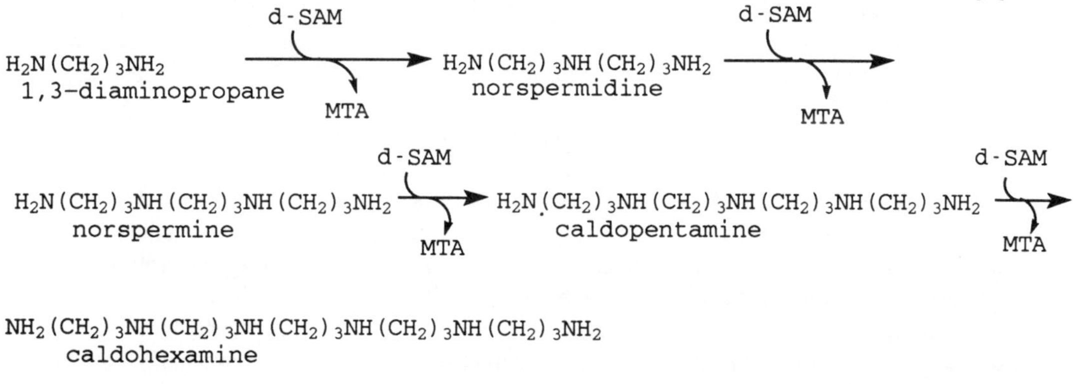

substrates which serve as aminopropyl acceptors. The enzyme has the capacity to transfer the aminopropyl moiety to a primary amino group linked to either a trimethylenic (1,3-diaminopropane or norspermidine) chain or to a tetramethylenic (putrescine or spermidine) chain. Thus, this property distinguishes the enzyme from all other aminopropyltransferases previously described which only recognize tetramethylenic structures as aminopropyl moiety acceptors.[91]

Except for the examples from *Clostridium*[83] and *Sulfolobus*,[90] there has not been any other formal report or further characterization of the putative pathway suggested by Scheme 6 in any biological source. Whether pathways analogous to those depicted earlier in Schemes 1 and 2 have a role in generating norspermidine or other trimethylenic polyamines, using 1,3-diaminopropane as the substrate instead of putrescine, is unknown.

The distribution of norspermidine throughout bacteria is seemingly restricted to organisms that are adapted to survival in extremely acidic and/or high-temperature environments. As noted earlier, this has included quantitation of this triamine first in *Caldariella*,[15] then in *Thermus*,[61,89] *Sulfolobus*, [90,92] and *Clostridium*[83] sp. It was also found in archaebacteria[92] *Thermodiscus maritimus*, *Pyrodictium occultum*, *Thermoproteus tenax*, *Thermofilum pendens*, *Desulfurococcus* sp., and *Thermococcus* sp. The occurrence of norspermidine, however, extends beyond this niche to algae,[13,84] mollusks,[20] and arthropods.[19,20] Numerous uncommon genera of algae within the phyllum Phaeophyta, and common algae[13] such as *Euglena gracilis*, *Trachelomonas* sp., *Chlamydomonas mowusii*, *Chlorella vulgaris*, *Scenedesmus acutus*, the red alga *Cyanidium caldarium*,[84] and others[13] have all been reported to produce norspermidine.

Norspermidine has been reported in trace amounts in mollusks such as snails (*Patella coerulea*), mussels (*Mytilis galloprovincialis*), and clams (*Pinna nobilis*).[20] It has also been detected in arthropods such as crab (*Carcinus mediterraneus*)[20] and shrimp (*Penaeus kerathurus*[20] and *Penaeus setiferus*[19]), as well as in insects[20,54,61] including the cockroach, cricket, fly, and midge. The occurrence of norspermidine in some multicellular organisms must be interpreted with caution, since it is not known whether these examples represent indigenous biosynthetic capabilities or whether the occurrence is a consequence of alimentary habits.

C. NORSPERMINE (THERMINE)

The tetraamine norspermine (1,11-diamino-4,8-diazaundecane), also called *sym*-norspermine and trivially named thermine, was first detected and identified in the extreme thermophilic eubacterium, *Thermus thermophilus*.[14] It was recently discovered concurrently with norspermidine in the higher plant, alfalfa (*Medicago sativa* L.), associated with genetic lines tolerant to drought.[9] This latter report is the only observation to date of the occurrence of norspermidine and norspermine in higher plants. Among lower plant forms, norspermine has been found in mosses[46,58] and in lichens.[58]

As in the case for norspermidine, the enzymes responsible for the biosynthesis of norspermine have been characterized from few sources. Aminopropyltransferase reactions were demonstrated in the anaerobic extreme thermophilic eubacterium, *Clostridium thermohydrosulfuricum*,[83] and the extreme thermophilic archaebacterium, *Sulfolobus solfataricus*,[90] which had the capacity to carry out the two-step conversion of 1,3-diaminopropane to norspermine using decarboxylated *S*-adenosylmethionine as the aminopropyl group donor. Thus, norspermine like norspermidine, was produced in these bacteria by reactions shown in Scheme 6 above. A preliminary report presented evidence for a similar synthetic capacity in extracts from alfalfa.[93]

The distribution of norspermine in nature nearly parallels that of norspermidine. The two compounds have been found to coexist in many genera, thus supporting the notion that norspermidine is a biosynthetic precursor of norspermine according to Scheme 6.[14] Other bacteria in which norspermine was reported have included thermophilic archaebacteria such as *Caldariella acidophila*,[15] and all of the archaebacteria listed in the previous section that contain norspermidine plus *Thermoplasma* sp.[92] It was found in unicellular eukaryotic algae including *Euglena gracilis*,[11] *Phaeodactylum tricornutum*,[94] and the thermoacidophilic red alga *Cyanidium caldarium*,[84] and in multicellular algae[13] including *Chlamydomonas mowusii*, *Chlorella ellipsoidea*, *Enteromorpha prolifera*, *Closterium* sp., *Aegagrophila sauteri*, *Spirogyra* sp., and *Nitella flexilis*. Although norspermine and norspermidine were not found in six different species of *Agrobacterium*, both polyamines could be synthesized by these bacteria when 1,3-diaminopropane was added to the medium. The mechanisms responsible for these biosyntheses involved the condensation of 1,3-diaminopropane and L-aspartic β-semialdehyde via a Schiff-base complex analogous to Scheme 1.[36]

Norspermine has also been reported in such diverse phylla as Mollusca and Arthropoda. A representative of the former is the octopus (*Octopus vulgaris*), which was found to contain substantial amounts of norspermine.[20] In Arthropoda, the tetraamine was found in such members as crab,[20] shrimp,[19,20] the cockroach,[54,61] cricket,[54] fly,[54] and the midge.[54]

D. CALDOPENTAMINE

Caldopentamine (1,15-diamino-4,8,12-triazapentadecane)[95] was initially identified from the extreme thermophilic eubacterium, *Thermus thermophilus* HB8.[61] It has not yet been reported in any other bacteria. *T. thermophilus* HB8 is capable of producing several uncommon polyamines as reviewed in this chapter. Interestingly, the particular complement of polyamines present in cultures of this thermophile varied with the growth temperature.[61,95] Cultures grown at 65°C produced predominantly the triamines homospermidine and norspermidine, and some thermospermine, but only trace amounts of caldopentamine. At optimum growth temperature of 75°C, the triamine complement was reduced and the titer of caldopentamine increased. Cultures grown at 80°C contained predominant titers of norspermine and caldopentamine and only trace amounts of triamines. The correlation between increased growth temperature and increased biosyntheses of tetra- , penta- , and hexaamines (see next section) prompted the proposal that the trimethylenic long-chained polyamines may be required for the thermostability of these bacteria.[17,61,83] Dependence upon these polyamines for *in vitro* protein synthetic activity at 70°C was observed in extracts of *T. thermophilus*, *Sulfolobus acidocaldarius*, and *S. solfataricus*.[96]

A pentaamine has not been conclusively identified as a product of plant metabolism. However, presumptive evidence for the occurrence of caldopentamine or some similar analogue has been found in polyamines extracted and analyzed by HPLC from drought-tolerant genetic lines of alfalfa and heat-tolerant genetic lines of cotton.[79] Final confirmation of these findings, however, awaits structural determination of the putative polyamines.

The long-chained trimethylenic polyamines may have other roles apart from supporting thermophily. Oshima[61] also found caldopentamine, norspermine, norspermidine, and thermospermine in extracts from larvae of the cockroach (*Periplaneta brunnea*). In fact, the polyamine composition of the cockroach was qualitatively similar to that of the extreme thermophilic bacterium, *T. thermophilus*. What potential role these uncommon polyamines might have in insects is unknown.

E. CALDOHEXAMINE

Caldohexamine (1,19-diamino-4,8,12,16-tetraazanonadecane) is the largest trimethylenic polyamine reported thus far from any organism.[97] The finding of this exceptional polyamine, as a quantitatively minor component in *Thermus thermophilus*, established that the entire set of trimethylenic polyamines of 1,3-diaminopropane, norspermidine, norspermine, caldopentamine, and caldohexamine occurred in a single bacterium. The occurrence of all five of these polyamines in one organism and the demonstrated occurrence of aminopropyltransferase enzymic activities in thermophilic bacteria with capacities to mediate the sequence of reactions in Scheme 6,[83,90] provides the likelihood that the trimethylenic polyamines are produced by d-SAM-dependent aminopropylation reactions. However, a more complete definition of this pathway remains to be established.

VII. HOMOCALDOPENTAMINE AND HOMOCALDOHEXAMINE

The long-chained polyamines, homocaldopentamine[98] (1,16-diamino-4,8,12-triazahexadecane) and homocaldohexamine[57] (1,20-diamino-4,8,12,16-tetraazaeicosane), have also been identified in *Thermus thermophilus*. The titer of homocaldopentamine varied with the growth temperature of this extreme thermophilic bacterium in a manner analogous to that of caldopentamine,[61] as discussed above. The occurrence of homocaldopentamine and homocaldohexamine has not been confirmed from any other organism to date. However, homocaldopentamine or a similar analogue has been observed in hplc analyses of drought-tolerant alfalfas and heat-tolerant cottons,[79] as discussed above. A comparison of their structures (Table 1) would suggest that homocaldopentamine and homocaldohexamine are derived by successive aminopropyl transfers to thermospermine. The singular occurrence of homocaldopentamine and homocaldohexamine, along with thermospermine, in the same organism that also exhibits the singular occurrence of caldopentamine and caldohexamine, along with the other trimethylenic polyamines, provides an appealing basis on which to suggest the following: a single aminopropyltransferase enzyme with broad substrate specificity could accomplish the reaction sequences presented in Scheme 6, as well as the analogous reactions leading to these other long-chained polyamines, in this extreme thermophilic bacterium. If this is the case, this enzyme would have properties similar to those of the enzyme described by Cacciapuoti et al.[90]

VIII. CONCLUSIONS

A perusal of the uncommon polyamines listed in Table 1 reveals that most of these compounds are formed by aminopropyl additions to appropriate acceptor amines. At this time, only the two Schemes 1 and 6 are known to catalyze these additions. However, Scheme 1 has been investigated and confirmed to exist in only one plant species.[42] Scheme 6 has not been confirmed in plants, although presumptive evidence for its existence was reported.[79,93] Even spermidine synthase (putrescine aminopropyltransferase), which catalyzes a reaction analogous to those in Scheme 6, but with high specificity for putrescine in common polyamine biosynthesis, has received only minimal investigative attention in plants.[99] Spermine synthase has not been purified or characterized from any plant source. Owing to the marked stabilizing properties of the higher molecular weight tetraamines and pentaamines on membranes,[100,101] nucleic acids,[102] and the protein synthesizing apparatus,[96] there is ample evidence to support the hypothesis that the long-chained uncommon polyamines may serve as drought-stress and/or heat-stress protectants from these conditions in plants. Definitive testing of this hypothesis must await further characterization and the availability of purified aminopropyltransferase and polyamine oxidase enzymes from plants.

ACKNOWLEDGMENTS

The authors wish to acknowledge the assistance of Tara L. Kuehn in the literature search conducted for the citations of this review. Research support from the Southwest Consortium on Plant Genetics and Water Resources CRSR-88-34186-3340 is also acknowledged.

REFERENCES

1. **Galston, A.W. and Sawhney, R.K.,** Polyamines in plant physiology, *Plant Physiol.*, 94, 406, 1990.
2. **Slocum, R.D., Kaur-Sawhney, R., and Galston, A.W.,** The physiology and biochemistry of polyamines in plants, *Arch. Biochem. Biophys.*, 235, 283, 1984.
3. **Galston, A.W. and Smith, T.A.,** Eds., *Polyamines in Plants*, Martinus-Nijhoff/Dr. W. Junk, Dorbrecht, 1985.
4. **Tabor, H. and Tabor, C.W.,** Eds., Polyamines, *Meth. Enzymol.*, Vol. 94, Academic Press, New York, 1983, 497.
5. **Caldarera, C.M., Clo, C., and Guarnieri, C.,** Eds., *Biomedical Studies of Natural Polyamines*, Editrice CLUEB, Bologna, 1986.
6. **Kuttan, R., Radhakrishnan, A.N., Spande, T., and Witkop, B.,** *Sym*-homospermidine, a naturally occurring polyamine, *Biochemistry*, 10, 361, 1971.
7. **Smith, T.A.,** Homospermidine in *Rhizobium* and legume root nodules, *Phytochemistry*, 16, 278, 1977.
8. **Fujihara, S., Nakashima, T., and Kurogochi, Y.,** Occurrence of a new polyamine, canavalmine, in the sword bean, *Canavalia gladiata*, *Biochem. Biophys. Res. Commun.*, 107, 403, 1982.
9. **Rodriguez-Garay, B., Phillips, G.C., and Kuehn, G.D.,** Detection of norspermidine and norspermine in *Medicago sativa* L. (alfalfa), *Plant Physiol.*, 89, 525, 1989.
10. **Rolle, I., Hobucker, H.E., Kneifel, H., Paschold, B., Riepe, W., and Soeder, C.J.,** Amines in unicellular green algae. 2. Amines in *Scenedesmus acutus*, *Anal. Biochem.*, 77, 103, 1977.
11. **Kniefel, H., Schuber, F., Aleksijevic, A., and Grove, J.,** Occurrence of norspermine in *Euglena gracilis*, *Biochem. Biophys. Res. Commun.*, 85, 42, 1978.
12. **Villanueva, V.R., Adlakha, R.C., and Calvayrac, R.,** Biosynthesis of polyamines in *Euglena gracilis*, *Phytochemistry*, 19, 787, 1980.
13. **Hamana, K. and Matsuzaki, S.,** Widespread occurrence of norspermidine and norspermine in eukaryotic algae, *J. Biochem.*, 91, 1321, 1982.
14. **Oshima, T.,** Thermine. A new polyamine from an extreme thermophile, *Biochem. Biophys. Res. Commun.*, 63, 1093, 1975.
15. **DeRosa, M., DeRosa, S., Gambacorta, A., Carteni-Farina, M., and Zappia, V.,** Occurrence and characterization of new polyamines in the extreme thermophile *Caldariella acidophila*, *Biochem. Biophys. Res. Commun.*, 69, 253, 1976.
16. **Oshima, T.,** A new polyamine, thermospermine, 1,12-diamino-4,8-diazadodecane, from an extreme thermophile, *J. Biol. Chem.*, 254, 8720, 1979.
17. **Oshima, T. and Baba, M.,** Occurrence of *sym*-homospermidine in extremely thermophilic bacteria, *Biochem. Biophys. Res. Commun.*, 103, 156, 1981.
18. **Tait, G.H.,** The formation of homospermidine by an enzyme from *Rhodopseudomonas viridis*, *Biochem. Soc. Trans.*, 7, 199, 1979.
19. **Stillway, L.W. and Walle, T.,** Identification of the unusual polyamines 3,3'-diaminodipropylamine and *N,N'*-*bis*(3-aminopropyl)-1,3-propanediamine in the white shrimp *Penaeus setiferus*, *Biochem. Biophys. Res. Commun.*, 77, 1103, 1977.
20. **Zappia, V., Porta, R., Carteni-Farina, M., DeRosa, M., and Gambacorta, A.,** Polyamine distribution in eukaryotes: Occurrence of *sym*-norspermine in arthropods, *FEBS Lett.*, 94, 161, 1978.
21. **Hamana, K. and Matsuzaki, S.,** Occurrence of *sym*-homospermidine in the Japanese newt, *Lynops pyrrhogaster*, *FEBS Lett.*, 99, 325, 1979.
22. **Pegg, A.E.,** Recent advances in the biochemistry of polyamines in eukarytoes, *Biochem. J.*, 234, 249, 1986.
23. **Tabor, C.W. and Tabor, H.,** Polyamines, *Annu. Rev. Biochem.*, 53, 749, 1984.
24. **Cohen, S.,** *Introduction to the Polyamines*, Prentice-Hall, Englewood Cliffs, New Jersey, 1971, 17.
25. **Smith, T.A. and Wilshire, G.,** Distribution of cadaverine and other amines in higher plants, *Phytochemistry*, 14, 2341, 1975.
26. **Moruzzi, G. and Caldarera, C.M.,** Occurrence of polyamines in the germs of cereals, *Arch. Biochem. Biophys.*, 105, 209, 1964.
27. **Umezu, M.,** On the quantitative changes of amino acids and amines in the process of sake′-brewing. VI. On the amines contained in rice and rice-koji, *Hakko Kogaku Zasshi*, 39, 247, 1961, *Chem. Abst.*, 58, 9596c, 1963.
28. **Smith, B.N. and Meeuse, B.J.D.,** Production of volatile amines and skatole at anthesis in some arum lily species, *Plant Physiol.*, 41, 343, 1966.
29. **Schiitte, H.R.,** Biosynthesis of lupine alkaloids, *Angew. Chem. [Engl.]*, 4, 991, 1965.
30. **Leistner, E. and Spenser, I.D.,** Biosynthesis of the piperidine nucleus. Incorporation of chirally labeled [1-^{3}H]cadaverine, *J. Am. Chem. Soc.*, 95, 4715, 1973.

31. **Hohlt, H.E., Maynard, D.N., and Barker, A.V.,** Ammonium tolerance of some cultivated Solanaceae, *J. Am. Soc. Hort. Sci.*, 95, 345, 1970.
32. **Tabor, C.W. and Tabor, H.,** Polyamines in microorganisms, *Microbiol. Rev.*, 49, 81, 1985.
33. **Dion, A.S. and Cohen, S.S.,** Polyamine stimulation of nucleic acid synthesis in an uninfected and phage-infected polyamine auxotroph of *Escherichia coli* K12, *Proc. Natl. Acad. Sci. USA*, 69, 213, 1972.
34. **Scherer, P. and Kneifel, H.,** Distribution of polyamines in methanogenic bacteria, *J. Bacteriol.*, 154, 1315, 1983.
35. **Hamana, K., Matsuzaki, S., Niitsu, M., and Samejima, K.,** Synthesis of novel polyamines in *Paracoccus, Rhodobacter*, and *Micrococcus, FEMS Microbiol. Lett.*, 67, 267, 1990.
36. **Hamana, K., Matsuzaki, S., Niitsu, M., and Samejima, K.,** Polyamine distribution and the potential to form novel polyamines in phytopathogenic agrobacteria, *FEMS Microbiol. Lett.*, 65, 269, 1989.
37. **Hamana, K. and Matsuzaki, S.,** Distribution of polyamines in actinomycetes, *FEMS Microbiol. Lett.*, 41, 211, 1987.
38. **Hamana, K., Niitsu, M., Samejima, K., and Matsuzaki, S.,** Occurrence of aminopropylcadaverine and its aminopropyl derivatives aminopentylnorspermidine and N,N′-bis(3-aminopropyl)cadaverine in *Halococcus acetoinfaciens, FEMS Microbiol. Lett.*, 50, 79, 1988.
39. **Tait, G.H.,** A new pathway for the biosynthesis of spermidine, *Biochem. Soc. Trans.*, 4, 610, 1976.
40. **Tait, G.H.,** Bacterial polyamines, structures and biosynthesis, *Biochem. Soc. Trans.*, 13, 316, 1985.
41. **Yamamoto, S., Hamanaka, K., Suemoto, Y., Ono, B., and Shinoda, S.,** Evidence for the presence of a novel biosynthesis pathway for norspermidine in *Vibrio, Can. J. Microbiol.*, 32, 99, 1986.
42. **Srivenugopal, K.S. and Adiga, P.R.,** Coexistence of two pathways of spermidine biosynthesis in *Lathyrus sativus* seedlings, *FEBS Lett.*, 112, 260, 1980.
43. **Yamamoto, S., Aoyama, Y., Kawaguchi, M., Iwada, A., and Makita, M.,** Polyamine oxidase in *Eichhornia crassipes* Solms, *Chem. Pharm. Bull.*, 31, 3315, 1983.
44. **Birecka, H., Dinolfo, T.E., Martin, W.B., and Frochlich, M.W.,** Polyamines and leaf scenescence in pyrrolizidine alkaloid-bearing heliotropium plants, *Phytochemistry*, 23, 991, 1984.
45. **Srivenugopal, K.S. and Adiga, P.R.,** Enzymic synthesis of *sym*-homospermidine in *Lathyrus sativus* (grass pea) seedlings, *Biochem. J.*, 190, 461, 1980.
46. **Yamamoto, S., Iwado, A., Hashimoto, Y., Aoyama, Y., and Makita, M.,** Gas chromatography-mass spectrometry of polyamines as their *N*-ethoxycarbonyl derivatives and identification of *sym*-homospermidine and *sym*-norspermine in mosses and ferns, *J. Chromatogr.*, 303, 99, 1984.
47. **Marshall, M., Russo, G., Van Etten, J., and Nickerson, K.,** Polyamines in dimorphic fungi, *Curr. Microbiol.*, 2, 187, 1979.
48. **Hamana, K. and Matsuzaki, S.,** Unusual polyamines in slime molds *Physarum polycephalum* and *Dictyostelium discoideum, J. Biochem.*, 95, 1105, 1984.
49. **Oshima, T.,** Novel polyamines of extremely thermophilic bacteria, in *Biochemistry of Thermophily*, Friedman, S.M., Ed., Academic Press, New York, 1978, 211.
50. **Hamana, K. and Matsuzaki, S.,** *sym*-Homospermidine is a major polyamine in a sulfur-oxidizing bacterium *Thiobacillus novellus, Biochem. Int.*, 13, 439, 1986.
51. **Hamana, K., Kamekura, M., Onishi, H., Akazawa, T., and Matsuzaki, S.,** Polyamines in photosynthetic eubacteria and extreme-halophilic archaebacteria, *J. Biochem.*, 97, 1653, 1985.
52. **Hamana, K., Miyagawa, K., and Matsuzaki, S.,** Occurrence of *sym*-homospermidine as the major polyamine in nitrogen-fixing cyanobacteria, *Biochem. Biophys. Res. Commun.*, 112, 606, 1983.
53. **Hamana, K. and Matsuzaki, S.,** Occurrence of homospermidine as a major polyamine in the authentic genus *Flavobacterium, Can. J. Microbiol.*, 36, 228, 1990.
54. **Hamana, K., Suzuki, M., Wakabayashi, T., and Matsuzaki, S.,** Polyamine levels in the gonads, sperm, and salivary gland of cricket, cockroach, fly and midge, *Comp. Biochem. Physiol.*, 92B, 691, 1989.
55. **Matsuzaki, S., Xiao, L.-P., Suzuki, M., Hamana, K., Niitsu, M., and Samejima, K.,** Occurrence of aminopropylhomospermidine and canavalmine in the hamster epididymis, *Biochem. Int.*, 15, 817, 1987.
56. **Kuttan, R. and Radhakrishnan, A.N.,** Studies on the biosynthesis of *sym*-homospermidine in sandal (*Santalum album* L.), *Biochem. J.*, 127, 61, 1972.
57. **Oshima, T.,** The genes and genetic apparatus of extreme thermophiles, in *Thermophiles: General, Molecular, and Applied Microbiology*, Brock, T.D., Ed., John Wiley & Sons, New York, 1986, 137.
58. **Hamana, K. and Matsuzaki, S.,** Distinct difference in the polyamine compositions of Bryophyta and Pteridophyta, *J. Biochem.*, 97, 1595, 1985.
59. **Jänne, J., Pösö, H., and Raina, A.,** Polyamines in rapid growth and cancer, *Biochim. Biophys. Acta*, 473, 241, 1978.
60. **Shirahata, A., Takeda, Y., Kawase, M., and Samejima, K.,** Detection of spermine and thermospermine by thin-layer chromatography, *J. Chromatogr.*, 262, 451, 1983.
61. **Oshima, T.,** Unusual polyamines in an extreme thermophile, *Thermus thermophilus*, in *Advances in Polyamine Research*, Vol. 4, Bachrach, U., Kaye, A., and Chayen, R., Eds., Raven Press, New York, 1983, 479.

62. **Weaver, R.H. and Herbst, E.J.,** Metabolism of diamines and polyamines in microorganisms, *J. Biol. Chem.*, 231, 637, 1958.
63. **Weaver, R.H. and Herbst, E.J.,** The oxidation of polyamines by *Neisseria perflava*, *J. Biol. Chem.*, 231, 647, 1958.
64. **Smith, T.A.,** Purification and properties of the polyamine oxidase of barley plants, *Phytochemistry*, 11, 899, 1972.
65. **Smith, T.A.,** Amine levels in mineral-deficient *Hordeum vulgare* leaves, *Phytochemistry*, 12, 2093, 1973.
66. **Smith, T.A., Croker, S.J., and Loeffler, R.S.T.,** Occurrence in higher plants of 1-(3-aminopropyl)-pyrrolinium and pyrroline: products of polyamine oxidase, *Phytochemistry*, 25, 683, 1986.
67. **Croker, S.J., Loeffler, R.S.T., Smith, T.A., and Session, R.B.,** 1,5-Diazabicyclo[4.3.0]nonane, the oxidation product of spermine, *Tetrahedron Lett.*, 24, 1559, 1983.
68. **Zoltewicz, J.A. and Bloom, L.B.,** The apparent pk_a of the oxidation of spermine or spermidine: a pk_a masked by a ring-chain tautomeric equilibrium, *Bioorganic Chem.*, 18, 85, 1990.
69. **Brandänge, S., Erikkson, L.-H., and Rodriguez, B.,** Ring-chain tautomerism of an *N,N*-acetal formed by enzymatic oxidation of spermine or spermidine, *Acta Chem. Scand.*, B38, 526, 1984.
70. **Bagni, N.,** Spermidine and spermine in seeds, *Giorn. Bot. Ital.*, 102, 67, 1968.
71. **Smith, T.A.,** Putrescine, spermidine and spermine in higher plants, *Phytochemistry*, 9, 1479, 1970.
72. **Morgen, D.M.L.,** Polyamine oxidases, *Biochem. Soc. Trans.*, 13, 322, 1985.
73. **Smith, T.A.,** The di- and polyamine oxidases of higher plants, *Biochem. Soc. Trans.*, 13, 319, 1985.
74. **Smith, T.A.,** Further properties of the polyamine oxidase from oat seedlings, *Phytochemistry*, 16, 1647, 1977.
75. **Smith, T.A.,** Polyamine oxidase from barley and oats, *Phytochemistry*, 15, 633, 1976.
76. **Suzuki, Y. and Yanagisawa, H.,** Purification and properties of maize polyamine oxidase: a flavoprotein, *Plant Cell Physiol.*, 12, 1085, 1980.
77. **Hirasawa, E., Watanabe, H., and Suzuki, Y.,** Polyamine oxidase of millet shoots, *Phytochemistry*, 25, 1739, 1986.
78. **Yanagisawa, H., Soto, A., Hushiai, S., Kamiya, A., and Torii, N.,** Polyamine oxidase from water hyacinth, *Plant Physiol.*, 85, 906, 1987.
79. **Kuehn, G.D., Bagga, S., Rodriguez-Garay, B., and Phillips, G.C.,** Biosynthesis of uncommon polyamines in higher plants and their relation to abiotic stress responses, in *Polyamines and Ethylene: Biosynthesis, Physiology and Interaction*, Flores, H.E., Arteca, R.N., and Shannon, J.C., Eds., American Society of Plant Physiology Press, Rockville, M.D., 1990, 190.
80. **DeRosa, M., DeRosa, S., Gambacorta, A., Carteni-Farina, M., and Zappia, V.,** The biosynthetic pathway of new polyamines in *Caldariella acidophila*, *Biochem. J.*, 176, 1, 1978.
81. **Bachrach, U., Persky, S., and Razin, S.,** Metabolism of amines. 2. The oxidation of natural polyamines by *Mycobacterium smegmatis*, *Biochem. J.*, 76, 306, 1960.
82. **Bachrach, U.,** Spermidine oxidase from *Serratia marcescens*, *J. Biol. Chem.*, 237, 3443, 1962.
83. **Paulin, L., Ruohola, H., Nykänen, I., and Pösö, H.,** The incorporation of 1,3-diaminopropane into thermine by an extreme thermophile: a novel route for the biosynthesis of polyamines, *FEMS Microbiol. Lett.*, 19, 299, 1983.
84. **Hamana, K., Matsuzaki, S., Niitsu, M., Samejima, K., and Nagashima, H.,** Polyamines of unicellular thermoacidophilic red alga *Cyanidium caldarium*, *Phytochemistry*, 29, 377, 1990.
85. **Nakajima, T., Wolfgram, F., and Clark, W.G.,** Identification of 1,4-methylhistamine, 1,3-diaminopropane, and 2,4-diaminobutyric acid in bovine brain, *J. Neurochem.*, 14, 113, 1967.
86. **Johnson, M.W. and Markham, R.,** Nature of the polyamines in plant viruses, *Virology*, 17, 276, 1962.
87. **Mitra, S. and Kaesberg, P.,** Interaction of polyamines with turnip yellow mosaic virus ribonucleic acid, *Biochem. Biophys. Res. Commun.*, 11, 146, 1963.
88. **Beer, S.V. and Kosuge, T.,** Spermidine and spermine — polyamine components of turnip yellow mosaic virus, *Virology*, 40, 930, 1970.
89. **Oshima, T.,** Novel polyamines in *Thermus thermophilus*: isolation, identification, and chemical synthesis, *Meth. Enzymol.*, 94, 401, 1983.
90. **Cacciapuoti, G., Porcelli, M., Carteni-Farina, M., Gambacorta, A., and Zappia, V.,** Purification and characterization of propylamine transferase from *Sulfolobus solfataricus*, an extreme thermophilic archaebacterium, *Eur. J. Biochem.*, 161, 263, 1986.
91. **Tabor, H. and Tabor, C.W.,** Eds., Polyamines, *Meth. Enz.*, 94, 1983.
92. **Kneifel, H., Stetter, K.O., Andreessen, J.R., Wiegel, J., König, H., and Schoberth, S.M.,** Distribution of polyamines in representative species of archaebacteria, *System. Appl. Microbiol.*, 7, 241, 1986.
93. **Bagga, S., Kuehn, G.D., and Phillips, G.C.,** Polyamine oxidase and aminopropyltransferase enzymes responsible for biosynthesis of uncommon polyamines in alfalfa, *Agron. Abstr.*, 1989, 173, 1989.
94. **Hamana, K. and Matsuzaki, S.,** Further study on polyamines in primitive unicellular eukaryotic algae, *J. Biochem.*, 97, 1311, 1985.
95. **Oshima, T.,** A pentaamine is present in an extreme thermophile, *J. Biol. Chem.*, 257, 9913, 1982.

96. **Friedman, S. M. and Oshima, T.,** Polyamines of sulfur-dependent archaebacteria and their role in protein synthesis, *J. Biochem.,* 105, 1030, 1989.
97. **Oshima, T. and Senshu, M.,** Unusual long polyamines in a thermophile, in *Polyamines: Basic and Clinical Aspects,* Imahori, K., Ed., VNU Science Press, New York, 1985, 113.
98. **Oshima, T. and Kawahata, S.,** Homocaldopentamine, a new naturally occurring pentaamine, *J. Biochem.,* 93, 1455, 1983.
99. **Yamanoha, B. and Cohen, S.S.,** *S*-adenosylmethionine decarboxylase and spermidine synthase from Chinese cabbage, *Plant Physiol.,* 78, 784, 1985.
100. **Schuber, G.,** Influence of polyamines on membrane functions, *Biochem. J.,* 260, 1, 1989.
101. **Meers, P., Hong, K., Bentz, J., and Papahadjopoulos, D.,** Spermine as a modulator of membrane fusion: interactions with acidic phospholipids, *Biochemistry,* 25, 3109, 1986.
102. **Basu, H.S. and Marton, L.J.,** The interaction of spermine and pentaamines with DNA, *Biochem. J.,* 244, 243, 1987.

Chapter 9

MOLECULAR ANALYSES OF POLYAMINE SYNTHESIS IN HIGHER PLANTS

Russell L. Malmberg and Erin Bell

TABLE OF CONTENTS

I. INTRODUCTION

Polyamines and the polyamine-conjugates in plants have been correlated with a wide array of phenomena in higher plants[1] (also, see other chapters in this volume), yet it is difficult to precisely pin down what their specific role is in any physiological process. In part this is due to a common difficulty with studies involving any putative growth regulator. How is one to prove that a given correlation of a growth response with the dramatic change in concentration of a compound is meaningful or part of the primary event, as opposed to coincidental, or a secondary consequence. How does one prove causation? One of the approaches being taken by several laboratories is an analysis of polyamine synthesis at the molecular level. At a minimum, this should allow a deeper analysis of the many phenomena associated with polyamines. The use of transgenic plants may eventually allow some of the hypotheses about polyamine function to be tested. While it is not clear that these molecular analyses will cut through the causation dilemma, they should provide a first step in that direction by permitting a more mechanistic approach.

In this article, we review two recent papers that describe some use of molecular genetic or transgenic plant methods to gain insights into the regulation or functions of polyamines in plants.

II. OVER EXPRESSING A YEAST ORNITHINE DECARBOXYLASE GENE IN TRANSGENIC ROOTS OF *NICOTIANA RUSTICA* CAN LEAD TO ENHANCED NICOTINE ACCUMULATION

The yeast gene encoding ornithine decarboxylase (ORNdc) has been cloned and well characterized.[2] Hammill et al.[3] have introduced this clone into *Nicotiana rustica* as a means of perturbing alkaloid synthesis. Since putrescine is a precursor of nicotine in tobacco species, introducing a heterologous ORNdc that is uncoupled from normal plant regulatory controls might significantly alter nicotine production, providing new insights into this pathway.

The yeast ORNdc gene contains no introns, so Hammill et al.[3] linked it directly, minus yeast promoter and terminator, to a CaMV 35S promoter and polyadenylation sequence. Some of the vectors made use of a doubled 35S enhancer sequence in the promoter. They then transformed roots of *Nicotiana rustica* with *Agrobacterium rhizogenes* strains containing these vectors. A number of independent transformed root cultures were obtained, and the presence of the introduced DNA was confirmed by southern blotting. Many of the lines had multiple inserts.

Some of the transformed lines had normal levels of ORNdc activity, but some showed higher enzyme activities. ORNdc activity varied with the stage of the growth cycle. In lines where the ORNdc activity was higher than the control, it was roughly three times higher at the stage of maximum enzyme activity during the middle of the growth cycle. At the end of the growth cycle, when the control lines had little ORNdc, the activity in the transformed lines was as much as 20 times higher. Thus some lines not only showed evidence of higher enzyme activity as a result of the yeast gene introduction, but also showed changes in regulation during the growth cycle. One root line with an elevated and constitutive ORNdc activity, was examined in more detail. The levels of other enzymes examined, including arginine decarboxylase (ARGdc), were nearly identical to the control at all stages of the growth cycle. Northern blots revealed an RNA of the correct size for the yeast gene, and showed no evidence of cross-hybridization of the yeast gene to the *Nicotiana rustica* RNA.

Amine levels in several lines were measured both at mid and late growth phases. Two lines had roughly twofold increases in the levels of putrescine. There were no significant differences in the levels of spermidine and spermine, suggesting that the concentrations of these compounds are strongly regulated at specific levels. Nicotine production in several lines with the introduced yeast gene was elevated about twofold over control levels. There was a difference between constructs with the duplicated 35S enhancer vs. the normal single enhancer; the increased nicotine was only found in those with the doubling. Comparing ORNdc enzyme activity with nicotine production across the entire set of control and transformed lines revealed a weak but statistically significant positive correlation ($r = 0.66$, $p < 0.05$).

Varying the levels of ORNdc is thus one method that can result in a corresponding change in nicotine levels. ORNdc levels are clearly not the only determining factor, since the correlation was not perfect. Arginine decarboxylase (ARGdc) may also be contributing to the putrescine precursor for nicotine, and this enzyme was not deregulated in these studies.

III. ANALYSIS OF A cDNA ENCODING ARGININE DECARBOXYLASE FROM OAT REVEALS SIMILARITY TO THE *ESCHERICHIA COLI* ARGININE DECARBOXYLASE AND EVIDENCE OF PROTEIN PROCESSING

One of the problems posed by polyamine synthesis in higher plants is the apparently simultaneous existence of two pathways leading to putrescine synthesis. Each pathway may have specific regulated roles in growth and development.[1] Recently, we reported the molecular cloning and analysis of the first enzyme of one pathway, arginine decarboxylase (ARGdc) from oats (*Avena sativa*).[4] We review those results here. Our initial strategy was to purify the active enzyme, then obtain an amino acid sequence, and construct a corresponding oligonucleotide probe for the cDNA.

Young oat leaves were chosen as the source of protein because of the high levels of ARGdc in this tissue. We partially purified ARGdc activity by preparative isoelectric focusing. We then used radiolabelled difluoromethylarginine[5] (DFMA) as a tag for the enzyme, since this inhibitor covalently binds to the active site of the enzyme. This permitted us to locate the ARGdc-DFMA complex following SDS-PAGE. One gel slice contained about twice the radioactivity as the other slices, migrating as a polypeptide with an apparent molecular weight of 24 kDa. Sequence analysis indicated that more than one polypeptide was present in the sample, so further purification on a C8 HPLC column was performed, again using radiolabelled DFMA as a marker to identify the fraction of interest. We were then able to obtain an amino acid sequence of this DFMA-binding polypeptide, construct a degenerate oligonucleotide, and identify cDNA clones from a λgt10 library of oat leaf mRNA.

Some of the cDNA clones were then sequenced. One clone, pADC2, contains 2124 base pairs (bp), close to the size of the mRNA as measured on northern blots of oat leaf polyA$^+$ mRNA. pADC2 contains an open reading frame of 1820 bp, encoding a protein of 66 kDa. This size was unexpectedly large, given that the apparent size of the isolated DFMA-binding polypeptide was 24 kDa. We used pADC2 to direct an *in vitro* transcription-translation reaction, and determined the product size by SDS-PAGE as 62 kDa, closer to the size predicted by the open reading frame than to the size isolated from oat extracts. Analysis of the derived amino acid sequence indicated extensive regions of homology to the *E. coli* ARGdc sequence of Moore and Boyle,[6] as shown in Figure 1. The regions of homology are scattered throughout the length of the sequence, and the blocks of homology are spaced similar distances apart in both the oat and *E. coli* enzymes.

The open reading frame of pADC2 contains the amino acid sequence determined from analysis of isolated ARGdc, but it is internal, in the carboxyl third of the putative protein. The predicted molecular weight of the protein beginning with the internal amino end that we identified by sequencing is approximately 21 kDa, which is close to the size of 24 kDa predicted from migration of the ARGdc-DFMA complex in SDS-PAGE (for clarity, we will refer to this protein as 24 kDa). This, plus the fact that there is no methionine at the internal amino end, suggests that the ARGdc-DFMA complex we isolated contained a fragment of the full length protein. Protein processing or proteolysis must have occurred prior to the final stages of purification. Since the 24 kDa band was the only one identified when gels were assayed for

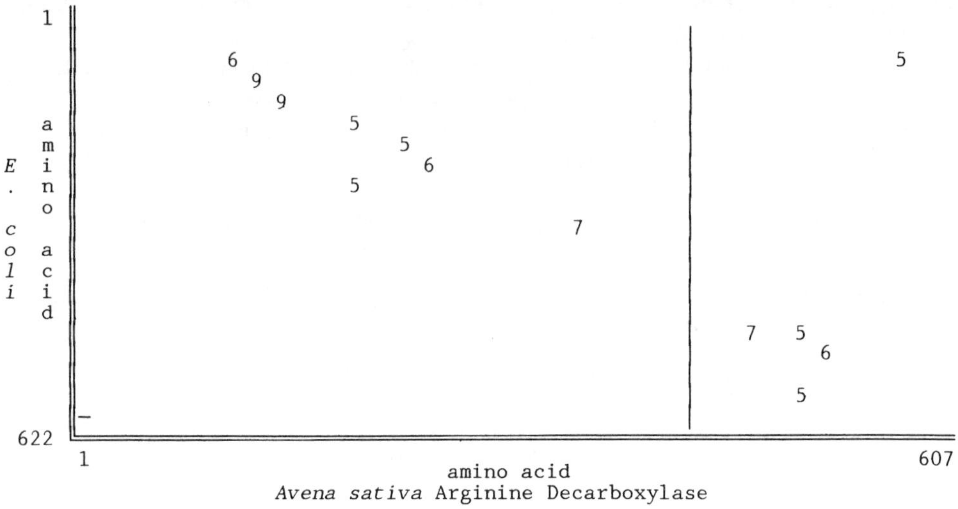


```
(104,93)
gsvYqgVfPv KVNQH kdVvqdmvHf
   |  | |  |||||   |     |
ngdYflVyPi KVNQH rrViesliHs

(125,112)
mvhfgydhsy GLEAGSK pELliamscl
           |||||||  ||
eslihsgepl GLEAGSK aELmavlaha

(151,136)
ltkakpgayl VCNGYKD saYvaLALaa
           |||||||   |  |||
ahagmtrsvi VCNGYKD reYirLALig

(249,234)
kalnkLhwLk LLHFH vGSmipttdi
     |  |  |||||  ||
reagrLdsLq LLHFH lGSqmanird

(293,276)
gvetmttlDc GGGLGVDY dGTRsgssdm
        |  ||||||||  |||
lgvniqcfDv GGGLGVDY eGTRqsdcsv

(343,324)
yhGvPHPvlc TESGRA masyHsmiil
  | |||    ||||||     |
enGlPHPtvi TESGRA vtahHtvlvs

(474,492)
PthkAtLvDv TCDSDG kvDkfIrdte
|   | | |  ||||||   |  |
PerrAvLlDi TCDSDG aiDhyIdgdg

(511,531)
ggyyvavllt GAYQE aLsNkHNLFG
           |||||  | | |||||
nppmlgffmv GAYQE iLgNmHNLFG

(521,541)
GAYQEaLsNk HNLFG gpslVrVvgt
||||| | |  |||||     | |
GAYQEiLgNm HNLFG dteaVdVfvf

(574,75)
kqdissviee RARE nkvwemvekL
           ||||          |
rlrsinaafk RARE sygyngdyfL
```

FIGURE 1. Comparison of the amino acid sequences of the *Avena sativa* and *E. coli* arginine decarboxylases. The Intelligenetics program Ddmatrix was used to generate this comparison. The higher the number in the chart, the more similar are the two regions. The regions with the ten best scores are indicated in detail below the graph. Numbers in parentheses represent the x,y coordinates of amino acid sequences being compared. A vertical line indicates the position of the first amino acid obtained during the amino acid sequencing of the DFMA-binding fragment.

radioactivity, either this processing event is very thorough or DFMA binds preferentially to the processed protein. DFMA binding is an enzyme-catalyzed process; preferential binding to the processed form of ARGdc might mean that proteolysis is required for maximal enzyme activity. Alternatively, it may be that the full-length protein contains structures which are sensitive to cleavage under the isolation conditions used.

We raised an antibody against a polypeptide derived from the carboxyl terminal sequence, and then used this antibody to detect ARGdc on western blots of oat leaf proteins. The protein extraction of the oat leaves was carried out under conditions designed to minimize the possibility of proteolysis.[7] Two major bands react with the antibody, one at 24 kDa and one at 34 kDa; the 24 kDa western blot band migrates at precisely the same R_f as does the DFMA-binding polypeptide. The presence of these two bands suggests that protein processing of ARGdc is an *in vivo* physiological event and not just an artifact of isolation.

Two features of the arginine decarboxylase protein from oat are noteworthy. First, there is the extensive amino acid sequence similarity to the *E. coli* enzyme, with nine distinct regions scattered throughout the full 66 kDa open reading frame. Do these regions have functional significance? Several regions are within the 24 kDa fragment we identified by DFMA binding, and thus may be associated with the active site of the enzyme. Second, ARGdc protein may be processed *in vivo*, yielding both the 24 kDa DFMA binding polypeptide, and another, unexpected, 34 kDA polypeptide. While pADC2 encodes a 66 kDa protein, and a form near this molecular weight can be detected after *in vitro* transcription-translation, western blot analysis indicates that in oat leaves two much smaller forms of the protein predominate. Since the antibody was raised to a fusion protein containing the carboxyl-end of the 24 kDa form, the 34 kDa and 24 kDa forms of ARGdc presumably share a common carboxyl-end epitope. They may represent alternative forms of processing of oat leaf ARGdc. Since the *E. coli* and oat enzymes have regions of homology throughout the sequence, most of the nascent polypeptide may be required enzyme activity. Perhaps the polypeptide cleavage event activates the enzyme, but the two consequent polypeptides remain associated with each other.

IV. CONCLUSIONS

The molecular analysis of plant polyamine synthesis is in its infancy. Compared to the technical sophistication that exists in the mammalian literature, the plant literature is embarrassingly backward. The two examples just discussed represent a beginning of a molecular analysis of the plant pathway, providing examples of both systematic perturbation of polyamine synthesis and of analysis of a specific enzyme. The relative ease with which transformed plants can be obtained should allow a deeper understanding of the regulation and functions of polyamines in plants.

ACKNOWLEDGMENTS

Research of the authors has been supported by National Science Foundation grant DMB-87-15799, by U.S. Department of Agriculture competitive research grant GAM-89-01056, and by a University of Georgia Biotechnology Fellowship.

REFERENCES

1. **Evans, P.T. and Malmberg, R.L.,** Do polyamines have roles in plant development?, *Annu. Rev. Plant Physiol. Plant Mol. Biol.*, 40, 235, 1989.
2. **Fonzi, W.A. and Sypherd, P.S.,** The gene and primary structure of ornithine decarboxylase from *Saccharomyces cerevisiae*, *J. Biol. Chem.*, 262, 10127, 1987.
3. **Hammill, J.D., Robins, R.J., Parr, A.J., Evans, D.M., Furze, J.M., and Rhodes, M.J.C.,** Over-expressing a yeast ornithine decarboxylase gene in transgenic roots of *Nicotiana rustica* can lead to enhanced nicotine accumulation, *Plant Mol. Biol.*, 15, 27, 1990.
4. **Bell, E. and Malmberg, R.L.,** Analysis of a cDNA encoding arginine decarboxylase from oat reveals similarity to the *Escherichia coli* arginine decarboxylase and evidence of protein processing, *Mol. Gen. Genet.*, 227, 431, 1990.
5. **Bitonti, A.J., Casara, P.J., McCann, P.P., and Bey, P.,** Catalytic irreversible inhibition of bacterial and plant arginine decarboxylase activities by novel substrate and product analogues, *Biochem. J.*, 242, 69, 1987.
6. **Moore, R.C. and Boyle, S.,M.,** Nucleotide sequence and analysis of the *speA* gene encoding biosynthetic arginine decarboxylase in *Escherichia coli*, *J. Bacteriol.*, 172, 4631, 1990.
7. **Cordonnier, M.M., Greppin, H., and Pratt, L.H.,** Monoclonal antibodies with differing affinities to the red-absorbing and far red-absorbing forms of phytochrome, *Biochemistry* 24, 3246, 1985.

Polyamines in Plant Growth and Development

Chapter 10

POLYAMINES IN CELL DIVISION, FRUIT SET AND DEVELOPMENT, AND SEED GERMINATION

Marcos Egea-Cortines and Yosef Mizrahi

TABLE OF CONTENTS

I. INTRODUCTION

The activity of polyamines as growth promoters was first shown in bacteria and yeasts in the pioneering work of Tabor and Tabor in the 1950s.[1] Research on this subject was extended to animals,[2] in which high levels of endogenous polyamines were found, and to plants.[3,4] Later, the possibility that polyamines also play a role in cell division in plants was raised. The effect of external polyamines in inhibiting senescence of oat protoplasts and enhancing incorporation of radioactive precursors [^{3}H]-uridine and [^{3}H]-leucine into RNA and proteins suggested an antisenescence or growth-promoting action for polyamines in plants.[5-7] The role of polyamines in cell division in plants has been studied in a wide range of systems, such as protoplasts cultured *in vitro*,[5,6,8,9] fruit set and development (see below), internodal growth,[10-12] and algae.[13-16]

In plants polyamines were first believed to be synthesized primarily from putrescine (Put) derived from arginine decarboxylase (ADC),[17] unlike the situation in animal cells, where polyamines are synthesized exclusively via ornithine decarboxylase (ODC), whose activity had been found to be related to cell division.[18,19] However, ODC was later found in rapidly dividing cells of plants of the Solanaceae, (e.g., tobacco and tomato),[20] and in the unicellular alga *Euglena gracilis*.[21] Further studies showed the similarity of the enzyme found in several plant species to that of animals,[22-24] indicating that ODC could be a key enzyme for polyamine synthesis during cell division in plant systems. In this respect, plants are similar to bacteria, in which both ADC and ODC participate in polyamine biosynthesis.

The function of polyamines in cell division or morphogenetic processes in plant systems has been studied using four kinds of experiments: (1) seeking for correlations between polyamine levels, activities of their biosynthetic enzymes and cell division or morphogenetic processes; (2) studying the effects of exogenous polyamines on cell division; (3) studying the effect of inhibitors of polyamine biosynthesis, with and without exogenous polyamines, on the morphogenetic process; and (4) using polyamine metabolism-related mutants. This four-pronged approach has been used to study polyamine metabolism in relation to flowering in tobacco[25] but has not been applied to studies of cell division in plants.

We are presently using this approach, focusing on two morphogenetic processes in which cell division is involved: fruit set and development and germination. This work will be reviewed below.

II. POLYAMINES IN CELL DIVISION

A. CELL DIVISION IN PLANT SYSTEMS

In plants, cell division occurs in the developing embryo, developing fruits, meristematic tissues in the adult plants, and in proliferating tissues in culture. These systems may be divided into two categories, according to the characteristics of the process: (1) triggered systems, where cell division is triggered by an external inducer (hormones, fertilization, nutrients, etc.), and (2) continuous systems, in which cell division is a nonperturbed state occurring under optimal conditions (meristematic tissues, algal cultures during the logarithmic phase, and cells cultured *in vitro*).

In some triggered systems (e.g., release of *Helianthus tuberosus* tuber tissue from dormancy with auxins), the mitotic activity seems to be synchronized,[26] facilitating correlations between the metabolism of polyamines and the different events of the cell division cycle (see Chapter 11). In others, an attempt to synchronize the cells might lead to artifactual results due to the synchronizing process.[27,28]

In systems that divide continuously, the cells are not synchronized and all the stages of the cell cycle may be found at any given time. Hence, protein and DNA synthesis occur all the time in the tissue at large. This fact makes it difficult to correlate polyamine metabolism to any specific stage of cell division, and only general conclusions may be drawn upon comparison with nondividing tissues.

B. POLYAMINE LEVELS IN CELL DIVISION

In plants, cell division was found to be correlated with increased polyamine biosynthesis activity, based on a comparison between dividing tissues and metabolically active tissues that are not dividing.[20] Many studies on polyamine metabolism were aimed at finding correlations between this high mitotic activity, increased polyamine biosynthesis, and the expected increased polyamine levels.

Most of the studies performed on polyamine metabolism were limited to the free polyamines (dansylated directly in plant extracts by the method of Smith[29]), the common ones being putrescine (Put), spermidine (Spd), and spermine (Spm). However, polyamines were also found in the form of conjugated and bound molecules.* The levels and distribution of polyamines among the various fractions in a variety of plants are presented in Table 1. The lowest levels were found in fascicles of *Cyperus rotundus,* and the highest were found in the young seeds of tobacco.

In triggered systems, the levels of free polyamines increase before the onset of cell division. There are several systems, such as algae,[21] developing apple fruits [30] and tomato fruits after fruit-set,[31] in which the polyamines seem to accumulate premitotically and then decrease sharply upon onset of mitosis. The fact that their levels are high before mitosis might suggest a role for them in the synthesis of protein,[32] RNA,[33] and DNA[13,14] that occurs during this period, but their direct involvement has not been demonstrated. In continuous systems, such as root tips of corn, the amount of free polyamines is lower in the root meristematic tissue than in the root base.[34] When the results are expressed in nanomoles per gram fresh weight, a decrease of spermidine is evident as the samples are taken further from the root tip. In tomato roots, in contrast, the levels of putrescine, spermidine, and spermine are higher in the meristem than in nondividing portions of the root.[35]

Levels of conjugated and bound polyamines are somewhat higher in plants of the Solanaceae[31,36] than in plants of other families. High levels of conjugated polyamines have been found during flowering in tobacco[37] and in citrus flowers,[38] although in tobacco their direct involvement is rather controversial.[39] Their level is high in tobacco ovaries,[40] tomato ovaries,[31] *Trigonella* leaves, and pepper.[41] In dividing protoplasts of *Vigna aconitifolia*, the amount of free and bound polyamines is 200-fold higher than in nondividing protoplasts (Table 1).[42]

Less is known about the chemistry, metabolism, and function of bound polyamines than about conjugated or free polyamines. In mung beans, the ratio of free polyamines to bound polyamines was high during the growth stages, decreasing with the aging of the tissue.[43] In corn roots, bound polyamines were higher in the root meristem than in the root base,[34] spermine being found only in conjugated forms. The application of labeled spermidine to tobacco[44] and *Petunia*[45] protoplasts yielded protein-bound spermidine, but in oats the labeled Spd was converted to diaminopropane, which was found bound to a protein.[45] There is accumulating evidence for the binding of polyamines to proteins in tobacco,[44] Jerusalem artichoke,[46-48] oats, and *Petunia,*[45] and to the enzyme ribulose bisphosphate carboxylase in alfalfa.[49]

In an experiment performed recently with Jerusalem artichoke, gamma-irradiated cells did not show substantial differences in polyamine levels, as compared with the control treatment, suggesting that there is no connection between cell division and polyamines.[50]

* We refer in the text to "conjugated" polyamines as those which are covalently linked to perchloric acid-soluble moieties, such as hydroxycinnamic acids. "Bound" polyamines are covalently linked to perchloric acid-insoluble molecules, such as proteins. Both classes of polyamines can be liberated by acid hydrolysis.

TABLE 1
Distribution of Polyamines Among the Different Fractions in Various Plants

Species	Tissue	Units	Total Polyamines	Polyamine Fraction % of Total			
				Free	Conjugated	Bound	Ref.
Petrosilium hortense	Radicle	nmol/gfw	281	25.6	N.D.[a]	74.4	118
Helianthus annuus	Radicle	nmol/gfw	615	67	2.4	30.6	118
Solanum tuberosum	Radicle	nmol/gfw	2339	37.6	24.6	7.8	118
Cyperus rotundus	Fascicle	nmol/gfw	15	N.D.	N.D.	100	118
Trigonella foenum-graenum	Leaves	nmol/mg.prot	10.85	26.8	21.4	51.7	41
Vigna aconitifolia	Div. protopl.	nmol/10^6 cells	325.5	45.6	N.D.	54.4	42
Vigna aconitifolia	Nondiv. protopl.	nmol/10^6 cells	1.6	50.46	N.D.	49.54	42
Nicotiana tabacum	9-Day old ovary	nmol/gfw	33716	1.36	97.92	0.7	40
Nicotiana tabacum	34-Day seeds	nmol/gfw	2013	23.24	80.62	9.68	40
Nicotiana tabacum	55-Day seeds	nmol/gfw	404.9	21.21	27.46	50.58	40
Lycoparsicon esculentum	0-Day old ovary	nmol/gfw	8006	12.21	63.7	24.08	86
Lycoparsicon esculentum	5-Day old ovary	nmol/gfw	3312	17.84	38.85	43.29	86
Lycoparsicon esculentum	Breaker fruit	nmol/gfw	233	35.19	2.5	62.23	86

[a] N.D. = not detected.

C. POLYAMINE BIOSYNTHETIC ENZYMES DURING CELL DIVISION

High polyamine biosynthetic enzyme activity characterizes rapidly proliferating tissues.[51] The main polyamine biosynthetic enzymes in plants are ODC, ADC, and *S*-adenosylmethionine decarboxylase (SAMDC). Putrescine is synthesized by either ADC or ODC, and its relative importance varies with different systems. In tomato fruits[20] and tobacco ovaries[40] ODC seems to be the main enzyme, whereas in carrot suspension cultures[52] and rice internodes[10] ADC is the prevalent enzyme.

The activities of ODC, ADC, and SAMDC are modulated by plant growth regulators that induce cell division in plant systems.[53] The application of ethylene or gibberellic acid to rice plants grown in deep water activated SAMDC and ADC triggering cell division.[10] In tubers of *Helianthus tuberosus*, the application of auxins induced growth that is based on cell expansion rather than on cell division, together with increased activity of ADC and SAMDC but not of ODC.[54] The application of gibberellin to dwarf peas enhanced ADC activity[11] and internodal growth. Smith et al.[12] related ADC activity to different endogenous gibberellins and growth rates in different genotypes of pea, the high level of ADC being related to growth due to cell elongation and not cell division. ODC activity was found to be enhanced by auxins and gibberellins in germinating barley seeds[55] and by auxins in tomato ovaries.[56] Auxins enhanced SAMDC in *Helianthus tuberosus* tubers[57] but inhibited the activity in carrot cell cultures in which auxins inhibit differentiation.[58]

The enzyme activity assayed *in vitro* is not always a good indicator of enzyme participation in polyamine synthesis during cell division. There is a difference between the activity found by extracting the enzymes from the tissue and that assayed by incorporation of radioactive precursors into polyamines *in vivo*. Because the activity of the extractable ADC in cell lines of maize was lower than that found *in vivo,* Hiatt proposed that ADC was inactivated by the extraction method.[59] The application of radioactive irreversible inhibitors is a valid system to assay the amount of enzyme-protein in the tissue.[60] Thus, Schwartz et al. localized ODC in the meristematic zone of *Zea mays* roots by the application of the radioactive ODC inhibitor DL-α-difluoromethylornithine (DFMO).[60]

TABLE 2
Effect of Polyamine Biosynthesis Inhibitors (20 m*M*) With and Without Putrescine (0.17 m*M*) on Cell Number in Heterotrophically Grown *Chlorella emersonii* Cultures[65]

Treatment	Cell Number 10 days After Inoculation. x 10^6 ml^{-1}	% of Control
Control	16.4 ± 0.28	100
Put	19.5 ± 0.35	119
DFMO	10.5 ± 0.35	64.5
DFMO + Put	15.8 ± 0.17	95
DFMA	14.2 ± 0.17	86.3
DFMA + Put	17.3 ± 0.53	107

D. EFFECT OF POLYAMINE BIOSYNTHESIS INHIBITORS ON CELL DIVISION

Several inhibitors of polyamine metabolism have been used in the study of the metabolism and biochemistry of polyamines in plants. Polyamine biosynthesis inhibitors interfere with several morphogenetic processes as well as with the division of various tissues.

In triggered systems, polyamine biosynthesis inhibitors had a strong effect on growth resulting from cell division. DFMO inhibited cell division-based growth of pollinated fruits of tobacco[61] and tomato,[62] and in germinating seedlings of *Acer saccharum.*[63]

In continuous systems, such as cultures of the red microalga *Porphyridium* sp.[64] grown under continuous light and the green alga *Chlorella emersonii*[65] grown heterotrophically (dark conditions with 1% glucose), the presence of DFMO in the medium reduced the growth rate of the algae. The application of DL-α-difluoromethylarginine (DFMA) had no effect on the growth rate of *Porphyridium* sp.[64] and a very slight effect on *Chlorella emersonii*[65] (Table 2). The addition of putrescine together with the inhibitor counteracted the effect of both DFMO and DFMA.

It is interesting, however, that in those systems in which growth is triggered or maintained by an external source of auxins, polyamine biosynthesis inhibitors do not affect cell division. Thus in explants of Jerusalem artichoke tubers grown in the presence of 0.2 ppm of 2,4-D and 5.0 ppm of BAP, the SAMDC inhibitor methylglyoxal *bis*-(guanylhydrazone) (MGBG), at concentration of 1 mM, did not affect cell division in the culture but rather the formation of tracheary elements.[66] On the other hand, in explants supplied with no auxins but with 1 m*M* Spd, cell number increased by 80%. Likewise, DFMO did not inhibit the growth of tomato fruits when set with NAA at concentration of 4000 ppm[31] (the recommended concentration for agricultural purposes). In this case, ADC activity increased by about 30% in the presence of DFMO.

In continuous systems, such as tobacco tissues cultured *in vitro* on a basal medium containing auxins, the application of DFMO did not inhibit cell division,[67] but even promoted a substantial increase in fresh and dry weight, whereas the application of DFMA had the opposite effect. The growth of carrot cell cultures was not inhibited by either DFMA or DFMO. These inhibitors had a strong effect on both ODC and ADC, respectively, but no effect on spermidine and spermine levels, whereas the putrescine level was significantly reduced.[68] It is suggested that the growth of the tissue is based on the spermidine and spermine levels that were not depleted by the inhibitors. It is worth mentioning that the carrot cells also grow in a medium supplemented with auxins. In seedling roots of *Phaseolus vulgaris*, 10^{-4} *M* canavanine inhibits growth by about 40%, this inhibition being totally overcome by the application of 10^{-5} *M* putrescine.[69]

It seems, therefore, that in systems triggered or maintained by external plant growth regulators, cell division is not inhibited by polyamine biosynthesis inhibitors. Furthermore, polyamines are not depleted by polyamine biosynthesis inhibitors in different plants such as tobacco explants *in vitro*,[67] and tomato fruits,[31] despite the inhibiting effect of these compounds on cell division.

E. FUNCTION OF POLYAMINES DURING CELL DIVISION IN PLANT SYSTEMS

Polyamines were found to enhance a number of cellular processes in plants, such as protein synthesis,[32] RNA synthesis,[33] and DNA methylation.[70] Other processes, such as DNA synthesis, seem to be polyamine dependent, although comprehensive experiments to support this hypothesis are necessary. The identification of spermidine bound to the eukaryotic initiation factor 4D in Chinese hamster ovary cells[71] suggests a direct role of bound spermidine in protein synthesis in eukaryotes. All these facts suggest that polyamines participate in general processes that enable a plant cell to grow normally and divide.

The finding that RUBISCO is a natural substrate for transglutaminase-mediated conjugation with polyamines in plants[49] suggests that polyamines might play a role in regulating not only protein and RNA synthesis but also photosynthesis. Growth resulting from cell expansion is not sensitive to polyamine biosynthesis inhibitors in several systems, indicating that the role of polyamines in cell division might be more specific than that of simple cofactors or activators of general cell processes. Further research in this direction is needed.

III. POLYAMINES IN FRUIT SET AND DEVELOPMENT

A. THE FRUIT SET PROCESS

The development of fruits is usually triggered by pollination. Fruit set can also be achieved by the application of plant growth regulators such as auxins (strawberry),[72] gibberellins (tomato),[73] and ethylene (pineapple).[74] The setting treatment induces rapid cell division, which usually lasts for several days up to two months, depending on the species and the variety. In avocado fruits, however, the cells start dividing at a slow rate after fruit set and continue dividing until the fruit is removed from the tree.[75]

The final cell number in the fruit depends on both the number of cells found in the ovary at anthesis and the mitotic process that occurs after anthesis. The relative importance of each period depends on the species and the variety.[75] Polyamines seem to play a regulatory role in morphogenetic processes preceding fruit set; that is, in the formation of the flowers.[76-79] In most fruits, the final cell number is reached shortly after fruit set (e.g., in the Rutgers variety of tomato 10 to 14 days after anthesis).[80] In mandarins, the mitotic process lasts about a month, with half of the period occurring before anthesis.[81] The mitotic process after anthesis is influenced by climatic factors, the setting treatment (pollination or treatment with plant growth regulators),[82,83] and fruit positioning in the cluster.[84,85]

Heimer and co-workers were the first to demonstrate that ODC is active in rapidly dividing plant cells. They later suggested the involvement of polyamines in the cell division phase of fruit development.[20] Thereafter, the role of polyamines in fruit development was studied from different points of view, e.g., cell division, fruit set, and ripening, as is discussed below.

B. POLYAMINE LEVELS DURING FRUIT SET AND DEVELOPMENT

Polyamine levels are high during the onset of fruit development, as compared with other stages of fruit development, or even as compared with other actively dividing plant structures. For instance, total polyamine levels found in tomato fruits at fruit set are close to 10^4 nmol/g fresh weight (free, conjugated, and bound).[86] High polyamine levels at full anthesis have been reported in apples,[30] *Phaseolus vulgaris*,[87] tobacco,[40] and mandarins,[88] whereas in avocado[89] their levels are somewhat low at anthesis and increase after full bloom. However, avocado is a special plant, regarding its pattern of fruit set, in that the percentage of flowers that set among the thousands of flowers in each cluster is very small.

After the initial stages of development, the polyamine levels decrease. A recent study performed with peas showed that the spermine level decreases when the fruits are set with gibberellin, whereas putrescine and spermidine levels increase.[90] The addition of spermidine and especially of spermine accelerates the senescence of the ovaries, suggesting that high levels of free polyamines can induce senescence in unpollinated ovaries. The development of pea fruits is due to cell expansion with only little cell division. Cell division also occurs in unpollinated peas,[91] indicating that it is not responsible for growth in pea fruits after set. During the development of the fruits after the cell division period, polyamine levels were low in all the fruits studied. In tomato fruits, no change was found in free, conjugated, and bound polyamines after cell division was terminated up to ripening.[86]

The onset of ripening of fruits starts usually when the fruits achieve their maximal size. This stage is known as 100% development. The time from anthesis until fruits are fully developed depends on the species and the variety, on climatic factors, on the position of the fruit in the cluster, etc. The burst of ethylene synthesis, which is considered to be both the trigger and the modulator of ripening,[92] sets in place a sequence of events which culminate in the production of a ripe fruit. The ripening process involves the formation of specific compounds, including proteins, pigments, flavor components, etc., and the degradation of others like chlorophyll, cell walls, and some alkaloids. It can be considered as a major morphogenetic event that ends with the senescence of the organ.[93]

The inhibition of ethylene formation by exogenous polyamines in midclimacteric apples[94] and in oat leaves[95] suggested that polyamines act as inhibitors of the ripening process. In the pericarp of the nonripening tomato mutant *Alcobaca*,[96] increased levels of free putrescine were found in the late stages of ripening, whereas spermidine levels did not differ significantly between the mutant and the normal tomato variety Rutgers. Spermine was not found in either of the genotypes. In three normal tomato varieties with different shelf-life lengths and ripening speeds, levels of polyamines were higher in the slow-ripening variety than in the fast-ripening variety, during the late stages of ripening (light-red to red stage).[97] The polyamine levels during ripening in normal tomato decrease slowly, or remain at low levels throughout the process.[86]

Ethylene inhibits polyamine biosynthetic enzymes in various plant tissues,[98,99] including ripening tomato fruits.[100] Thus, the increased polyamine levels in nonripening tomato mutants or slow-ripening varieties could be the result of the lack of ethylene, whereas in normal varieties ethylene would inhibit polyamine synthesis, leading to low polyamine levels in the final stages of ripening. Indeed, in the *nor* nonripening tomato mutant, which was found to be deficient in ethylene production, the activities of ADC and ODC remained remarkably high during ripening.[100] In contrast, ODC and ADC activities decreased during ripening in normal tomatoes having normal climacteric production of ethylene.[100] Inoculation of the fruits with the fungus *Rhizopus stolonifer* inhibited the activity of ODC and ADC in the fruits of normal and *rin* and *nor* nonripening mutant varieties, indicating that the ethylene formed by the fungus inhibited polyamine biosynthesis in tomato pericarp. In fact, the application of ethylene alone or together with the fungus had the same quantitative effect on the enzymes as the fungus alone. The treatment of *nor* tomato fruits with ethylene does not trigger ripening or endogenous ethylene formation, indicating that the inhibition of the enzymes is a direct effect of ethylene. In agreement with these results, in mandarin, a nonclimacteric fruit, the polyamine levels are comparatively high during the final stages of development and ripening.[88] Since ethylene production in nonclimacteric fruits is very low, polyamine synthesis is expected to be uninhibited during the ripening process.

The high polyamine content found in the ovaries of plants with a high set percentage, e.g., in tobacco,[40] mandarin,[88] *Phaseolus vulgaris*,[87] or tomato,[31] and the low polyamine content in ovaries of plants with low fruit set capacity, such as avocado,[89] suggests a direct correlation between high polyamine levels and high setting potential. This is further supported by the works in apple, in which low polyamines are related to fruit drop.[30]

C. POLYAMINE BIOSYNTHETIC ENZYMES DURING FRUIT SET AND DEVELOPMENT

The polyamine biosynthetic enzymes, ODC and ADC, show high activities during the early stages of development in several fruits. In tomato[20] and tobacco ovaries[40] the activity of ODC is low during anthesis and increases during the mitotic stages of the ovaries. In auxin-induced tomatoes, the pattern of ODC and ADC activities is identical to that in pollinated fruits, although the activity of ODC is slightly lower.[56] After the cell division period, ODC activity is similar to that found in the ovaries at full anthesis. Concomitantly, ADC activity increases, but to a much lesser extent than ODC.[20] There are other fruits in which ADC shows higher activity than ODC. Mandarins, for example, have an increased activity of ADC and ODC during anthesis and a small peak immediately after pollination. The activity of ODC decreases after fruit set faster than that of ADC, effecting a sharp increase of the ADC/ODC ratio.[88]

In avocado fruits, ADC and ODC activities were found to differ among the varieties and tissues studied. In the Fuerte variety, Winer et al.[101] found high ADC activity in the seed coat during the initial stages of fruit development, while the activity of this enzyme in the mesocarp was low throughout fruit development. In the Simmonds variety, both ADC and ODC were found. However, ODC activity was higher and it increased after fruit set, as in other fruits described before, decreasing as final fruit size was achieved.[89]

The activities of both ODC and ADC are high during the initial stages of fruit development in *Phaseolus vulgaris*.[87] In this case, ADC activity is higher than ODC, the activities of both enzymes decreasing dramatically as fruit size increases. In this species, activities of ADC, ODC and SAMDC are higher before anthesis than after fruit set.[87]

From all the cited above, it seems that the activity of the polyamine biosynthetic enzymes after the cell division period is very low as compared with the initial stages of development of the fruits, and it remains low throughout the rest of fruit development, including ripening, with the exception of mandarins.[88]

D. EFFECT OF EXOGENOUS POLYAMINES ON FRUIT SET

The application of exogenous putrescine before anthesis or at full anthesis caused an increase in the final size of the fruit and improved set in apples.[102,103] In a similar experiment performed with pears, putrescine was found to improve fruit set, to delay senescence of the ovules and to enhance pollen germination and fertilization by two days.[104] Putrescine did not affect the ethylene formation that occurs during the setting process in pears, suggesting that its effect as a senescence inhibitor, in this case, is not mediated by ethylene. In pears, maximum set was found at concentration of 10^{-3} M of putrescine, whereas the same treatment in apples had an optimum set with concentrations two orders of magnitude lower. Putrescine increased nitrogen levels in the flowers but not in the leaves in pears.[104] The fact that the application of nitrogen to apples before anthesis improves fruit set[105] suggests that the effect of exogenous polyamines on fruit set might be nonspecific, and that they might merely serve as an exogenous nitrogen source.

E. EFFECT OF POLYAMINE BIOSYNTHESIS INHIBITORS ON FRUIT DEVELOPMENT

The application of DFMO to tomato fruits during their initial stages of development had a strong inhibitory effect on their growth. This inhibition was overcome by the application of

TABLE 3
Effect of DFMO and Putrescine (1 m*M* Each) on Tomato Fruit Fresh Weight (mg/fruit) When Given Immediately, 1 Day, or 10 Days After Pollination[62]

	Time of Application[b]/Time of Measurement (Days after Pollination)			
Treatment[a]	**0/5**	**0/10**	**1/6**	**10/17**
Control (H_2O)	42.09[z] (100)[cd]	1017.4[z] (100)	130.0[z] (100)	6493[z] (100)
DFMO	12.52[y] (29.7)	178.5[y] (17.5)	29.2[y] (22.5)	6345[z] (97.7)
Put	48.78[z] (115.9)	1245.6[z] (122.4)	116.0[z] (89.2)	7281[z] (112.1)
DFMO + Put	39.94[z] (94.5)	1047.2[z] (102.9)	94.4[z] (72.6)	6197[z] (95.4)

[a] The concentrations of the chemicals refer to the concentration of the solutions applied by the split stem technique.
[b] 0, immediately after pollination; 1, 24 hours after pollination.
[c] The difference between values marked "z" and those marked "y" in the same column is statistically significant.
[d] Figures in parentheses represent percent of control.

putrescine (Table 3).[62] Polyamines apparently play a role in pollen germination,[106] since pollen germination can be inhibited by DFMO.[107] These results indicate that the inhibition of fruit growth could be merely an inhibition of pollen germination and the fertilization process. The fact that DFMO also inhibited growth when it was applied one day after pollination indicates that the effect is based on the inhibition of cell division rather than of pollen germination and fertilization process (Table 3).[62] Treatment with DFMO had no effect on growth when it was applied after the cell division period (ten days after pollination). These results were verified by determining the effect of DFMO on the increase of cell number in the developing fruit.[80] Inhibition of growth by DFMO in tomato fruits is due to the inhibition of cell division, and this can be overcome by the application of putrescine (Table 4). In a similar study performed with tobacco ovary, DFMO inhibited growth, also effecting lower protein and polyamine levels. DFMA also inhibited growth and, paradoxically, ODC activity, suggesting that the inhibitor was converted to DFMO by the high arginase activity present in the tissue.[61] This last possibility was later empirically supported by the study of Slocum et al.[108]

The experiments described above indicate that the major enzyme for polyamine synthesis during fruit set and development in tobacco and tomato is ODC, but ADC activity in other fruits might also be important.[87,88,101] It is, therefore, important to use specific inhibitors in order to identify the enzyme responsible for the synthesis of polyamines. The hypothesis that ODC is the major synthetic enzyme during cell division and ADC during cell expansion[54] is supported by the data available on fruit set and development, but further research is still required to substantiate it.

The application of DFMO to auxin-induced tomato fruits does not inhibit growth as it does in pollinated ovaries.[31] This could be due to the presence of exogenous auxins that seem to decrease the sensitivity of plant tissues to polyamine biosynthesis inhibitors. The involvement of polyamines in embryogenesis has been studied in different plants. Since auxin-induced tomato fruits lack seeds, it is possible that in pollinated fruits DFMO exerts its effect by inhibition of embryogenesis. Thus, fruits induced to develop parthenocarpically would not be sensitive to DFMO. The formation of the seeds is a prerequisite for fruit growth, since they establish, by an auxin-driven flux, the metabolic sink in the newly forming fruit.[109] On the other hand, seed formation continues after the cell division process is completed. During this period, fruit development is not sensitive to DFMO.[62] Thus, the question whether the lack of inhibition of growth by DFMO in auxin-induced tomato fruits is due to the presence of auxins or to the lack of embryogenesis remains open.

TABLE 4
The Effects of DFMO, DFMA, and Putrescine on Fresh Weight and Cell Number in Tomato Fruits Grown *In Vivo* Five Days After Pollination[a80]

Treatment	Fresh weight of tomato fruits (mg)[b]	Number of cells per fruit (x10^6)
Control (H_2O)	418[z]	7.15[z]
1 mM DFMO	126[y] (30)[c]	2.59[y] (36)
1 mM DFMO + 1 mM Put	383[z] (91)	6.55[z] (91)
1 mM Put	440[z] (105)	8.00[z] (111)
1 mM DFMA	382[z] (91)	6.23[z] (87)

[a] The results are the averages of two experiments. Three ovaries per treatment were used in each experiment.
[b] Different letters in the same column represent a significant difference at the 1% level.
[c] The numbers in parentheses represent percent of the control.

IV. POLYAMINES IN SEED GERMINATION

A. THE GERMINATION PROCESS

After a dormancy period, if required, the seed germinates, provided appropriate environmental conditions are present. Germination starts after water has been imbibed by the seed. During the imbibition period, the seed increases in size due to the increase in the water content without increasing its net dry matter. Later the seed becomes metabolically active, digesting the storage material to be used for the morphogenetic events until the seedling develops and becomes self-sufficient in terms of nutrient uptake and photosynthesis. Early in germination, new proteins are synthesized. These proteins are encoded by mRNA that had been accumulated during seed formation. The protein synthesis during the early stages of germination is not required for emergence of the radicle,[110] which seems to be a purely biophysical event, involving only cell expansion. The emergence of the radicle is considered to be the end of the germination process. There is a lag period between the emergence of the radicle and the onset of cell division. The length of this period depends upon the species and can be as long as 50 days from the beginning of water uptake by the seeds.[111]

B. POLYAMINE LEVELS DURING GERMINATION

The fact that polyamines were found to activate protein synthesis in ribosomal preparations from *Helianthus tuberosus*[32] suggest that polyamines act as activators of the protein synthesis occurring early in germination. Working on *Phaseolus vulgaris*, Bagni studied the correlation between the level of polyamines and the level of protein and RNA during germination.[112] Spermidine and spermine were found to increase during germination in the embryo, together with RNA and proteins, whereas both polyamines and macromolecules decreased their levels in the cotyledons. A correlation between free polyamine levels RNA and protein content during seed germination was found in *Phaselous mungo, Pisum sativum, Tragopogon porrifollium, Triticum aestivum*, and *Zea mays*.[113]

The polyamine content in seeds before germination depends upon the species and variety. In maize seeds, low germination potential has been correlated with low protein and free polyamine contents.[114,115] In rice seeds, the situation seems to be the opposite, with the content of free polyamines being one order of magnitude higher in low-germinating seeds compared to those exhibiting a high germination potential.[116] Application of plant growth regulators influenced free polyamine levels. Kinetin increased ADC activity and polyamine levels in rice seeds with high and low germinating vigor, whereas abscisic acid had the opposite effect. Anguillesi et al.[117] found good temporal correlation between the onset of mitosis in the radicle meristem and polyamine levels in several plant seedlings tested. It is important to note that unless studies on cell division in seedlings are done after radicle emergence, the onset of cell division might not be found.[63] After a period of stratification in maple seedlings, there was a sharp increase in polyamine levels that was correlated with the onset of mitosis.[63] In a recent study examining large numbers of plants, no relation was found between polyamine levels and mitotic activity.[118] There is still a possibility that polyamines might be metabolized to other compounds that might influence cell division.[119] This possibility should be investigated further.

C. POLYAMINE BIOSYNTHETIC AND CATABOLIC ENZYMES DURING GERMINATION

Tissues of germinating plant seedlings are commonly used for biochemical studies of enzymes involved in polyamine metabolism. ADC was studied in seedlings of oat, barley, rice, maize, wheat, and rye,[120] ODC in barley seeds,[55] and diamine oxidase (DAO) and polyamine oxidase (PAO) in a variety of plants.[116-127] There are, however, very few studies in which the activity of any of these enzymes has been related to the germination process.

In barley seeds, the activity of ODC was found to increase soon after termination of water uptake.[55] Gibberellic acid and β-indolylacetic acid (10^{-5} *M* and 10^{-3} *M* respectively) induced a fourfold increase in the ODC activity, whereas 10^{-3} *M* putrescine and spermidine together inhibited the activity of this enzyme. Cadaverine, a natural diamine, enhanced the activity of ODC, whereas 1,8-diaminooctane, an artificial diamine, had no effect. In *Phaseolus mungo* seeds, ADC activity increased during the first 18 h of germination, after water imbibition, and then decreased again to basal levels.[128] The activity was temporally correlated with protein and RNA content.

The catabolism of polyamines has received special attention in germinating seeds. In maize, the activity of PAO increased dramatically during germination.[122] The activity in etiolated seedlings was double that in light-grown seedlings, but the levels of endogenous polyamines were identical in the two light treatments, indicating that the activity of the enzyme was not correlated with polyamine levels. In fact, the enzyme was localized in the cell wall whereas the polyamines were found in the cytoplasm, suggesting that the enzyme and substrate are in different compartments.

PAO responds to plant growth regulators. Application of ABA to rice seeds enhanced PAO activity and inhibited ADC activity in the embryos, whereas kinetin had the opposite effect.[116]

D. EFFECT OF POLYAMINE BIOSYNTHESIS INHIBITORS ON GERMINATION

Felix and Harr[129] studied a wide variety of inhibitors of ADC, ODC, and SAMDC as a means of determining the role of polyamines in germination. They found that the inhibitors neither depleted polyamine levels nor inhibited germination. Free polyamines were actually higher in the presence of DFMO. DFMO, D-Arg, and dicyclohexylamine, applied to germinating seeds of *Acer saccharum*, inhibited cell division but not cell elongation and radicle growth. The latter two processes were inhibited by the application of 10 m*M* MGBG.[63] Since germination is not dependent upon cell division, a correlation should not be expected between seed germination and polyamine levels.

V. CONCLUDING REMARKS

Many of the processes reviewed above and in other chapters of this volume are cell division-dependent morphogenetic events. Therefore, further research is needed to clarify the specific action of polyamines during cell division. The only system in which most aspects of polyamine metabolism (e.g., synthesis, catabolism, and polyamine levels) has been studied is *Helianthus tuberosus*. Thus, more systems should be comprehensively studied in order that the metabolism of polyamines and their possible role in the different physiological processes can be elucidated. The characterization of free, conjugated, and bound polyamine pools should be a major priority. Studies of polyamine metabolites (e.g., oxidation products and conjugates) by means of radioactive precursors might also identify compounds that are susceptible to depletion under conditions in which cell division is inhibited. This information would be instructive.

ACKNOWLEDGMENTS

Special thanks to Ms. Dorot Imber for her valuable remarks as well as for editing the manuscript.

REFERENCES

1. **Tabor, C.W. and Tabor, H.,** Polyamines, *Annu. Rev. Biochem.*, 53, 749, 1984.
2. **Bachrach, U.,** Metabolism and function of spermine and related polyamines, *Annu. Rev. Microbiol.*, 24, 109, 1970.
3. **Bertossi, F., Bagni, N., Moruzzi, G., and Caldarera, C. M.,** Spermine as a new growth-promoting substance for *Helianthus tuberosus* (Jerusalem artichoke) *in vitro*, *Experientia*, 21, 80, 1965.
4. **Bagni, N.,** Aliphatic amines and a growth factor of coconut milk as stimulating cellular proliferation of *Helianthus tuberosus* (Jerusalem artichoke) *in vitro*, *Experientia*, 22, 732, 1966.
5. **Altman, A., Kaur-Sawhney, R., and Galston, A.W.,** Stabilization of oat protoplasts through polyamine-mediated inhibition of senescence, *Plant Physiol.*, 60, 570, 1977.
6. **Galston, A.W., Altman, A., and Kaur-Sawhney, R.,** Polyamines, ribonuclease and the improvement of oat leaf protoplasts, *Plant Sci. Lett.*, 11, 69, 1978.
7. **Altman, A. and Bachrach, U.,** Involvement of polyamines in plant growth and senescence, in, *Advances in Polyamine Research*, Vol. 3, Caldarera, C.M., Zappia, V., and Bachrach, U., Eds., Raven Press, New York, 1981, 365.
8. **Eilers, R.J., Sullivan, J.G., and Skirvin, R.M.,** Analyzing the effects of exogenous polyamines and growth regulators on plating efficiency of sweet potato protoplasts using a central composite test design, *Plant Cell Rept*, 7, 216, 1988.
9. **Kaur-Sawhney, R., Flores, H.E., and Galston, A.W.,** Polyamine-induced DNA synthesis and mitosis in oat leaf protoplasts, *Plant Physiol.*, 65, 368, 1980.
10. **Cohen, E. and Kende, H.,** The effect of submergence, ethylene and gibberellin on polyamines and their biosynthetic enzymes in deepwater-rice internodes, *Planta*, 169, 498, 1986.
11. **Dai, Y., Kaur-Sawhney, R., and Galston, A.W.,** Promotion by gibberellic acid of polyamine biosynthesis in internodes of light-grown dwarf peas, *Plant Physiol.*, 69, 103, 1982.
12. **Smith, M.A., Davies, P.J., and Reid, J.B.,** Role of polyamines in gibberellin-induced internode growth in peas, *Plant Physiol.*, 78, 92, 1985.
13. **Adlakha, R.C., Villanueva, V.R., Calvayrac, R., and Edmunds, L.N.J.,** Arrest of cell division blocks the utilization of polyamines in synchronized cultures of photoautotrophically grown *Euglena*, *Arch. Biochem. Biophys.*, 201, 60, 1980.
14. **Cohen, E., Arad (Malis), S., Heimer, Y.H., and Mizrahi, Y.,** Polyamine biosynthetic enzymes in the cell cycle of *Chlorella*, *Plant Physiol.*, 74, 385, 1984.

15. **Kneifel, H. and Hegewald, E.,** Norspermidine and other polyamines in the genus *Scenedesmus*, *Arch. Hydrobiol.*, 60, 87, 1980.
16. **Maiss, B., Kneifel, H., and Soeder, C.J.,** Amines in algae.7. Concentration changes of polyamines in synchronous *Chlorella emersonii*: kinetic relationship to DNA and RNA content, *Z. Pflanzenphysiol.*, 106, 213, 1982.
17. **Montague, M.J., Koppenbrink, J.W., and Jaworski, E.G.,** Polyamine metabolism in embryogenic cells of *Daucus carota*, *Plant Physiol.*, 62, 430, 1978.
18. **Mamont, P.S., Duchesne, M., Grove, J., and Bey, P.,** Anti-proliferative properties of DL-α-difluoromethyl ornithine in cultured cells. A consequence of the irreversible inhibition of ornithine decarboxylase, *Biochem. Biophys. Res. Commun.*, 81, 58, 1978.
19. **Pegg, A.E. and McCann, P.P.,** Polyamine metabolism and function, *Am. J. Physiol.*, 243, C212, 1982.
20. **Heimer, Y.M., Mizrahi, Y., and Bachrach, U.,** Ornithine decarboxylase activity in rapidly proliferating plant cells, *FEBS Lett.*, 104, 146, 1979.
21. **Aleksijevic, A., Grove, J., and Schuber, F.,** Studies on polyamine biosynthesis in *Euglena gracilis*, *Biochim. Biophys. Acta.*, 565, 199, 1979.
22. **Heimer, Y.M. and Mizrahi, Y.,** Characterization of ornithine decarboxylase of tobacco cells and tomato ovaries, *Biochem. J.*, 201, 373, 1982.
23. **Cohen, E., Arad (Malis), S., Heimer, Y.M., and Mizrahi, Y.,** Polyamine biosynthetic enzymes in *Chlorella*: characterization of ornithine and arginine decarboxylase, *Plant Cell Physiol.*, 24, 103, 1983.
24. **D'Orazi, D. and Bagni, N.,** Ornithine decarboxylase activity in *Helianthus tuberosus*, *Physiol. Plant.*, 71, 177, 1987.
25. **Malmberg, R.L. and McIndoo, J.,** Ultraviolet mutagenesis and genetic analysis of resistance to methylglyoxal-*bis* (guanylhydrazone) in tobacco, *Mol. Gen. Genet.*, 196, 28, 1984.
26. **Serafini-Fracassini, D., Bagni, N., Cionini, P.G., and Bennici, A.,** Polyamines and nucleic acids during the first cell cycle of *Helianthus tuberosus* tissue after the dormancy break, *Planta*, 148, 332, 1980.
27. **Kubitschek, H.E.,** Evidence for the generality of linear growth, *J. Theor. Biol.*, 28, 15, 1970.
28. **Okuda, A. and Cooper, S.,** The continuum model: an experimental and theoretical challenge to the G1 model of cell cycle regulation, *Exp. Cell Res.*, 185, 1, 1989.
29. **Smith, T.A. and Best, G.R.,** Polyamines in barley seedlings, *Phytochem.*, 16, 841, 1977.
30. **Biasi, R., Bagni, N., and Costa, G.,** Endogenous polyamines in apple and their relationship to fruit set and fruit growth, *Physiol. Plant.*, 73, 201, 1988.
31. **Egea-Cortines, M., Cohen, E., Arad, S., and Mizrahi, Y.,** Effect of difluoromethylornithine on the growth of pollinated and naphthalene acetic acid-induced tomato ovaries, in *Polyamines and Ethylene: Biosynthesis, Physiology and Interactions*, Flores, H.E., Arteca, R.N., and Shannon, J.C., Eds., American Soc. Plant Physiologists, Rockville, MD, 1990, 325.
32. **Cocucci, S. and Bagni, N.,** Polyamine-induced activation of protein synthesis in ribosomal preparation from *Helianthus tuberosus* tissue, *Life Sci.*, 7, 113, 1968.
33. **Guylfoyle, T.J. and Hanson, J.B.,** Increased activity of chromatin-bound ribonucleic acid polymerase from soybean hypocotyl with spermidine and high ionic strength, *Plant Physiol.*, 51, 1022, 1973.
34. **Dumortier, F.M., Flores, H.E., Shekhawat, N.S., and Galston, A.W.,** Gradients of polyamines and their biosynthetic enzymes in coleoptiles and roots of corn, *Plant Physiol.*, 72, 915, 1983.
35. **Cohen, E.,** Possible involvement of polyamines and their biosynthetic enzymes in cell division in plants, Ph.D thesis, Ben-Gurion University of Negev, in Hebrew, 1984.
36. **Tiburcio, A.F., Kaur-Sawhney, R., Ingersoll, R.B., and Galston, A.W.,** Correlation between polyamines and pyrrolidine alkaloids in developing tobacco callus, *Plant Physiol.*, 78, 323, 1985.
37. **Martin-Tanguy, J.,** The occurrence and possible function of hydroxycinnamoyl acid amides in plants, *Plant Growth Regul.*, 3, 381, 1985.
38. **Kushad, M.M., Orvos, A.R., and Yelenosky, G.,** Relative changes in polyamines during citrus flower development, *Hort. Sci.*, 25, 946, 1990.
39. **Wyss-Benz, M., Streit, L., and Ebert, E.,** Feruloylputrescine and caffeoylputrescine are not involved in growth and floral bud formation of stem explants from *Nicotiana tabacum* L. var Xanthi nc, *Plant Physiol.*, 92, 924, 1990.
40. **Slocum, R.D. and Galston, A.W.,** Changes in polyamine biosynthesis associated with postfertilization growth and development in tobacco ovary tissues, *Plant Physiol.*, 79, 336, 1985.
41. **Tiburcio, A.F., Masdeu, M.A., Dumortier, F.M., and Galston, A.W,** Polyamine metabolism and osmotic stress. I. Relation to protoplast viability, *Plant Physiol.*, 82, 369, 1986.
42. **Kaur-Sawhney, R., Shekhawat, N.S., and Galston, A.W,** Polyamine levels as related to growth, differentiation and senescence in protoplast-derived cultures of *Vigna aconitifolia* and *Avena sativa*, *Plant Growth Regul.*, 3, 329, 1985.
43. **Goldberg, R. and Perdrizet, E.,** Ratio of free to bound polyamines during maturation in mung bean hypocotyl cells, *Planta.*, 161, 531, 1984.

44. **Apelbaum, A., Canellakis, Z.N., Applewhite, P.B., Kaur-Sawhney, R., and Galston, A.W.,** Binding of spermidine to a unique protein in thin-layer tobacco tissue culture, *Plant Physiol.*, 88, 996, 1988.
45. **Mizrahi, Y., Applewhite, P.B., and Galston, A.W,** Polyamine binding to proteins in oat and petunia protoplasts, *Plant Physiol.*, 91, 738, 1989.
46. **Serafini-Fracassini, D., Del Duca, S., and D'Orazi, D.,** First evidence for polyamine conjugation mediated by an enzymic activity in plants, *Plant Physiol.*, 87, 757, 1988.
47. **Serafini-Fracassini, D., Del Duca, S., D'Orazi, D., and Mossetti, U.,** Conjugation of polyamines by an enzyme activity in dividing cells of *Helianthus tuberosus*, in *Perspectives in Polyamine Research*, Perin, A., Scalabrino, G., Sessa, A., and Ferioli, M.E., Eds., Wichtig Editore, Milano, 1988, 89.
48. **Serafini-Fracassini, D., Del Duca, S., and Torrigiani, P.,** Polyamine conjugation during the cell cycle of *Helianthus tuberosus*: nonenzymatic and transglutaminase-like binding activity, *Plant Physiol. Biochem.*, 27, 659, 1989.
49. **Margosiak, S.A., Dharma, A., Bruce-Carver, M.R., Gonzales, A.P., Louie, D., and Kuehn, G.D.,** Identification of the large subunit of ribulose 1,5-bisphosphate carboxylase/oxygenase as a substrate for transglutaminase in *Medicago sativa* L. (Alfalfa) *Plant Physiol.*, 92, 88, 1990.
50. **Phillips, R., Press, M.C., Bingham, L., and Grimmer, C.,** Polyamines in cultured artichoke explants: effects are primarily on xylogenesis rather than cell division, *J. Exp. Bot.*, 39, 473, 1988.
51. **Galston, A.W.,** Polyamines as modulators of plant development, *BioScience*, 33, 382, 1983.
52. **Feirer, R.P., Mignon, G., and Litvay, J.D.,** Arginine decarboxylase and polyamines required for embryogenesis in the wild carrot, *Science*, 223, 1433, 1984.
53. **Altman, A.,** Polyamines and plant hormones, in *The Physiology of Polyamines.*, Bachrach, U. and Heimer, Y.M., Eds., CRC Press, Inc., Boca Raton, FL, 1989, 121.
54. **Bagni, N., Barbieri, P., and Torrigiani, P.,**Polyamine titer and biosynthetic enzymes during tuber formation of *Helianthus tuberosus*, *J. Plant Growth Regul.*, 2, 177, 1983.
55. **Kyriakidis, D.A.,** Effect of plant growth hormones and polyamines on ornithine decarboxylase activity during the germination of barley seeds, *Physiol. Plant.*, 57, 499, 1983.
56. **Mizrahi, Y. and Heimer, Y.H.,** Increased activity of ornithine decarboxylase in tomato ovaries induced by auxin, *Physiol. Plant.*, 54, 367, 1982.
57. **Torrigiani, P., Serafini-Fracassini, D., and Bagni, N.,** Polyamine biosynthesis and effect of dicyclohexylamine during the cell cycle of *Helianthus tuberosus* tuber, *Plant Physiol.*, 84, 148, 1987.
58. **Fienberg, A.A., Choi, J.H., Lubich, W.P., and Sung, Z.R.,** Developmental regulation of polyamine metabolism in growth and differentiation of carrot culture, *Planta*, 162, 532, 1984.
59. **Hiatt, A.,**Polyamine synthesis in maize cell lines, *Plant Physiol.*, 90, 1378, 1989.
60. **Schwartz, M., Altman, A., Cohen, Y., and Arzee, T.,** Localization of ornithine decarboxylase and changes in polyamine content in root meristems of *Zea mays*, *Physiol. Plant.*, 67, 485, 1986.
61. **Slocum, R.D. and Galston, A.W.,** *In vivo* inhibition of polyamine biosynthesis and growth in tobacco ovary tissues, *Plant Cell Physiol.*, 26, 1519, 1985.
62. **Cohen, E., Arad, S., Heimer, Y.M., and Mizrahi, Y.,** Participation of ornithine decarboxylase in early stages of tomato fruit development, *Plant Physiol.*, 70, 540, 1982.
63. **Walker, M A., Roberts, D.R., Shih, C., and Dumbroff, E.B.,** A requirement for polyamines during the cell division phase of radicle emergence in seeds of *Acer saccharum*, *Plant Cell Physiol.*, 26, 967, 1985.
64. **Dubinsky, O.,** The involvement of polyamines and their biosynthetic enzymes in the cell division of the unicellular red alga *Porphyridium* sp, M.SC. thesis, Ben-Gurion University of The Negev, Israel, in Hebrew, 1985.
65. **Feler, E.,** The involvement of polyamines and their biosynthetic enzymes during heterotrophic growth of *Chlorella emersonii*, Ms.C thesis, Ben-Gurion University of the Negev, Israel, in Hebrew, 1989.
66. **Phillips, R., Press, M.C., and Eason, A.,** Polyamines in relation to cell division and xylogenesis in cultured explants of *Helianthus tuberosus*, *J. Exp. Bot.*, 38, 164, 1987.
67. **Burtin, D., Martin-Tanguy, J., Paynot, M., and Rossin, N.,** Effects of the suicide inhibitors of arginine and ornithine decarboxylase activities on organogenesis, growth, free polyamine and hydroxycinnamoyl putrescine levels in leaf explants of *Nicotiana xanthi* n.c. cultivated *in vitro* in a medium producing callus formation, *Plant Physiol.*, 89, 104, 1989.
68. **Fallon, K.M. and Phillips, R.,** Polyamines in relation to growth in carrot cell cultures, *Plant Physiol.*, 88, 224, 1988.
69. **Palavan-Unsal, N.,** Polyamine metabolism in the roots of *Phaseolus vulgaris*. Interaction of the inhibitors of polyamine biosynthesis with putrescine in growth and polyamine biosynthesis, *Plant Cell Physiol.*, 28, 565, 1987.
70. **Cano, C., Herrera-Estrella, L., and Ruiz-Herrera, J.,** DNA methylation and polyamines in regulation of development of the fungus *Mucor rouxii*, *J. Bacteriol.*, 170, 5946, 1988.

71. **Park, M.H.,** The identification of an eukaryotic initiation factor 4D precursor in spermidine-depleted chinese hamster ovary cells, *J. Biol. Chem.*, 263, 7447, 1988.
72. **Nitsch, J.P.,** Growth and morphogenesis of the strawberry as related to auxin, *Amer. J. Bot.*, 37, 211, 1950.
73. **Mapelli, S., Torti, G., Badino, M., and Soressi, G.P.,** Effect of GA_3 on flowering and fruit-set in a mutant of tomato, *Hort. Sci.*, 14, 736, 1979.
74. **Bartholomew, D.P. and Paull, R.E.,** Pineapple, in, *Handbook of Fruit Set and Development*, Monselise, S. P., Ed.,CRC Press, Boca Raton, FL, 1986, 371.
75. **Coombe, B.G.,** The development of fleshy fruits, *Ann. Rev. Plant Physiol.*, 27, 507, 1976.
76. **Bendeck de Cantu, L. and Kandeler, R.,** Significance of polyamines for flowering in *Spirodela punctata*, *Plant Cell Physiol.*, 30, 455, 1989.
77. **Kaur-Sawhney, R., Tiburcio, A.F., and Galston, A.W.,** Spermidine and flower-bud differentiation in thin layer explants of tobacco, *Planta*, 173, 282, 1988.
78. **Kaur-Sawhney, R., Kandpal, G., McGonigle, B., and Galston, A.W.,** Further experiments on spermidine-mediated floral-bud formation in thin-layer explants of Wisconsin 38 tobacco, *Planta*, 181, 212, 1990.
79. **Malmberg, R.L., McIndoo, J., Hiatt, A.C., and Lowe, B.A.,** Genetics of polyamine synthesis in tobacco: developmental switches in the flower, *Cold Spring Harbor Symp. Quant. Biol.*, L, 475, 1985.
80. **Teitel, D.C., Cohen, E., (Malis) Arad, S., Birnbaum, E., and Mizrahi, Y.,** The possible involvement of polyamines in the development of tomato fruits *in vitro*, *Plant Growth Regul.*, 3, 309, 1985.
81. **Talon, M., Zacarias, L., and Primo-Lillo, E.,** Hormonal changes associated with fruit set and development in mandarins differing in their parthenocarpic ability, *Physiol. Plant.*, 79, 400, 1990.
82. **Bunger-Kibler, S. and Bangerth, F.,** Relationship between cell number, cell size and fruit size of seeded fruits of tomato (*Lycopersicon esculentum* Mill.), and those induced parthenocarpically by the application of plant growth regulators, *Plant Growth Regul.*, 1, 143, 1983.
83. **Mapelli, S., Frova, C., Torti, G., and Soressi, G.P.,** Relationship between set, development and activities of growth regulators in tomato fruits, *Plant Cell Physiol.*, 19, 1281, 1978.
84. **Bangerth, F. and Ho, L.C.,** Fruit position and fruit set sequence in a truss as factors determining final size of tomato fruits, *Ann. Bot.*, 53, 315, 1984.
85. **Bangerth, F.,** Dominance among fruits/sinks and the search for a correlative signal, *Physiol. Plant.*, 76, 608, 1989.
86. **Egea-Cortines, M., Cohen, E., Arad, S., Bagni, N., and Mizrahi, Y.,** unpublished results.
87. **Palavan, N. and Galston, A.W.,** Polyamine biosynthesis and titer during various developmental stages of *Phaseolus vulgaris*, *Physiol. Plant.*, 55, 438, 1982.
88. **Nathan, R., Altman, A., and Monselise, S.P.,** Changes in activity of polyamine biosynthetic enzymes and in polyamine contents in developing fruit tissues of 'Murcott' mandarin, *Sci. Hort.*, 22, 359, 1984.
89. **Kushad, M.M., Yelenosky, G., and Knight, R.,** Interrelationship of polyamine and ethylene biosynthesis during avocado fruit development, *Plant Physiol.*, 87, 463, 1988.
90. **Carbonell, J. and Navarro, J.L.,** Correlation of spermine levels with ovary senescence and with fruit set and development in *Pisum sativum* L., *Planta*, 178, 482, 1989.
91. **Vercher, Y., Molowny, A., and Carbonell, J.,** Gibberellic acid effects on the ultrastructure of endocarp cells of unpollinated ovaries of *Pisum sativum*, *Physiol. Plant.*, 71, 302, 1987.
92. **Yang, S.,** Biosynthesis and action of ethylene, *Hort. Sci.*, 20, 41, 1985.
93. **Brady, C.J.,** Fruit ripening, *Annu. Rev. Plant. Physiol.*, 38, 155, 1987.
94. **Apelbaum, A., Burgoon, A.C., Anderson, J.D., Lieberman, M., Ben-Arie, R., and Matoo, A.K.,** Polyamines inhibit biosynthesis of ethylene in higher plant tissue and fruit protoplasts, *Plant Physiol.*, 68, 453, 1981.
95. **Fuhrer, J., Kaur-Sawhney, R., Shih, L., and Galston, A. W.,** Effect of exogenous 1,3-diaminopropane and spermidine on senescence of oat leaves. II. Inhibition of ethylene and possible mode of action, *Plant Physiol.*, 70, 1597, 1982.
96. **Dibble, A.R.G., Davies, P.J., and Mutschler, M.A.,** Polyamine content of long-keeping *alcobaca* tomato fruit, *Plant Physiol.*, 86, 338, 1988.
97. **Saftner, R.A. and Baldi, B.G.,** Polyamine levels and tomato fruit development: possible interaction with ethylene, *Plant Physiol.*, 92, 547, 1990.
98. **Icekson, I., Bakhanashvili, M., and Apelbaum, A.,** Inhibition by ethylene of polyamine biosynthetic enzymes enhanced lysine decarboxylase activity and cadaverine accumulation in pea seedlings, *Plant Physiol.*, 82, 607, 1986.
99. **Icekson, I., Goldlust, A., and Apelbaum, A.,** Influence of ethylene on *S*-adenosylmethionine decarboxylase activity in etiolated pea seedling., *J. Plant Physiol.*, 119, 335, 1985.
100. **Bakanashvili, M., Barkai-Golan, R., Kopeliovitch, E., and Apelbaum, A.** Polyamine biosynthesis in *Rhizopus*-infected tomato fruits: possible interaction with ethylene, *Physiol. Mol. Plant Pathol.*, 31, 41, 1987.

101. **Winer, L. and Apelbaum, A.,** Involvement of polyamines in the development and ripening of avocado fruits, *J. Plant Physiol.*, 126, 223, 1986.
102. **Costa, N. and Bagni, N.,** Effect of polyamines on fruit-set of apple (c.v. Rubyspur), *Acta Hort.*, 120, 239, 1981.
103. **Costa, G. and Bagni, N.,** Effects of polyamines on fruit-set of apple, *Hort. Sci.*, 18, 59, 1983.
104. **Crisosto, C.H., Lombard, P.B., Sugar, D., and Polito, V.S.,** Putrescine influences ovule senescence, fertilization time, and fruit set in "Comice" pears, *J. Amer. Soc. Hort. Sci.*, 113, 708, 1988.
105. **Dennis, F.G.,** Apple, in, *Handbook of Fruit Set and Development*, Monselise, S.P., Ed., CRC Press, Boca Raton, FL, 1986, 1.
106. **Bagni, N., Adamo, P., Serafini-Fracassini, D,. and Villanueva, V.R.,** RNA, Proteins and polyamines during tube growth in germinating apple pollen, *Plant Physiol.*, 68, 727, 1981.
107. **Rajam, M.V.,** Restriction of pollen germination and tube growth in lily pollen by inhibitors of polyamine metabolism, *Plant Sci.*, 59, 53, 1989.
108. **Slocum, R.D., Bitonti, A.J., McCann, P.P., and Feirer, R.P.,** DL-a-Difluoromethyl [3,4-^{3}H]arginine metabolism in tobacco and mammalian cells. Inhibition of ornithine decarboxylase activity following arginase-mediated hydrolysis of DL-a-difluoromethylargine to DL-a-difluoromethylornithine, *Biochem. J.*, 255, 197, 1988.
109. **Varga, A. and Bruinsma, J.,** Role of seeds and auxins in tomato fruit growth, *Z. Pflanzenphysiol. Bd.*, 80, 95, 1976.
110. **Galli, M.G., Levi, M., and Sparvoli, E.,** Lack of correlation between overall protein synthesis and the onset of cell elongation in germinating embryos of *Haplopappus gracilis*, *Physiol. Plant.*, 51, 321, 1981.
111. **Shih, C.Y., Dumbroff, E.B., and Peterson, C.A.,** Developmental studies of the stratification-germination process in sugar maple syrup, *Can. J. Bot.*, 63, 903, 1985.
112. **Bagni, N.,** Metabolic changes of polyamines during the germination of *Phaseolus vulgaris*, *New Phytol.*, 69, 159, 1970.
113. **Villanueva, V.R., Adlakha, R.C., and Cantera-Soler, A M.,** Changes in polyamine concentration during seed germination, *Phytochemistry*, 17, 1245, 1978.
114. **Anguillesi, M.C., Grilli, I., and Floris, C.,** Polyamines and protein metabolism in maize inbreds differing in seed protein content, *J. Exp. Bot.*, 33, 10, 1982.
115. **Sepulveda, G. and Sanchez de Jimenez, E.,** Polyamine distribution among maize embryogenic tissues and its relation to seed germination, *Biochem. Biophys. Res. Commun.*, 153, 881, 1988.
116. **Mukhopadhyay, A., Choudhuri, M.M., Sen, K., and Ghosh, B.,** Changes in polyamines and related enzymes with loss of viability in rice seeds, *Phytochemistry,* 22, 1547, 1983.
117. **Anguillesi, M.C., Grilli, I., and Floris, C.,** Rate of synthesis of spermine and spermidine in germinating seeds of *Glycine*, *Helianthus* and *Triticum Planta*, 148, 24, 1980.
118. **Felix, H. and Harr, J.,** Association of polyamines to different parts of various plant species, *Physiol. Plant.*, 71, 245, 1987.
119. **Christ, M., Harr, J., and Felix, H.,** Transport of polyamines in sugar beet seedlings, *Z. Naturforsch.*, 44c, 59, 1989.
120. **Smith, T.A.,** Arginine decarboxylase of oat seedlings, *Phytochemistry,* 18, 1447, 1979.
121. **Federico, R., Alisi, C., and Forlani, F.,** Properties of the polyamine oxidase from the cell wall of maize seedlings, *Phytochemistry*, 28, 45, 1989.
122. **Torrigiani, P., Scoccianti, V., and Bagni, N.,** Polyamine oxidase activity and polyamine content in maize during seed germination, *Physiol. Plant.*, 74, 427, 1988.
123. **Angelini, R., Manes, F., and Federico, R.,** Spatial and functional correlation between diamine-oxidase and peroxidase activities and their dependence upon de-etiolation and wounding in chick-pea stems, *Planta*, 182, 89, 1990.
124. **Federico, R. and Angelini, R.,** Occurrence of diamine oxidase in the apoplast of pea epicotyls, *Planta*, 167, 300, 1986.
125. **Federico, R. and Angelini, R.,** Distribution of polyamines and their related catabolic enzyme in etiolated and light-grown Leguminosae seedlings, *Planta*, 173, 317, 1988.
126. **Nadeu, P., Delaney, S., and Chouinard, L.,** Effects of cold hardening on the regulation of polyamine levels in wheat (*Triticum aestivum* L.) and alfalfa (*Medicago sativa* L.), *Plant Physiol.*, 84, 73, 1987.
127. **Torrigiani, P., Serafini-Fracassini, D., and Fara, A.,** Diamine oxidase activity in different physiological stages of *Helianthus tuberosus* tuber, *Plant Physiol.*, 89, 69, 1989.
128. **Chatterjee, S., Choudhuri, M.M., and Ghosh, B.,** Changes in polyamine content during root and nodule growth of *Phaseolus mungo*, *Phytochemistry*, 22, 1553, 1983.
129. **Felix, H. and Harr, J.,** Influence of inhibitors of polyamine biosynthesis on polyamine levels and growth of plants, *Z. Naturforsch.*, 44c, 55, 1989.

Chapter 11

CELL CYCLE-DEPENDENT CHANGES IN PLANT POLYAMINE METABOLISM

Donatella Serafini-Fracassini

TABLE OF CONTENTS

I. HOW TO STUDY THE INVOLVEMENT OF POLYAMINES IN CELL GROWTH

Since the pioneering work of Herbst and Snell in 1948,[1] it has been known that the growth of all living organisms requires polyamines (PA) and that their content is proportional to cellular growth rates. Since PA are normal constituents of all cells, their function in growth processes cannot be studied by supplying them exogenously, which can sometimes cause inhibition, but rather by depleting the cells of PA or by supplying them to PA-depleted cells. PA-depleted cells can be obtained by inducing mutants which are deficient or lacking in their biosynthetic enzymes, by inhibiting the activity of these enzymes, or by utilizing naturally deficient organs.

In higher plants, the occurrence of multiple pathways of putrescine (PU) biosynthesis causes great difficulties in obtaining PA-depleted mutants. Similarly, the use of various inhibitors of PA biosynthesis can be unreliable, especially in short-term experiments, because they are sometimes scarcely or only transitorily active on plant enzymes *in vivo* and a complete depletion of PA cannot be achieved owing to the following facts: (1) there are alternative and possibly compensatory pathways for their biosynthesis,[2] (2) there are normally large amounts of intracellular PA, which have slow turnover rates and may be "stored" in a bound form, undetectable by the usual extraction methods and only "released" when necessary. For these reasons, organs which are temporarily PA-deficient, like dormant tubers of *Helianthus tuberosus*, offer a useful model system for studies on the role of PA in the regulation of cellular growth.

II. A POLYAMINE-DEFICIENT SYSTEM: THE DORMANT TUBER OF *HELIANTHUS TUBEROSUS*

Plants of *H. tuberosus*, propagated by vegetative reproduction, provide a homogeneous tissue: the parenchyma of the tuber. The latter, formed in late summer (end of August), slows down its growth during flowering (September to October), then temporarily resumes growth and finally becomes dormant in early winter (November to December). Its medullary parenchyma cells cease multiplying early on, enlarge and slowly accumulate reserve substances until February,[3,4] they can be considered in G_0 phase or arrested in G_1 phase. The cells are morphologically homogeneous and are characterized by a large vacuole, a thin layer of cytoplasm, adherent to the thin cell wall, and including heavily stained nuclei, with small nucleoli.[5] Their nongreen plastids, unlike the usual form,[5] contain tubular complexes, phytoferritin, and protein crystals (Figure 1). The endoplasmic reticulum and polysomes are scarce; the latter are unable to incorporate amino acids in a cell-free extract if exogenous PA are not supplied.[6] These parenchyma cells contain very small amounts of several growth substances such as indol-3-yl-acetic acid (IAA), cytokinins, and gibberellins, but a considerable amount of inhibitors such as abscisic acid-like substances;[3] PA also are present in amounts insufficient to sustain growth ($< 10\ \mu M$) (Figure 2). All the PA biosynthetic enzyme activities[7] (Figure 3) as well as diamine oxidase[8] (DAO) and transglutaminase-like[9] (TGase-like) activities (Figure 4) are low or practically absent, while arginase[7] (ARGase), on the contrary, is active (Figure 3). Tuber dormancy must be considered as a period of slow metabolism which ends, from a physiological point of view, in early February, when the utilization of reserve substances begins.[3,10] In March, the PA content increases abruptly.[3] The parenchyma cells are programmed to become totally depleted and die, when the translocation of nutrients to growing roots and sprouts is completed.

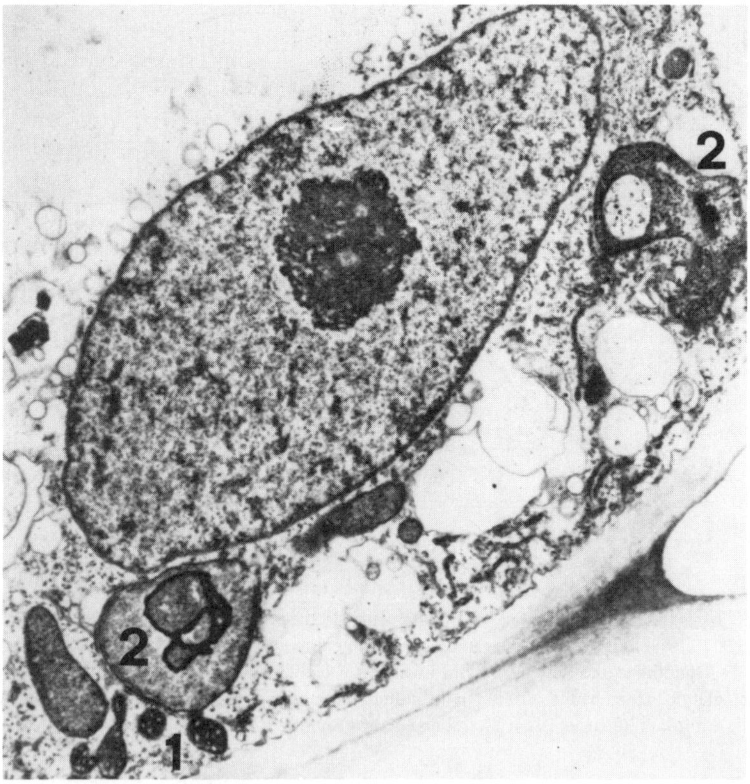

FIGURE 1. Electron micrograph of *Helianthus tuberosus* parenchyma cells in culture. Note the association of mitochondria (1) with the plastids which contain tubular complexes (2). Magnification x 7500. Courtesy of Professor A. Favali.

III. ACTIVATED SLICES OF THE TUBER

Dormancy can be broken by excising slices of tuber parenchyma and by treating them with IAA, 2,4-D, or PA added to the sterile growth medium.[11-13] The cells are particularly responsive if they are explanted during deep dormancy (January to early March). Owing to the low amount of endogenous hormones and PA, they cannot grow if these substances, together with minerals and sugars, are not exogenously supplied. The cells abruptly change their fate and function and this allows one to identify the metabolic and morphological parameters involved in this switch. Tissue slices can be cultured on agarized medium for long-term experiments (30 to 40 days) or in liquid medium for studies on the cell cycle.

A. LONG-TERM CULTURES

The first evidence that PA act as growth substances was obtained by observing that the weight increment of tuber explants in a culture medium containing PU, spermidine (SD), spermine (SM), or cadaverine (CD), was similar to that of explants treated with IAA.[13,14] Within one day, some internal cells located in a ring four to five layers deep just beneath the surface of the explant, start to divide[12] and repeatedly form internal walls thus becoming smaller and smaller.[15] Cell divisions proceed synchronously for two to three cycles; subsequently, these cells begin to differentiate, whereas the walls of the superficial cells of the explant start to suberify. At the end of the culture period, the explants, both those treated with PA or with IAA, reflect the typical

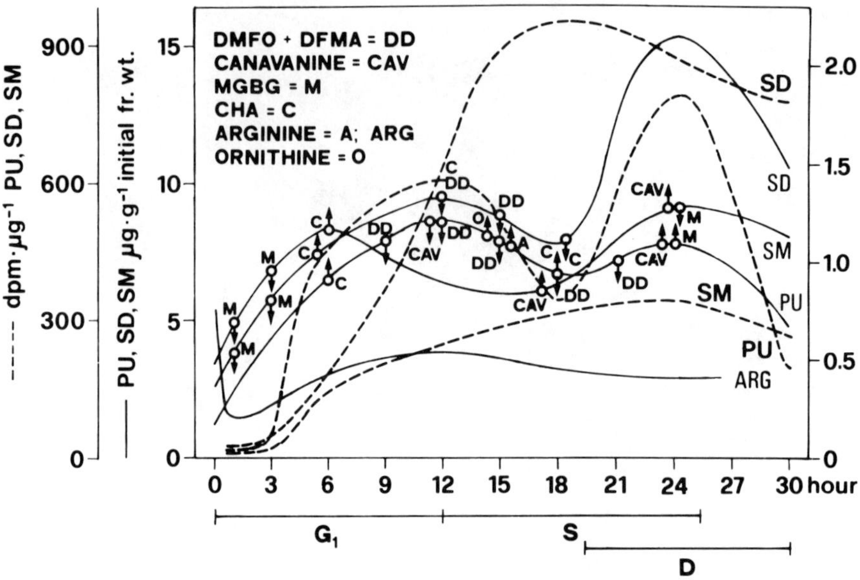

FIGURE 2. Spermine (SM), spermidine (SD), putrescine (PU), and arginine (ARG) content (—) during the cell cycle of *Helianthus tuberosus* grown in Bonner and Addicott's medium supplemented with 10 μ*M* 2,4-D.[17] Incorporation of 740 kBq L-[5-^{3}H] arginine into PU, SD and SM (---). Arrows indicate significant stimulation (↑) or inhibition (↓) of the PA accumulation caused by the addition in the culture medium of several PA inhibitors or precursors. The data reported were taken from references 7, 14, 17, 20, and 22 and replotted.

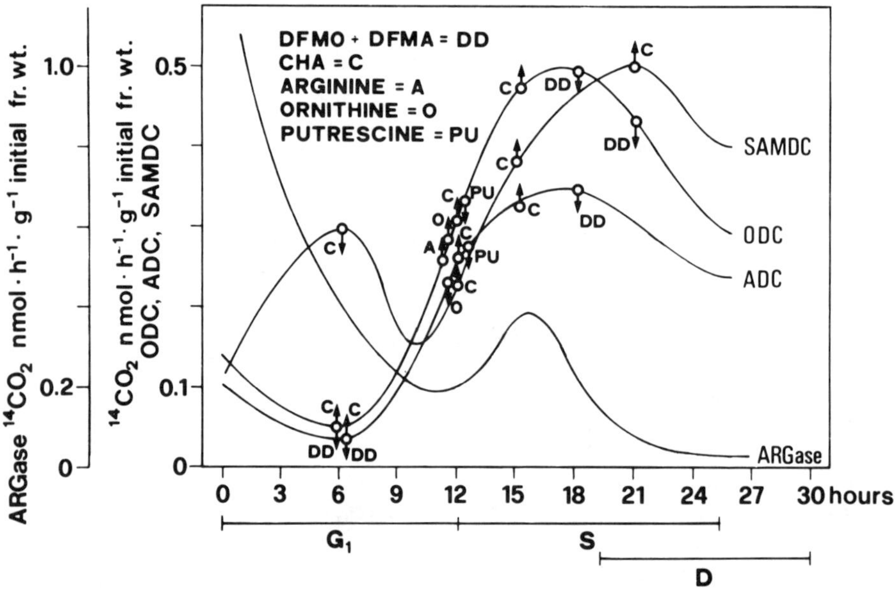

FIGURE 3. Arginine decarboxylase (ADC), ornithine decarboxylase (ODC), S-adenosylmethionine-decarboxylase (SAMDC), and arginase (ARGase) activities during the cell cycle of *Helianthus tuberosus*. Arrows indicate significant stimulation (↑) or inhibition (↓) of the activities caused by the addition in the culture medium (the same as in Figure 2) of PU or of several PA inhibitors and precursors. The data were taken from references 7, 20, and 36 and replotted.

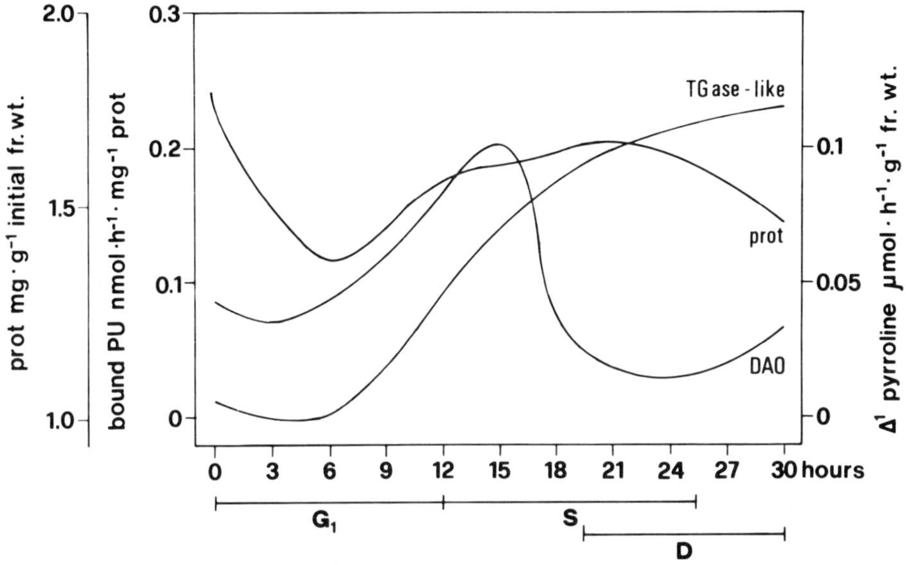

FIGURE 4. Protein content, transglutaminase (TGase)-like and diamine oxidase (DAO) activities during the cell cycle of *Helianthus tuberosus* (grown in the same conditions as described in Figure 2). The data reported were taken from references 8, 9, and 24 and replotted.

anatomy of the intact tuber, supporting the hypothesis that these treatments might cause the switching on of the same differentiation program.[15]

Treatment of these explants with inhibitors of PA biosynthesis have confirmed the PA requirement for growth. In fact, canavanine (CAV) and canaline (CAL), as well as DL-α-difluoromethylornithine (DFMO) and DL-α-difluoromethylarginine (DFMA) alone or in combination, cause a very strong inhibition of growth.[16] Studies on long-term cultures[13,15] suggest that PA, differently from IAA, are mainly active on cell division rather than on cell expansion.

Cell division is a very complex phenomen; the physiology and biochemistry is not completely understood. Many primary events occur more a less simultaneously. For this reason, it is necessary to separate cell division into periods having predominant characteristics that can be analyzed separately. In addition, it is useful to subdivide the cell itself into its compartments since each of them is a microcosm with its own particular behavior during the cell cycle.

B. THE CELL CYCLE

A multicellular system consisting of homogeneous cells, to a considerable extent dividing synchronously, is well approximated by the tuber slices of *Helianthus tuberosus* during the first cell cycle after activation in a growth medium containing 2,4-D.[17] The sudden switch from a state of slow metabolic activity occurring during dormancy to one which is characterized by the induction of the first cell cycle is evidenced by the appearance of morphological and physiological modifications.

1. Morphological Modifications

After only 3 h of activation, the nuclei, no longer adherent to the cell wall, begin to enlarge and later move to the center of the cell. The nucleoli also enlarge and, before nuclear division, fuse together and become more stainable and vacuolated; they are occasionally extruded.[5,15] The extrusion of nucleoli has also been observed in other freshly explanted tissues, in cells probably having a nurse function for the adjacent dividing cells.[18] At about 20 h an increasing number

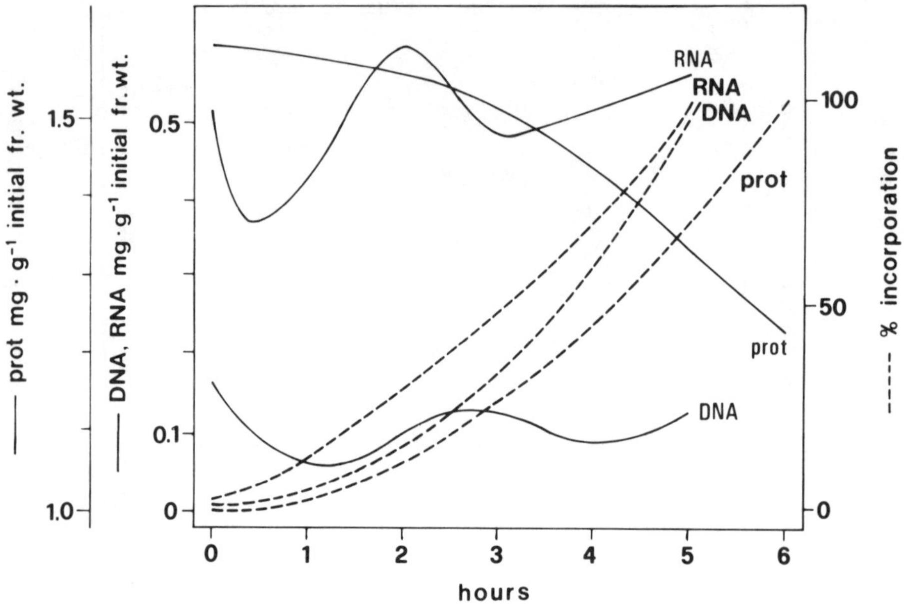

FIGURE 5. DNA, RNA, and protein content during the first half of G_1 phase (—). Incorporation of [6-^{3}H] thymidine, [5-^{3}H] uridine, and [^{14}C] methionine (---) added to the growth medium (the same as in Figure 2). 100 = maximum incorporation at the end of the experiments. The data reported were taken from references 23, 24, and 25 and replotted.

of cells contain two less densely stained nuclei, with several nucleoli; the nuclei are separated by a thin wall apparently growing centripetally.[15] Although mitoses have been reported to occur,[11,12,17] studies performed using colchicine have shown that in the first cycle the nuclei divide prevalently amitotically; in the subsequent cycles, normal mitoses become detectable.[15] Recently, amitotic divisions were observed in other freshly explanted tissues.[18] Cell divisions cease around 27 to 30 h.[15-17] Autoradiographic studies have indicated that the S phase begins after the twelfth hour and ends at 24 to 26 h, thus the G_2 phase is virtual.[17] The length of the cycle is constant when the cells are excised during deep dormancy, otherwise G_1 can be longer.[12]

2. Physiological Modifications

a. Early G_1 Phase

Polyamine biosynthesis is probably one of the earliest metabolic events induced by activation, since an increase in PA content can be measured after 15 min.[3,17] The level of PA attained after 1 h is sufficient to sustain protein synthesis. The synthesis of RNA and nonchromosomal DNA becomes detectable after 20 to 30 min.[17] (Figure 5). Within 1 h, many reserve substances are utilized: arginine content is halved within 30 min.[17] (Figure 2), asparagine, glutamine, aspartic acid, as well as other free amino acids or amides also decrease.[19] The synthesis of PA was studied *in vivo* by adding [^{3}H] arginine in the growth medium, but uptake was slow due to the large amount of endogenous arginine; the label appeared in PU, SD, and SM only at the sixth hour[17] (Figure 2). However, at that time, neither ODC nor ADC activities are particularly high in cell-free extracts[7] (Figure 3) and DFMO and DFMA do not cause a substantial inhibition of PU synthesis, even though they significantly reduce both enzyme activities.[20]

If a third biosynthetic pathway for PU through citrulline, detected in aged *H. tuberosus* slices,[21] is also present in this phase, it is not distinguishable from ODC or ADC activities, using the $^{14}CO_2$ entrapping method. On the contrary, SAMDC appears active and is considerably inhibited by cyclohexylamine (CHA) at the sixth hour[7] (Figure 3). Spermine and SD content is lowered, within 1 h, to 40% by MGBG, but the inhibition disappears by the sixth hour[22] (Figure 2).

In conclusion, the very early initiation of PU synthesis does not appear to be clearly correlated with the ODC and ADC activities assayed *in vitro*. However, caution is necessary in evaluating the enzyme activities *in vitro* and the efficiency of inhibitors *in vivo*. Free PA might be released from their bound form, as has been shown to occur mainly for SM.[10] At any rate, free PA continue to increase throughout the G_1 phase, this increase also being favored by the very low activities of enzymes that reduce the free PA pool size, either by oxidizing them or by covalently binding them to proteins, namely DAO and TGase-like activities[8,9] (Figure 4).

What is the physiological role of this PA accumulation? To try to understand it, other cellular events must be taken into account. Some information is now available: DNA, RNA, proteins, and also polysomes are very rapidly degraded during early G_1 phase, but all of these are also newly synthesized, as assessed by the incorporation of their labelled precursors.[17,23] The synthesis of enzymatic proteins involved in degradative phenomena, e.g., proteases can be inhibited by cycloheximide in the first 6 h, and probably RNAase and DNAase, no longer appear to be active after 6 h.[17,23,24] Polyamines have been found tightly-bound and loosely-bound to tRNA, rRNA, and to a poly(A)$^+$RNA, and the molar ratio PA/RNA increases in early G_1 phase (particularly PU/RNA). This suggests that PA can be directly involved in the activation of RNAs or in protecting them from nucleases.[25] It is known that SM bound to phe-tRNA, maintains the biological conformation of the latter, and that PA enhance the efficiency and fidelity of translation.[26] Moreover, ribosomes extracted from dormant tuber cells require the addition of PA and Mg^{2+} to be active. To the contrary, ribosomes extracted from dividing cells are active, since they already contain PA in sufficient amounts.[6]

Another early metabolic event is DNA synthesis which peaks around 3 to 6 h (depending on the extent of tuber dormancy and on the nutritional conditions in the field).[17,23] This synthesis occurs long before that of nuclear DNA, which starts after the twelfth hour. Other papers have failed to point out these early DNA changes in activated tuber slices probably because the observations were not made as early or as frequently or because the incubation times with thymidine were too short. By supplying [6-^{3}H] thymidine to the slices for 3 h and carrying out electron microscopy autoradiography or isolating organelles, mitochondria and plastids were shown to be responsible for this early DNA synthesis.[5,23] Thus, organelle and nuclear DNA replication are independent. DNA and protein synthesis in mitochondria are probably related to the division of the organelles themselves, occurring some time later,[5] or to changes in their metabolic functions. The observed increase in respiration (in aged slices, cyanide-insensitive respiration is also present)[27] might be necessary for the enhanced energy requirement connected with subsequent cell divisions and with the development of other organelles, such as bioenergetically incompetent plastids with their tubular complex,[5] as also suggested by the close association of the two organelles in the cytoplasm (Figure 1). Similarly, ptDNA synthesis may be a preparatory event for the division of plastids, observed some hours later; such plastids in long-term culture in the light, can differentiate thylakoids and become photosynthetically competent.[15] The increasing number of Golgi complexes and the numerous ER vesicles are indicative of a strong metabolic activity, that also must be supplied with energy.[5]

After 3 h, all three PA are also present in mitochondria.[23] A redistribution of free PA may occur during the extraction procedure but their presence in organelles in addition to that of their precursors and of their biosynthetic enzymes[28] can be confirmed also by the energy-dependent uptake of SD, driven by the membrane potential,[29] in isolated mitochondria of dormant tubers. The presence of PA in mitochondria, concomitant with mtDNA synthesis, but also with enhanced respiration, might be connected to either or both metabolic events. Newly synthesized DNA might require PA for its stabilization and/or PA might exert regulatory effects on respiration by enhancing the respiratory control ratio and the rate of respiration during ADP-ATP conversion.[30]

In chloroplasts of Chinese cabbage, Cohen et al.[31] found SD synthase activity, suggesting that PA can be synthesized therein; in chloroplasts of *Pinus radiata* ODC and ADC activities were detected.[28]

Another interesting cation, for which a role in controlling the phosphorylation of proteins involved in M and S phases has been proposed, is Ca^{2+}. The total endogenous Ca^{2+} content in *Helianthus tuberosus* tuber cells decreases sharply from 8 mM at 0 h to 4.8 mM at 6 h, thereafter remaining constant.[9] This decline in Ca^{2+} content can be attributed, in part, to its release during the excision of slices, but its amount is still rather high (possibly because the Ca^{2+} located in the wall, and not only cytoplasmic Ca^{2+} was measured). The cell cycle in preprogrammed cells seems to be induced by an increase in cytoplasmic Ca^{2+} in the 1 to 10 µM range due to an increased permeability of the plasma membrane.[32] Certain steps in the cell cycle may be regulated by Ca^{2+}-dependent protein kinases.

b. Late G_1 Phase

After the sixth hour, all the degradative processes observed in the first half of G_1 phase seem to end. By contrast, the biosynthesis of proteins, free PA and arginine continues (Figures 2 and 4).[17,24] The ODC, ADC, DAO, and TGase-like activities all increase considerably; the latter one is probably due to a newly-synthesized enzyme[9] (Figures 3 and 4).

c. Remarks on the Entire G_1 Phase

The G_1 phase can be divided into two parts: the first from early to mid G_1, the second from mid to late G_1. In the early part, the cell is involved simultaneously in eliminating all the markers of its dormant storage phase and in assuming the new characteristics linked to its meristematic (or repair) state, fully expressed in the second part. The first part of G_1 perhaps can also be considered a period of adaptation to an artificial environment (i.e., culture medium): water availability increases (the partial pressure of CO_2 decreases, that of O_2 increases); temperature increases (from 4°C to 24°C) as does hormone and nutrient availability (mineral and organic); light gets through and the pH changes during excision. The mechanical stress due to wounding of the peripheral cells and the change in pressure in the internal cells are also very important factors.

This two-period hypothesis can also be supported by the fact that subsequent cycles have a shorter G_1 phase[12] and that some of the early reaction phenomena take place even if the cells cannot undergo division.[17,19,33] In the latter case, perhaps, the cell initiates a new program that it simply cannot complete because of the unavailability of some substances.

d. S Phase

When the synthesis of nuclear and nucleolar DNA begins, a decrease in PA content takes place (Figure 2). However, the synthesis of PU, SD, and SM still continues as shown by the *in vivo* incorporation of arginine[17] (Figure 2), by the *in vitro* reactivation of arginase, and by the increasing activities of ADC, ODC and SAMDC; ODC and ADC reach their maximum activity in mid S phase when measured in cell-free extracts[7] (Figure 3). If the same occurs *in vivo*, this again raises the question as to why two biosynthetic pathways are activated at the same time and possibly in the same cell compartments. In this phase, both ADC and ODC activities, as well as that of arginase, are localized both in organelles and in the cytosol, whereas SAMDC is mainly cytosolic.[28] The decline in PU content is probably due to the considerable DAO activity occurring up to the early-S phase, which thereafter decreases sharply so that PA starts to accumulate again by the end of the S phase[8] (Figures 2 and 4).

Thus, DAO may have a central role in regulating the cellular level of PU and, consequently, of SD and SM. In order to verify this hypothesis, attempts have been made to block DAO activity *in vivo* by using aminoguanidine. However, such data are not reliable insofar as the inhibitor also blocks the uptake of the labelled PU supplied. PAO activity is not detectable in *Helianthus tuberosus*. The concomitant activity of DAO and ODC has also been observed in rapidly growing animal systems.[34] In all growing tissues, protein synthesis and respiration are enhanced, therefore PU may be also produced to furnish, through GABA, acids for the Krebs cycle and amino acids for proteins. DAO might also be involved in the biosynthesis of cell wall polymers, such as lignin.[35]

The *in vivo* effect of inhibitors on PA biosynthesis is not very marked or is transitory, so that scarce information is available on the consequences of PA depletion on the cell cycle. In addition to the reduction of SD, DNA, and RNA contents are also slightly lowered by CHA (Figure 2); in contrast, ODC, ADC, and SAMDC activity as well as PU accumulation are slightly enhanced (Figures 2 and 3). This effect is, however, only transitory since at 21 or 24 hours, all PA levels or enzyme activities are restored due to a sharp decline in intracellular free CHA content.[7]

Whereas during early G_1, MGBG severely inhibits SM and SD and also RNA and DNA accumulation, during early S phase it has no effect on PA content but causes an enhancement in RNA and DNA content[22] (Figure 2).

Canavanine (CAV) progressively inhibits the incorporation of [^{3}H] arginine into PU, SD, and SM, but an unexpected accumulation of these three compounds also occurs (Figure 2). Nucleic acids are also affected by this inhibitor.[22]

The combined effect of DFMO and DFMA is much more clear-cut (Figures 2 and 3): a 27% inhibition of ADC but only at 18 h, a 32% and 59% inhibition of ODC at 18 and 21 h respectively, a 60% decline in PU content between 16 and 21 h. However during cell division, enzyme activities and PA levels are restored.[20] Purified ODC, however, was more sensitive to DFMO than the crude extract.[36] In long-term *Helianthus tuberosus* cultures, DFMO and DFMA, alone or in combination, drastically or totally reduced growth, as do CAV, CAL, and CHA; all these inhibitors are active in *in vitro* assays.[7,16] Thus, their transitory effect during the cell cycle might be due to their rapid inactivation via a defense mechanism set up, but not maintained, by the *Helianthus tuberosus* cells.

Phillips et al.,[37,38] working on explants of a different cv. of *Helianthus tuberosus,* found SD, DAP, and only once CD, but never PU and SM. After 1 d in culture, SD content was slightly reduced by MGBG without causing any changes in cell number: however at a later stage xylogenesis was inhibited.[37] After 3 d, DFMO and DFMA, added separately, had little effect on cell division but inhibited tracheid differentiation.[38] DFMO and MGBG supplied for 3 d caused an elevation in SD levels and DFMA decreased SD levels;[37,38] the addition of PU and SD increased or inhibited cell division depending on the growth substances present in the medium.[37] These authors deny the possibility that PA affect cell division but rather suggest that they affect xylogenesis. This latter effect, however, might also be interpreted as a consequence of an incorrect DNA duplication, caused by the presence of these inhibitors, during the repeated cell divisions which precede differentiation. In this case too, since the inhibitors did not affect SD levels within the first 24 h it is not surprising that there was no effect on the first cell divisions.

Some of the experiments reported above suggest that the inhibition of PA synthesis also causes some disturbance to the nucleic acids. The effect of SD directly on DNA was studied by examining its competition with actinomycin D. In activated *Helianthus tuberosus* slices, SD was able to partially remove the *in vivo* inhibition of actinomycin D on DNA and RNA synthesis.[39] The direct interaction between DNA, actinomycin D, and PA was confirmed by *in vitro* studies which showed that the K_{app} of actinomycin D binding to DNA is significantly lowered by SM.[40]

Clearer data were obtained in bacteria and animals, which show that PA are essential for DNA, RNA, and protein synthesis of which they enhance the efficiency and fidelity.[41] For example, in animal systems ODC is inhibited by factors such as colchicine and vinblastine that delay or block the S phase, whereas the increase in PU content, by DAO inhibition, stimulates DNA synthesis and cell division. Again, various inhibitors of PA biosynthesis block DNA synthesis and the cell cycle and their effect seems to be specific. However, in animals, often only a partial depletion of PA can be achieved, and the block can be seen only after more than one doubling time period.[41] This was clearly demonstrated by Cohen et al.[42] also in *Chlorella vulgaris*: if ODC, which peaks just before S phase, is inhibited by DFMO, 84% of the cells still divide, but in the second cycle only 50% of autospore release was observed.

In *Helianthus tuberosus* the TGase-like activity *in vitro* continues to increase during the S phase becoming seven- to eightfold higher than the initial value[9] (Figure 4). This binding activity is partly due to a true TGase, as very recently ascertained by Beninati et al.[43] after the identification of glutamyl-PA, detected upon supplying endogeneous or artificial substrates to the plant extract. Some of its characteristics are different from those of the better known TGases of animal systems.[9,44,45] Conjugates with PU and SM, but mainly SD, are formed in *Helianthus tuberosus* cell-free extracts.[45] PA can form intramolecular or cross-linked conjugates and the length of the PA involved determines the length of the bridges.

Data on TGase activity during the cell cycle is also limited in animal cells. After 2.5 hours of stimulation, Korner and Bachrach[46] observed a decrease in TGase activity and antigenicity in C_6 glioma cells, suggesting that the enzyme was previously bound to the cell membrane and was released upon cell stimulation into the cytoplasm, where it was subject to proteolytic digestion. Thereafter a *de novo* synthesis compensated for this deficiency and a second peak in activity resulted.

In the activated cells of *Helianthus tuberosus*, the concentration of labelled PU, which is conjugated *in vivo* to relatively high molecular mass proteins during the cell cycle, is of the order of 0.2 μM. Thus, even taking the dilution of the internal PU into account, TGase contribution to the decrease in the free PA pool is not significant. Different protein-conjugated PA have been found *in vivo* during the *Helianthus tuberosus* cell cycle. Low molecular mass conjugates (18 kDa) appear in early G_1 phase, while high and very high molecular mass conjugates were formed during the S phase and up to the time of cell division.[9,47,48] The latter conjugates (formed both *in vivo* and in cell-free extracts) do not enter the stacking gel, when analyzed by SDS-PAGE.[9] In animals, similar high molecular mass conjugates were found by Beninati et al.[49] and others. Common substrates for this enzyme appear to be extracellular structural proteins.[50] In S phase, fibrillar material is released in the walls of *Helianthus tuberosus* cells[5] and wall proteins are synthesized in *Catharanthus roseus*.[51] Intracellular structural proteins, like actin and myosin, are also substrates of TGase in animals.[50] A functional dependence on PA for the formation of cytoskeletal structures, like actin filaments and microtubules, has been shown in PA-deficient CHO cells.[52] In addition, PA depletion causes a high incidence of binucleate fibroblasts.[53] It is reasonable to assume that, also in plants, PA may be involved in the formation of supramolecular cell structures. On the other hand, during S phase, many enzymes involved in DNA synthesis form multiprotein complexes associated with DNA[54] in which PA may possibly be involved.

With regard to PA bound to low molecular mass compounds, which are TCA-soluble, peptide-PA conjugates were also found in *Helianthus tuberosus* whereas hydroxycinnamoyl-PA[55] were not detectable.

e. Cell Division

The cell division (D) phase, which begins around the 20 h, largely overlaps with the S phase; thus, some changes in PA metabolism can be related to events linked to either or both phases. The second peak in PA accumulation and in PU synthesis from arginine, with SD synthesis remaining high, occurs in the middle of the division phase[17] (Figure 2). ADC and ODC activities are still high, though declining[7] (Figure 3) and DFMO and DFMA become practically ineffective[20] (Figures 2 and 3). Whereas MGBG inhibits SM accumulation, thus causing an accumulation of PU[20] (Figure 2), CAV inhibits PA synthesis but overall PA content increases[22] (Figure 2). Possibly, the latter inhibitor disturbs DAO activity which is still low, but tends to increase again (Figure 4). The conjugation of PU continues to increase and gives rise mainly to high molecular mass conjugates[9] (Figure 4). Protein accumulation, active until early D phase, begins to decline or proteins become insoluble and thus more difficult to detect[24] (Figure 4).

During the last part of this phase it appears that PA biosynthesis is no longer necessary, as though the PA levels attained are now sufficient to allow further progression of this phase. Cell division is, in fact, one of the phases that do not have restriction points. It is known that PAs form ionic linkages with pectic substances[56] and probably hydrogen bonds with neutral wall polysaccharides;[57] PAs have been found in cell walls.[58,59] Cell wall formation, one of the main events of late D phase, probably requires PA, thus causing a depletion of the free pool, assuming that some wall-bound PA are also TCA-insoluble. During this phase, TGase-like activity is not relevant (Figure 4). The sharp decline in PA content might also result from a degradative metabolism (totally or partly due to the resumption of DAO activity) as well as a slowing down of their synthesis (Figures 2 to 4). The main metabolic features occurring in dormant tubers and in activated slices during the first cell cycle are summarized in Figure 6.

IV. THE CELL CYCLE IN OTHER PLANTS

Variations in PA levels correlated with cell cycle events have been very recently found also in *Nicotiana tabacum* cell suspension cultures synchronized with aphidicoline, a specific inhibitor of DNA polymerase.[60] This system differs from that of *Helianthus tuberosus* in the sense that, at the onset of the cycle, the inhibited *N. tabacum* cells are not deficient in PA, which continue to increase (both free and TCA-insoluble PU) until the onset of the S phase, caused by the removal of aphidicoline. Suddenly, a drop in free PU and SD content occurs, as in *Helianthus tuberosus* , and in TCA-insoluble PU as well. Thereafter, in late S/G_2 phase, free PU increases again, followed by TCA-insoluble PU, and decreases during cell division again, as in *Helianthus tuberosus*. TCA-soluble bound PA appears to be neither related to any phase of the cycle nor are the oscillations reproducible. Also, in *N. tabacum*, the accumulation of PU during the prolonged G_1 phase, can be preparative for the subsequent S phase, because the cells are inhibited but still committed to dividing. In fact, within 3 h from the removal of the inhibitor, the cells already reach their maximum label index.[60] The specificity of the inhibitor and the strict coincidence between DNA synthesis and the decline in PA content, leads one to conclude that PA are "used" in the DNA synthesizing machinery and then "disposed of" after use, possibly by DAO-mediated catabolism. A similar decrease in PU and SD content was not observed in *Chlorella*, where PA titers increase up to the S phase even though ODC activity was very high only in G_1 phase.[42]

The occurrence in the S phase of a decrease in TCA-insoluble PU in *N. tabacum*[60] and the synthesis of labelled conjugates, also TCA-insoluble, observed *in vivo* and *in vitro* in *Helianthus tuberosus*[9] (Figure 4) are not conflicting. Both types of metabolism could coexist at different rates or involve different proteins.

The similarity between the bimodal trend of PA content in *Helianthus tuberosus* and *N. tabacum* tissues with that found by Heby et al.[61] in animal cells is very impressive. The behaviour of SAMDC (two maxima of activity in G_1 and S phases) and of ODC (a peak in early S) are very similar in *Helianthus tuberosus* and in animal tissues.[41,61] The main difference is the presence of a maximum in G_1 of ODC activity observed only in some animal cells, which, however, is not a constant feature, even though the increase in PU content in G_1 is constant.[41]

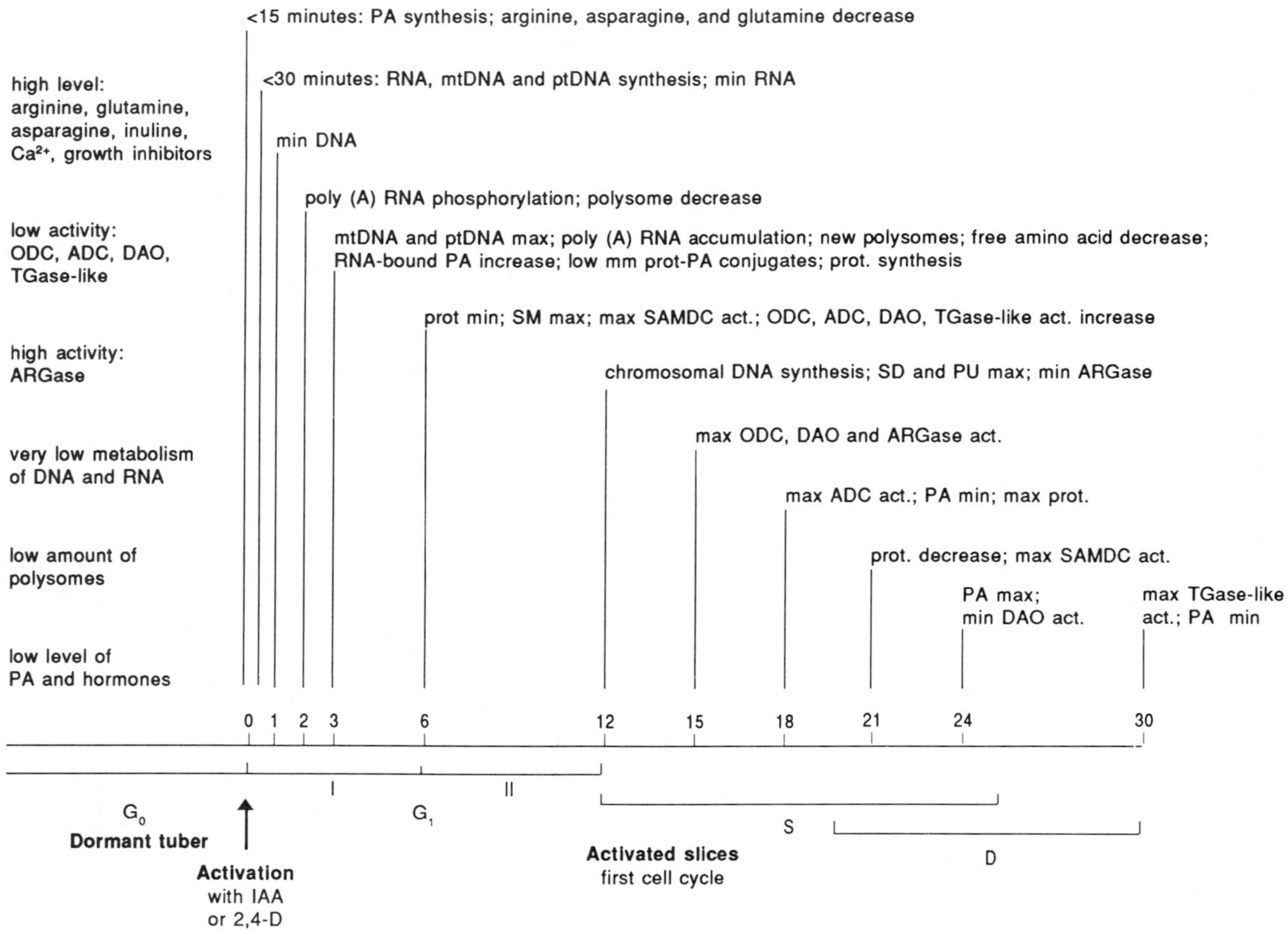

FIGURE 6. The main physiological features of parenchyma cells of *Helianthus tuberosus* tubers during dormancy and during the first cell cycle after the break of dormancy.

V. CONCLUDING REMARKS

Most of the experiments performed with *Helianthus tuberosus* have shown that inhibitors of PA synthesis also exert an effect on DNA levels or replication. In addition, PA also appear to be implicated in RNA and protein synthesis or function. PA synthesis reaches its maximum level just prior to DNA synthesis and cell division. Both events are affected or blocked by PA deprivation, as the numerous *in vitro* and *in vivo* studies performed with bacteria and animals indicate.[41]

It is now possible to propose that correlations also exist between PA levels and certain events of the cell cycle in plants. These correlations are convincing with regard to DNA synthesis, but less convincing when cell division as a whole is considered. The latter type of correlation is supported mainly by changes in the total amount of PA and enzyme activities. It is sometimes difficult to attribute a meaning to a certain level of endogenous PA since they can be stored, for example PU in the vacuole, whereas SD seems to be present in appreciable amounts in the walls (discussed in Chapter 7, this volume), where they cannot participate in intracellular metabolism. The isolation of PA-deficient mutants in plants will perhaps clarify this point. It should also be important to identify various physiological roles of the different PA. PA in the cell should not be considered simply as a unity but as distinct categories: free, nontightly-bound, tightly-bound, or covalently linked. To date, most of the studies have been performed only on free PA, but their exact role in the cell has not been clarified. Among the functions that PA might fulfill, free hydrogen or ionically-bound PA perhaps regulate the activity of some enzymes which work at a particular ionic strength or pH, by creating a suitable micro-environment around them. While such interactions can be studied *in vitro*, it is difficult to assess their importance *in vivo*. Covalently-bound PA are more easily studied *in vivo* because the conjugates are stable. However, in general, these conjugates are present in low amounts and are heterogeneous. In addition, the analysis of very high molecular mass conjugates is rather difficult. The identification of the protein partner of some of these conjugates should be very useful in understanding the function of PA and might offer further direct insight into the probable multiple roles of these cations in the cell cycle. Studies on the direct interactions between PA and specific molecules *in vivo* are only in their infancy but are very promising.

ACKNOWLEDGMENTS

The author acknowledges the collaboration offered over the years by all the past and present members of the lab and of Dr. Stefania Biondi for her help in reviewing the manuscript.

REFERENCES

1. **Herbst, E.J. and Snell, E.E.,** Putrescine as a growth factor for *Hemophilus parainfluenzae*, *J. Biol. Chem.*, 176, 989, 1948.
2. **Smith, T.A.,** Polyamines, *Annu. Rev. Plant Physiol.*, 36, 36, 1985.
3. **Bagni, N., Malucelli, B., and Torrigiani, P.,** Polyamines, storage substances and abscisic acid-like inhibitors during dormancy and very early activation of *Helianthus tuberosus* tuber tissue, *Physiol. Plant*, 49, 341, 1980.
4. **Duranton, H. and Maille, M.,** The metabolism of proline and the synthesis of the pigments in the Jerusalem artichoke, *Photochem. Photobiol.*, 2, 111, 1963.
5. **Favali, A., Serafini-Fracassini, D., and Sartorato, P.,** Ultrastructure and autoradiography of dormant and activated parenchyma of *Helianthus tuberosus*, *Protoplasma*, 123, 192, 1984.

6. **Cocucci, S. and Bagni, N.,** Polyamine-induced activation of protein synthesis in ribosomal preparation from *Helianthus tuberosus* tissue, *Life Sci.*, 7, 113, 1968.
7. **Torrigiani, P., Serafini-Fracassini, D., and Bagni, N.,** Polyamine biosynthesis and effect of dicyclohexylamine during the cell cycle of *Helianthus tuberosus* tuber, *Plant Physiol.*, 84, 148, 1987.
8. **Torrigiani, P., Serafini-Fracassini, D., and Fara, A.,** Diamine oxidase activity in different physiological stages of *Helianthus tuberosus* tuber, *Plant Physiol.*, 89, 69, 1989.
9. **Serafini-Fracassini, D., Del Duca, S., and Torrigiani, P.,** Polyamine conjugation during the cell cycle of *Helianthus tuberosus*: nonenzymatic and transglutaminase-like binding activity, *Plant Physiol. Biochem.*, 27, 659, 1989.
10. **Serafini-Fracassini, D. and Mossetti, U.,** Free and bound polyamines in different physiological stages of *Helianthus tuberosus* tuber, in *Recent Progress in Polyamine Research*, Selmeci, L., Brosnan, M., and Seiler, N., Eds., Akademiai Kiado, Budapest, 1985, 551.
11. **Adamson, D.,** Expansion and division in auxin-treated plant cells, *Can. J. Bot.*, 40, 719, 1962.
12. **Yeoman, M.M. and Aitchison, P.A.,** Growth patterns in tissue (callus) cultures, in *Plant Tissue and Cell Culture*, Street, H.E., Ed., Blackwell, Oxford, 1973, 240.
13. **Bertossi, F., Bagni, N., Moruzzi, G., and Caldarera, C.M.,** Spermine as a new promoting substance for *Helianthus tuberosus* (Jerusalem artichoke) *in vitro*, *Experientia*, 21, 80, 1965.
14. **Bagni, N., Serafini-Fracassini, D., and Torrigiani, P.,** Polyamines and growth in higher plants, in *Advances in Polyamine Research*, Vol. 3., Caldarera, C.M., Zappia, V., and Bachrach, U., Eds., Raven Press, New York, 1981, 337.
15. **Serafini-Fracassini, D. and Alessandri, M.,** Polyamines and morphogenesis in *Helianthus tuberosus* explants, in *Advances in Polyamine Research*, Vol. 4, Bachrach, U., Kaye, A.M., and Chayen, P., Eds., Raven Press, New York, 1983, 419.
16. **Bagni, N., Serafini-Fracassini, D., and Torrigiani, P.,** Polyamines and cellular growth processes in higher plants, in *Plant Growth Substances 1982*, Wareing, P.F., Ed., Academic Press, London, 1982, 473.
17. **Serafini-Fracassini, D., Bagni, N., Cionini, P.G., and Bennici, A.,** Polyamines and nucleic acids during the first cell cycle of *Helianthus tuberosus* tissue after the dormancy break, *Planta*, 148, 332, 1980.
18. **Altamura, M.M., Capitani, F., Serafini-Fracassini, D., Torrigiani, P., and Falasca, G.,** Root histogenesis from tobacco thin cell layers, *Protoplasma*, 161, 31, 1990.
19. **Scoccianti, V.,** Variation in the content of free and bound amino acids during dormancy and the first cell cycle in *Helianthus tuberosus* tuber explants, *G. Bot. Ital.*, 117, 237, 1983.
20. **Torrigiani, P. and Scoccianti, V.,** Inhibition of putrescine synthesis during the cell cycle in *Helianthus tuberosus* tuber explants, *Plant Physiol. Biochem.*, 28, 779, 1990.
21. **Speranza, A. and Bagni, N.,** Products of L-^{14}C-carbamoyl citrulline metabolism in *Helianthus tuberosus* activated tissue, *Z. Pflanzenphysiol.*, 88, 163, 1978.
22. **Bagni, N., Torrigiani, P., and Barbieri, P.,** Effect of various inhibitors of polyamine synthesis on the growth of *Helianthus tuberosus*, *Med. Biol.*, 59, 403, 1981.
23. **Torrigiani, P. and Serafini-Fracassini, D.,** Early DNA synthesis and polyamines in mitochondria from activated parenchyma of *Helianthus tuberosus*, *Z. Pflanzenphysiol.*, 97, 353, 1980.
24. **Serafini-Fracassini, D., Grandi, B., Dinnella, C., and Del Duca, S.,** Transglutaminase-like activity during the cell cycle of *Helianthus tuberosus*, presented at 2nd Int. Conf. on Transglutaminases and Protein Cross-linking Reactions, Cannes, June 24 to 28, 1990, 73.
25. **Serafini-Fracassini, D., Torrigiani, P., and Branca, C.,** Polyamines bound to nucleic acids during dormancy and activation of tuber cells of *Helianthus tuberosus*, *Physiol. Plant*, 60, 351, 1984.
26. **Quigley, G.J., Teeter, M.M., and Rich, A.,** Structural analysis of spermine and magnesium ion binding to yeast phenylalanine transfer RNA, *Proc. Natl. Acad. Sci. U.S.A.*, 75, 64, 1978.
27. **Arrigoni, O., De Santis, A., Arrigoni-Liso, R., and Calabrese, G.,** The increase of hydroxyproline-containing proteins in Jerusalem artichoke mitochondria during the development of cyanide-insensitive respiration, *Biochem. Biophys. Res. Commun.*, 74, 1637, 1977.
28. **Torrigiani, P., Serafini-Fracassini, D., Biondi, S., and Bagni, N.,** Evidence for the subcellular localization of polyamines and their biosynthetic enzymes in plant cells, *J. Plant Physiol.*, 124, 23, 1986.
29. **Pistocchi, R., Antognoni, F., Bagni, N., and Zannoni, D.,** Spermidine uptake by mitochondria of *Helianthus tuberosus*, *Plant Physiol.*, 92, 690, 1990.
30. **Chaffee, R.R.J., Arine, R.M., and Rochelle, R.H.,** The possible role of intracellular polyamines in mitochondrial metabolic regulation, *Biochem. Biophys. Res. Commun.*, 86, 293, 1979.
31. **Cohen, S.S., Marcu, D.E., and Balint, R.F.,** Light-dependent fixation of polyamines into chloroplasts of chinese cabbage, *Febs Lett.*, 141, 93, 1982.
32. **Trewavas, A.J.,** Growth substances, calcium and the regulation of cell division, in *The Cell Division Cycle in Plants*, Bryant, J.A. and Francis, D., Eds., Cambridge Univ. Press, Cambridge, 1985, 133.

33. **Serafini-Fracassini, D. and Filiti, N.,** Effects de la gibbérelline A_7 le métabolisme des acides nucléiques après la levée de la dormance du tubercule d'*Helianthus tuberosus*, *C. R. Acad. Sc.*, Paris, Serie III, 294, 645, 1982.
34. **Perin, A., Sessa, A., and Desiderio, M.A.,** Diamine oxidase in regenerating and hypertrophic tissues, in *Structure and Function of Amine Oxidases*, Mondovì, B., Ed., CRC Press, Boca Raton, FL, 1985, 179.
35. **Angelini, R., Manes, F., and Federico, R.,** Spatial and functional correlation between diamine-oxidase and peroxidase activities and their dependence upon de-etiolation and wounding in chick-pea stem, *Planta*, 182, 89, 1990l.
36. **D'Orazi, D. and Bagni, N.,** Ornithine decarboxylase activity in *Helianthus tuberosus*, *Physiol. Plant.*, 71, 177, 1987.
37. **Phillips, R., Press, M.C., and Eason, A.,** Polyamines in relation to cell division and xylogenesis in cultured explants of *Helianthus tuberosus*: lack of evidence for growth-regulatory action, *J. Exp. Bot.*, 38, 164, 1987.
38. **Phillips, R., Press, M.C., Bingham, L., and Grimmer, C.,** Polyamines in cultured artichoke explants: effects are primarily on xylogenesis rather than cell division, *J. Exp. Bot.*, 39, 473, 1988.
39. **Bagni, N., Corsini, E., and Serafini-Fracassini, D.,** Growth-factors and nucleic acid synthesis in *Helianthus tuberosus*. I. Reversal of actinomycin D inhibition by spermidine, *Physiol. Plant.*, 24, 112, 1971.
40. **D'Orazi, D., Serafini-Fracassini, D., and Bagni, N.,** Polyamine effects on the stability of DNA-actinomycin D complex, *Biochem. Biophys. Res. Commun.*, 90, 362, 1979.
41. **Heby, O.,** Role of polyamines in the control of cell proliferation and differentiation, *Differentiation*, 19, 1, 1981.
42. **Cohen, E., (Malis) Arad, S., Heimer, Y.H., and Mizrahi, Y.,** Polyamine biosynthetic enzymes in the cell cycle of *Chlorella*. Correlation between ornithine decarboxylase and DNA synthesis at different light intensities, *Plant Physiol.*, 74, 385, 1984.
43. **Beninati, S., Piacentini, M., Di Rao, M., Del Duca, S., and Serafini-Fracassini, D.,** in preparation, 1991.
44. **Folk, J.E.,** Translutaminases, *Annu. Rev. Biochem.*, 49, 517, 1980.
45. **Serafini-Fracassini, D., Del Duca, S., and D'Orazi, D.,** First evidence for polyamine conjugation mediated by an enzymic activity in plants, *Plant Physiol.*, 87, 757, 1988.
46. **Korner, G. and Bachrach, U.,** Intracellular distribution of active and inactive transglutaminase in stimulated cultured C_6 glioma cells, *J. Cell. Physiol.*, 130, 44, 1987.
47. **Mossetti, U., Serafini-Fracassini, D., and Del Duca, S.,** Conjugated polyamines during dormancy and activation of tuber of Jerusalem artichoke, in *Conjugated Plant Hormones. Structure, Metabolism and Function*, Schreiber, K., Schütte, H.R., and Sembdner, G., Eds., Deutscher Verlag der Wissenschaften, Berlin, 1986, 369.
48. **Serafini-Fracassini, D., Del Duca, S., D'Orazi, D., and Mossetti, V.,** Conjugation of polyamines by an enzyme activity in dividing cells of *Helianthus tuberosus*, in *Perspectives in Polyamine Research*, Perin, A., Scalabrino, D., Sessa, A., and Ferioli, L., Eds., Wichtig Editore, Milano, 1988, 89.
49. **Beninati, S., Piacentini, M., Cocuzzi, E.T., Autuori, F., and Folk, J.E.,** Covalent incorporation of polyamines as γ-glutamyl derivatives in CHO cell protein, *Biochim. Biophys. Acta*, 952, 325, 1988.
50. **Lorand, L. and Conrad, S.M.,** Transglutaminases, *Mol. Cell. Biochem.*, 58, 9, 1984.
51. **Kodama, H., Ito, M., and Komamine, A.,** Molecular aspects of the cell cycle in *Catharanthus roseus* synchronous cell division cultures, in *Progress in Plant Cellular and Molecular Biology*, Nijkamp, H.J.J., Van der Plas, L.H.W., and Van Aartrijk, J., Eds., Kluwer Acad. Publ., Dordrecht, 1990, 532.
52. **Pohjanpelto, P., Virtanen, I., and Hölttä, E.,** Polyamine starvation causes disappearance of actin filaments and microtubules in polyamine-auxotrophic CHO cells, *Nature*, 293, 475, 1981.
53. **Sunkara, P.S., Rao, P.N., Nishioka, K., and Brinkley, B.R.,** Role of polyamines in citokinesis of mammalian cells, *Exp. Cell Res.*, 119, 63, 1979.
54. **Dunham, V.L. and Bryant, J.A.,** Enzymic controls of DNA replication, in *The Cell Division Cycle in Plants*, Bryant, J.A. and Francis, D., Eds., Cambridge Univ. Press, Cambridge, 1985, 37.
55. **Martin-Tanguy, J.,** The occurrence and possible function of hydroxycinnamoyl acid amides in plants, *Plant Growth Regul.*, 3, 381, 1985.
56. **D'Orazi, D. and Bagni, N.,** *In vitro* interactions between polyamines and pectic substances, *Biochem. Biophys. Res. Commun.*, 148, 1259, 1987.
57. **Scoccianti, V., Bagni, N., Dubinsky, O., and Arad (Malis), S.,** Interaction between polyamines and cells of marine unicellular red alga *Porphyridium* sp., *Plant Physiol. Biochem.*, 27, 899, 1989.
58. **Goldberg, R. and Pedrizet, E.,** Ratio of free to bound polyamines during maturation in mung-bean hypocotyl cells, *Planta*, 161, 531, 1984.
59. **Mariani, P., D'Orazi, D., and Bagni, N.,** Polyamines in primary walls of carrot cells: endogenous content and interactions, *J. Plant Physiol.*, 135, 508, 1989.
60. **Pfosser, M., Königshofer, H., and Kandeler, R.,** Free, conjugated, and bound polyamines during the cell cycle of synchronized cell suspension cultures of *Nicotiana tabacum*, *J. Plant Physiol.*, 136, 574, 1990.
61. **Heby, O., Gray, J.W., Lindl, P.A., Marton, L.J., and Wilson, C.B.,** Changes in L-ornithine decarboxylase activity during the cell cycle, *Biochem. Biophys. Res. Commun.*, 71, 99, 1976.

Chapter 12

POLYAMINES AND PLANT MORPHOGENESIS

Arthur W. Galston and Hector E. Flores

TABLE OF CONTENTS

I. INTRODUCTION

In addition to their much studied roles in the regulation of protein[1,2] and RNA[3] synthesis and in the initiation and control of cell division,[4] polyamines are also known to affect patterns of cell differentiation and morphogenesis in plants, animals and microorganisms.[4-8] In this chapter, we will briefly survey the evidence linking polyamines to the control of several important morphogenetic processes in plants. These include the following:

1. The organization of the shoot apex, especially the determination of vegetative vs. reproductive pathways of development, the morphogenetic patterns of the individual floral organs, and the growth and maturation of the ovary and fruit.
2. The germination and growth patterns of the spores and protonemata of the gametophyte generation of the fern *Adiantum capillus-veneris*.
3. The formation of somatic embryos in culture, especially in *Daucus* and *Brassica*.
4. The formation of tubers.
5. The formation of embryonic, lateral, and adventitious roots.

II. POLYAMINES AND MORPHOGENESIS

A. MORPHOGENESIS AT THE SHOOT APEX

Most stem apices start out by producing purely vegetative stem tissue containing buds that go on to produce leaf-bearing nodes and internodes. Depending on the type of plant, the meristem eventually gives rise to flowers, either as a result of built-in developmental patterns, as in the tomato, or following photoperiodic induction by either a short-day (as in *Xanthium*, *Pharbitis*, and *Chenopodium*) or long-day (as in *Hyoscyamus* or *Lolium*) stimulus.[9] In the process of photoperiodic induction, the leaves perceive and integrate the duration of light and dark exposure, which they then transduce into a mobile florigenic stimulus, transported from leaf to apex by way of living cells, presumably through the phloem. Once having received the floral stimulus, some apices are permanently transformed, giving rise to reproductive organs indefinitely without a further environmental stimulus. Other plants require repeated photoperiodic induction, reverting to the vegetative habit in the absence of such signals.

Before they are ready to receive and respond to a photoperiodic stimulus, plants must achieve "ripeness to flower" (competence), an ill-defined physiological state representing the attainment of the capacity for sexual maturation. In *Pharbitis nil*, the Japanese morning glory, the newly germinated seedling is already ripe-to-flower, and the cotyledons can be induced by a single exposure to one long night. In other plants, such as short-day cultivars of tobacco, the plant must grow for several weeks and attain considerable size before it can respond to induction by long night treatment. It is not known whether this failure to respond results from inability of the leaf to sense or transduce the photoperiodic stimulus, or of the bud to respond to the translocated florigenic message.

The nature of the endogenous floral stimulus is not known, despite decades of intensive investigation. Several of the plant hormones, especially gibberellins and ethylene, are known to affect or induce flowering in particular cases, but no substance has yet been found to satisfy all criteria for the floral stimulus, which must include the ability to pass between different photoperiodic types via grafts. Other candidates for the native floral stimulus have been proposed, mainly on the basis of correlations between changes in their titer and the reproductive behavior of the plant, and also because application of the substance in question can result in flowering. The polyamines fall into this class of candidate substances.

For many years, Martin-Tanguy[10] has been accumulating evidence linking various hydroxycinnamic acid amides of the common polyamines putrescine (Put) and spermidine

(Spd), hereafter called PA conjugates, to the onset of flowering in tobacco (see Chapter 4). In vegetative plants younger than about seven weeks, there are no marked patterns of accumulation of the PA conjugates. In older vegetative plants, PA conjugates accumulate in the most recently initiated leaves, but when the plant initiates flowering, the conjugates accumulate in the floral organs, and practically disappear from the leaves. How can we account for these opposite trends in leaves and flowers? The simplest hypothesis would be that the conjugates move from leaves to the young floral buds (as the floral stimulus does); an alternative, and less attractive hypothesis would involve massive metabolic destruction of PA conjugates in the leaves together with massive synthesis in the floral organs.

Malmberg and colleagues have regenerated plants from single protoplasts of tobacco and petunia resistant to various inhibitors of polyamine biosynthesis.[11,12] In all cases, the "mutant" plants showed aberrations in both polyamine metabolism and floral development. Unfortunately, many of these mutants were sterile, thus preventing proper genetic analysis of the mutation(s) involved, but the authors conclude that the polyamines and/or their conjugates appear to significantly affect floral development. Other evidence correlating polyamine metabolism and reproductive development comes from *Spirodela punctata*, a quantitative long-day plant,[13] *Zea mays*,[14] and various members of the *Araceae*.[15]

With thin-layer tissue cultures derived from surface explants of fruiting tobacco plants,[16] it is possible to regenerate either vegetative or reproductive buds, depending on the auxin/cytokinin concentration. When cytokinin content is equal to the auxin concentration at 1 μM, floral buds are produced, but raising the cytokinin level to 10 μM yields only vegetative buds. In such a system derived from Wisconsin-38 tobacco, three lines of evidence connect polyamines and flowering:[17] (1) floral buds contain much higher levels of polyamines, especially Spd, than do vegetative buds, (2) inhibition of spermidine synthase activity by cyclohexylamine produces parallel declines in Spd level and flowering, and increases the number of vegetative buds formed, and (3) the addition of exogenous Spd to vegetative cultures causes some flowers to appear. To be effective, the Spd must be present for at least three weeks, during the crucial period when buds are determined.[18] In Samsun tobacco,[19] spermine, rather than spermidine is the most effective polyamine for the formation of flowers in the thin layer culture system.

It should be clearly understood that these tissue cultures are derived from plants that have already flowered and fruited. Thus, the above results do not reflect on floral initiation in the strict sense, but rather on evocation.[20] Thus, the fact that thin layer tobacco cultures will also form flowers in response to oligosaccharides[21] and ethylene[22] may indicate that they and polyamines are acting on a common pathway leading to floral evocation. One can sense a common thread in these three examples: ethylene. Polyamines are known to affect the production[23] and action of ethylene,[24] and among the many effects of oligosaccharides is the elicitation of ethylene biosynthesis.[25] Thus, if either the initiation or the evocation of flowers were sensitive to ethylene concentration, and if the critical concentration range differed in different plants, one could formulate an integrated theory of flowering in thin layer cultures of tobacco based on ethylene. Such a theory might also pertain to flowering in general.

In *Xanthium strumarium*, a short-day plant, exposure of leaves to successive inductive nights results in a rise in conjugated polyamines, especially of spermine,[7] per unit protein nitrogen. Spermine conjugates rise sharply after one inductive long night, remain high during the second cycle, then decline precipitously during the third and fourth inductive cycles. Spermidine, and to a lesser extent, putrescine, follow the same trend. About eight days after the initial inductive cycle, a rise is noted in conjugated polyamines in the terminal buds. Superficially, this also correlates well with the behavior expected of a floral stimulus. It is certainly essential to probe this question further, including analysis of vascular translocation sap.

Polyamines are also involved in the postfertilization events leading to the formation of the mature fruit. In tomato, application of α-difluoromethylornithine (DFMO), a "suicide" inhibitor of the putrescine-synthesizing enzyme ornithine decarboxylase, to a young fruit leads to an inhibition of cell division and therefore of fruit growth.[26] Application of putrescine to the inhibited system results in a partial reversal of both inhibitions, indicating that polyamines are essential components of the complex system involved in fruit growth. Whether polyamine levels normally limit the growth of fruits is not yet established, although some evidence[27,28] supports that view. For a more detailed discusion of polyamine involvement in fruit growth and development, the reader is referred to Chapter 10.

B. POLYAMINES AND MOPHOGENESIS IN FERN GAMETOPHYTES

For many years, Furuya[28]and his colleagues have studied the role of phytochrome, the blue light receptor and various plant growth regulators in the growth and morphogenesis of the gametophyte generation of the maidenhair fern, *Adiantum capillus-veneris*. The spores are dormant and will not germinate unless exposed to gibberellin or red light that converts P_r phytochrome to the P_{fr} form. The emerging cylindrical protonemal tube grows rapidly and straight toward the unilateral red light source, being phototropically oriented by phytochrome in the membrane. The single large nucleus in the protonemal tube is located about 20 μm behind the apex of the growing protonemal tip; it is locked in the G_1 stage of the cell cycle and does not divide unless directly exposed to blue light. Blue light causes a swelling at the tip and the prompt initiation of repeated nuclear division and cytokinesis, giving rise to the two dimensional flat prothallus on which the sex organs are formed. In red light, there is no nuclear division, and the single nucleus migrates into one of the two branches; upon blue light stimulation, the nucleated branch divides and gives rise to a prothallus, while the anucleate branch does not develop further. Particular gibberellins can cause the premature initiation of antheridia on the protonema, long before normal formation on the prothallus.

We applied approximately 0.1 to 1 m*M* DFMO and α-difluoromethylarginine (DFMA), the analogous inhibitor of arginine decarboxylase (ADC), the alternative enzyme whose action initiates the biosynthesis of putrescine.[29] To our surprise, while DFMO did not affect germination and inhibited protonemal growth, 0.5 to 1 μ*M* DFMA strongly promoted both processes. Since DFMA presumably inhibits putrescine biosynthesis via ADC, we added putrescine to the inhibited cultures to see whether it could reverse the effects of DFMA. In addition to being moderately inhibitory to both germination and protonemal extension, 1 to 5 m*M* Put had bizarre morphogenetic effects in the DFMA-treated protonemata. Protonemata were frequently greatly enlarged in diameter, and were covered with bubble-like swellings, especially just behind the apex, in the normally elongating zone.[30] Some of these bubbles swelled considerably during subsequent growth, and a few gave rise to unique, nonseptate branches of the protonemal tubes.

While it has not yet been possible to analyze these minute protonemata for polyamines, we assume that DFMA acts to lower cellular putrescine levels by inhibiting ADC. If so, the fact that the lowering of putrescine leads to increased germination and protonemal growth leads to the hypothesis that putrescine normally limits both processes. This could be rationalized if polyamines were located in the wall and served as a stiffening agent, like Ca^{2+} in higher plant systems.[31] Lowering of putrescine levels would lower wall resistance to osmotically-driven extension and, if this occurred in limited regions because of localized inhibition of ADC by DFMA, then bubbling and the subsequent observed events could occur. The exogenous putrescine essential to these phenomena could stabilize the system, although only in limited regions and within an unusual morphogenetic context. Recent reports indicate that both free and conjugated polyamines may be present in the plant cell wall,[32,33] consistent with the mechanism proposed here for fern protonemata.

In a few protonemata, approximating 0.5 to 1.0% of all those observed, antheridium-like cells were observed on the protonemata treated with 0.5 m*M* DFMA + 5 m*M* putrescine. While we could not definitely ascertain that they contained sperms, these unusual cells were never observed on the control protonemata, and are thus a consequence of the DFMA + putrescine treatment.

C. POLYAMINES AND SOMATIC EMBRYOGENESIS

Since the classic studies of F. C. Steward and co-workers at Cornell in the late 1950s-early 1960s,[34] several plant cell culture systems have been developed which allow the formation of large numbers of somatic embryos under fairly reproducible conditions. Although somatic embryogenesis can now be obtained in a large number of mono- and dicotyledonous plant species, most studies of the process at the physiological/biochemical/molecular level have been carried out with the classic system, phloem-derived cell cultures of wild carrot (Queen Ann's Lace). We will therefore emphasize this system in our discussion. The initial studies on carrot cell embryogenesis involved the "induction" of this process through culture in fairly complex media, notably upon addition of liquid endosperms such as coconut milk. It was later found, however, that carrot cell cultures competent to undergo embryogenesis could be maintained in chemically defined media supplemented with 2,4-dichlorophenoxyacetic acid (2,4,-D) at 0.1 to 10 μg/l. Embryogenesis would occur when the cells where transferred to fresh medium devoid of exogenous auxin. It is using this basic experimental protocol that most studies on the relation of polyamines and somatic embryogenesis have been carried out.

The first studies on polyamines and somatic embryogenesis were reported by Montague and co-workers.[35-36] They observed that the level of Put increased approximately twofold within 24 h after transfer of carrot cell cultures to a medium lacking 2,4,-D. A higher incorporation of label from ^{14}C-arginine paralleled this early rise in Put, and was also correlated with a transient increase in the activity of ADC.[36] Ornithine decarboxylase (ODC), the alternative putrescine biosynthetic enzyme, was not followed in these studies.

The availability of specific irreversible polyamine biosynthetic inhibitors in the early 1980s made it possible to further probe the correlations between polyamine levels and somatic embryogenesis. Feirer et al.[37] reported that addition of DFMA (1 mM) but not DFMO, inhibited embryo formation in carrot cells by up to 50%. Embryo production could be restored to control levels by addition of 1 to 5 mM Put, Spd, or spermine (Spm). Other inhibitors of polyamine biosynthesis produced similar effects; MGBG treatment resulted in lower levels of Spd and Put, accompanied by a reduction in embryogenesis and growth.[38] The above results were confirmed independently by Fienberg et al.[39] As reported previously,[35-36] ADC activity and the levels of Put, Spd, and Spm increased significantly in embryogenic carrot cells over the undifferentiating controls. Activity of *S*-adenosylmethionine decarboxylase (SAMDC) also showed a peak preceding embryogenesis, but no change in control cultures. A variant carrot cell line (W001), selected for resistance to 5-methyltryptophan and unable to undergo embryogenesis in the same conditions as for the wild type cell line (W001C)[40] failed to show changes in the above enzyme activities or in polyamine titers. The inhibitory effects of DFMA could also be reversed by Spd; furthermore, cyclohexylamine (CHA), an inhibitor of spermidine synthase, was also able to inhibit embryogenesis at 0.25 to 5.0 m*M*. Until very recently in both the animal and plant polyamine literature, CHA has been referred to as dicyclohexylamine (DCHA). It is now well established that DCHA does not inhibit spermidine synthase,[41] and what was being marketed as DCHA was actually CHA (see Chapter 2). A recent report is also consistent with the earlier picture, but also suggests that the situation is more complex than previously thought. When DFMO at 5 mM is added to carrot cells uring the preembryogenic phase, Put accumulation is drastically inhibited, but subsequent embryo formation is not affected.[42] L-Canavanine (CAV), a competitive inhibitor of ADC, blocks growth in the preembryogenic phase as well as latter embryo formation. DFMO appears to prolong the commitment of cells to embryogenesis; embryos treated with DFMO for 36 hours eventually developed into larger plants at a higher frequency, and grew better than the control plantlets. Using a similar approach, Fobert and Webb have suggested[43] that high levels of free Spd are important for embryogenesis of eggplant (*Solanum melongena*).

In addition to experiments involving polyamine biosynthetic inhibitors, somatic embryogenesis has also been studied in relation to organic nitrogen nutrition. It has been shown that when carrot cells are transferred to an embryogenic medium in the presence of 40 mg/l arginine, the cell clusters did not develop past the globular stage.[44] Exogenous Put also arrests the embryos at an early stage. However, when arginine is withdrawn from the medium, synchronous development of the arrested embryos occurs. It thus appears that under normal conditions, the endogenous Put derived from arginine may have a dramatic effect on cells competent to undergo embryogenesis.

Embryogenesis *in vitro* can also be obtained from immature pollen grains (androgenetic embryos), as in the classic reports on *Datura innoxia* and *Nicotiana tabacum*.[45-46] In microspores and anthers cultured *in vitro*, Put and Spd levels rise during embryogenesis, and androgenetic induction is preceded by a peak in arginine levels.[47] DFMA, but not DFMO, can inhibit embryogenesis of barley microspores,[48] and the inibition can be partially reversed by exogenous Spd. When Spd and Spm synthesis, and consequently embryogenesis, are inhibited in this system, endogenous Put accumulates.

The results discussed above, especially those involving specific inhibitors and reversibility by added polyamines, have led to the suggestion that polyamines are causally involved in somatic embryogenesis.[37-39] While the data are strongly suggestive, we must urge caution in their interpretation. In all cases studied, the inhibition of somatic embryogenesis is accompanied by a reduction in growth which can be very significant. As is typical for plant morphogenesis, it is very difficult to separate the effects of the various experimental treatments on growth (cell division and elongation) from those on morphogenesis per se. It may be useful to revisit these systems and attempt to define conditions which affect growth minimally while still inhibiting embryogenesis. Caution is also suggested by recent reports. For example, polyamine levels undergo significant changes in nucellar tissues of mango (*Mangifera indica*) as well as embryogenic callus.[49] However, Put in particular can reach very high levels even in nonembryogenic callus. Furthermore, addition of polyamines to the culture medium had no effect on callus formation or somatic embryogenesis; Spd addition actually suppressed callus growth in monoembryonic cultures. The relevance of polyamines to somatic embryogenesis has also been called into question in alfalfa (*Medicago sativa*) tissue cultures.[50] In two highly regenerable alfalfa genotypes, Put levels increased up to 32-fold during exposure to "induction" medium containing 2,4-D and kinetin, and fell dramatically after transfer to hormone-free differentiation medium. The rapid decline in Put coincided with the initiation of rapid tissue growth during early embryogenesis. Addition of DFMO, DFMA, CHA, and MGBG led to significant decreases in Put, Spd, and Spm titers in both genotypes, but inhibited somatic embryogenesis in only one of them. It was concluded that the pronounced changes observed in Put metabolism are not specifically related to embryogenesis, because similar changes in Put are also observed in response to abiotic stress (see review by Flores[51]) as well as during induction of root, shoot or callus formation.[52-54] A further complication in the interpretation of these results is the change in our conceptual approach to somatic embryogenesis competence and induction borne out by recent biochemical and molecular studies. While early work suggested that withdrawal of 2,4-D from the culture medium had an inductive effect on somatic embryo formation, it is currently accepted that, at least in carrot cultures, the cells can actually undergo development along an embryogenetic pathway while 2,4-D is still present in the medium, organizing into preglobular structures.[55] In this view, deletion of this growth regulator has a permissive rather than an inductive effect. The implication of this recent work is that, in attempting to identify biochemical markers and/or causal events for somatic embryogenesis, in the future we will have to look at earlier culture stages than have been examined so far.

D. POLYAMINES AND TUBER FORMATION

As discussed above and in Chapter 4, free and conjugated polyamines have been implicated in flower induction/initiation. There is also classic physiological evidence indicating that flower induction and tuberization may be linked.[56] As is well known, basic amides accumulate in the top shoots of tobacco during floral initiation.[10] In the potato interspecific hybrid SDA 6, tuberization is induced by short days, while flowering requires long days.[57] Paralleling these requirements for differential morphogenesis under and above ground, respectively, hydroxycinnamic acid amides (HCAs) accumulated in the stolon under short days whereas similar compounds appeared in stems and leaves under long days. The levels of HCAs in nontuberizing stolons were consistently low. Nitsch has shown that the tuberization signal in Jerusalem artichoke is graft-transmissible.[58] Chailakyan et al.[56] and Martin et al.[59] have reported that when a tobacco plant induced to flower by exposure to the appropriate photoperiod is grafted to a control (noninduced, nontuberizing) potato plant, the latter is induced to form tubers. These striking results suggest that the flowering and tuberizing stimulus are one and the same, are graft transmissible in accordance with the classic florigen hypothesis, and that the target meristems can be affected in dramatically different ways according to their developmental competence.

The above observations suggest that polyamines may be involved in tuberization. The problem has been recently approached by making use of *in vitro* tuberization systems. One such system utilizes explants obtained from etiolated stolons[60] and the other uses node cuttings derived from light-grown shoot cultures.[61] In both cases, tuberization occurs in a medium containing high levels of sucrose; cytokinin addition can enhance the rate of tuber formation, but does not seem to be an essential requirement, at least in the shoot culture system. In fact, Leubner and Amrhein[62] reported that potato tubers can form *in vitro* with or without benzyladenine (BA) addition. The growth response was similar in both conditions; however, after 30 days of culture in BA-containing medium, the HCA content of the explants was up to tenfold higher than in the controls. Inhibition of HCA synthesis by blocking the formation of the cinnamic acid precursors with an inhibitor of phenylalanine ammonia lyase (PAL), had little or no effect on tuberization. Based on these results, they suggested that it is unlikely that HCAs are causally involved in tuber formation. In our laboratory, we have adapted the shoot culture system, and confirmed several of these previous findings.[63] The axillary bud in our nodal explants develops into a minituber within 1 to 2 weeks of culture in high sucrose medium; exogenous cytokinin has little effect on tuberization. Addition of 1 mM aminooxyphenylpropionic acid (AOPP), a competitive inhibitor of PAL, has no effect on tuber formation. However, addition of DFMO completely blocks tuberization, while DFMA has no effect. Our results suggest that conjugated polyamines are not directly involved in tuber formation *in vitro*, but that Put derived via the ODC pathway is necessary for the process to take place. Whether this is because Put is necessary for cell proliferation as opposed to further morphogenesis, is not yet known.

E. POLYAMINES AND ROOT FORMATION

As in the other morphogenetic processes discussed in this chapter, the initial evidence relating polyamines to rhizogenesis is correlative. As is discussed in detail in Chapter 10, the correlation between high polyamine titers and high rates of cell division is well established in microbial, animal and plant cells. In plant systems, however, one must be careful in applying these correlations indiscriminately to both di- and polyamines. When polyamine titers are monitored in shoot and root apices and analyzed on a fresh weight basis, Spd and Spm show a clear gradient, being very high in the meristematic regions, and declining rapidly as one moves into the maturation (cell elongation) region.[47,64] When expressed on a per unit protein basis, Spd shows a relatively constant distribution, but the Spm gradient is still present; in fact, Spm is not detectable below the subapical region of the third internode in etiolated pea seedlings. Very similar gradients are observed in pea and corn roots.[65] In contrast, Put (and in the case of peas, also cadaverine) show completely opposite gradients, increasing sharply toward the region of cell elongation. In organs such as corn coleoptiles,[65] which lack a meristematic region, no significant gradients of di- or polyamines are present. The existence of these reverse gradients has been confirmed by other workers.[66] While we still do not know how these gradients arise, it is clear fom these studies that Spd and Spm are the only polyamines positively correlated with meristematic activity in shoot and root meristems.

Adventitious root formation induced by indolebutyric acid (IBA) has been studied in mung bean hypocotyls.[67-69] IBA-induced root formation is accompanied by large increases in Put, Spd, and Spm titers; these changes in polyamines were IBA concentration-dependent. Addition of polyamines and/or their precursors had no noticeable effect on root formation, but rooting was inhibited by CAV, canaline (another competitive ADC inhibitor), and MGBG.[67] Canaline is an aminooxy analog of ornithine, and is often used as an inhibitor of ODC. However, it can also inhibit many pyridoxal phosphate-dependent enzymes quite potently, so the results obtained with this inhibitor, in this and other systems, could involve significant side effects (see Chapter 2). The increases in Put, Spd, and Spm were observed prior to any detectable development of root primordia. MGBG affected both root formation and growth, reducing the levels of Spd/Spm while causing Put accumulation. This is consistent with MGBG acting in this case as an inhibitor of SAMDC. In this latter study,[68] Spm addition enhanced both adventitious root formation and elongation. Consistent with these findings, the incorporation of ^{14}C-arginine and ^{14}C-ornithine into Put and Spd correlated closely to the pattern of IBA-induced root formation. A first peak in label incorporation was detectable at 0 to 24 hours, corresponding to root primordial development; a second peak in Put biosynthesis was correlated with root elongation.[69] The above results are consistent with the hypothesis that auxin-induced root formation involves the promotion of polyamine biosynthesis, but the evidence is mostly correlative. Further work along these lines has been reported recently by the same group, using corn roots.[70-71] As we have previously observed,[64-65] Altman and co-workers have shown that Spd and Spm levels, as well as the incorporation of label from ^{14}C-ornithine, are highest in the root meristem and decrease toward the base. In contrast, incorporation of label from either ornithine or arginine into Put is at least several-fold higher than incorporation into Spd or Spm, and increases dramatically toward the base.[71] These changes are paralleled by gradients in ADC and ODC. The label from 5-^{14}C-DFMO also shows heavy localization in both terminal and lateral root meristems, supporting the association of active polyamine biosynthesis with regions of cell division. Because the meristem is the ultimate source of cells for both primary and lateral roots, one would expect that root growth and elongation are directly dependent on the cell division activity in the meristem. It is therefore surprising that addition of DFMO to primary roots of corn had little effect on root elongation.[71] However, other inhibitors (MGBG, CHA, and CAV) did cause a decrease in root length. This group has also shown that the reactivation of the quiescent center obtained by excision of the root cap brought about an increase in Spd content, which paralleled the renewed DNA synthetic activity as measured by ^{3}H-thymidine incorporation.

Results obtained with tree species (*Prunus avium*) also supports the contention that polyamines may play a significant role in early stages of root formation.[72] Addition of DFMO, CHA, and MGBG resulted in an inhibition of rooting, which could be partially reversed by the addition of the corresponding polyamine. Root initiation promoted by growth retardants (ancymidol, chlormequat, and paclobutrazol) is also paralleled by considerable increase in polyamine levels.[73]

III. CONCLUDING REMARKS

The experimental work we have summarized in this chapter is a rather heterogeneous assemblage relating polyamine titer and polyamine biosynthesis to several important plant morphogenetic processes. A large body of correlative evidence is available, but at the present time little progress has been made beyond this stage, even taking into account experiments with specific inhibitors of polyamine biosynthesis (notably DFMO and DFMA). In our opinion, the main difficulty arises not from lack of experimental systems or adequate experimental design, but from the fact that growth (cell division and elongation) and morphogenesis in plant cells are tightly linked. We must also deal with the very real possibility that for most, if not all of the processes we have examined, polyamines are involved only during the active phases of cell division, and have litle to do directly with morphogenesis per se. If polyamines are causally involved in any type of morphogenesis, however, it may be possible in the future, by judicious timing and dosage of inhibitors and/or other treatments, to separate the processes. The availability of clones for the major genes involved in polyamine biosynthesis (see Chapter 9) will make possible in the near future to manipulate the endogenous levels of polyamines in transgenic plants expressing those genes under the control of tissue-specific or inducible promoters. These additional tools will provide a new stimulus and a new angle for probing the involvement of polyamines in plant growth and development.

ACKNOWLEDGMENTS

The authors would like to thank Ms. Paula Sgrignoli for help with typing and editing this manuscript.

REFERENCES

1. **Barbiroli, B., Mezzetti, G., and Moruzzi, M. S.,** Polyamines and the initiation of protein synthesis in eukaryotes, in *The Physiology of Polyamines*, Vol. I, Bachrach, U. and Heimer, Y.M., Eds., CRC Press, Boca Raton, FL, 1989, 125.
2. **Algranati, I.D. and Goldenberg, S.H.,** Effects of polyamines and antibiotics on the structure and function of ribosomes, in *The Physiology of Polyamines*, Vol. I, Bachrach, U. and Heimer, Y.M., Eds., CRC Press, Boca Raton, FL, 1989, 143.
3. **Jacob, S.T. and Stether, D.A.,** Polyamines and RNA synthesis, in *The Physiology of Polyamines*, Vol. I, Bachrach, U. and Heimer, Y.M., Eds., CRC Press, Boca Raton, FL, 1989, 133.
4. **Bagni, N.,** Polyamines in plant growth and development, in *The Physiology of Polyamines*, Vol. II, Bachrach, U. and Heimer, Y.M., Eds., CRC Press, Boca Raton, FL, 1989, 107.
5. **Heby, O.,** Polyamines and cell differentiation, in *The Physiology of Polyamines*, Vol. I, Bachrach, U. and Heimer, Y.M., Eds., CRC Press, Boca Raton, FL, 1989, 83.
6. **Altman, A.,** Polyamines and plant hormones, in *The Physiology of Polyamines*, Vol. II, Bachrach, U. and Heimer, Y.M., Eds., CRC Press, Boca Raton, FL, 121, 1989.
7. **Galston, A.W., Kaur-Sawhney, R., Tiburcio, A.F., Hamasaki, N., Oshima, T., and Furuya, M.,** The control of morphogenesis by polyamines, in *Polyamines and Ethylene: Biochemistry, Physiology, and Interactions*, Flores, H.E., Arteca, R.N., and Shannon, J.C., Eds., American Society of Plant Physiologists, Rockville, MD, 1990, 224.
8. **Tabor, C.W. and Tabor, H.,** Polyamines, *Annu. Rev. Biochem.*, 53, 749, 1984.
9. **Vince-Prue, D.,** *Photoperiodism in Plants*, McGraw Hill, New York, 1975.

10. **Martin-Tanguy, J.,** The occurrence and possible significance of hydroxycinnamoyl acid amides in plants, *Plant Growth Regul.*, 3, 381, 1985.
11. **Malmberg, R.L. and McIndoo, J.,** Abnormal floral development of a tobacco mutant with elevated polyamine levels, *Nature*, 305, 623, 1983.
12. **Malmberg, R.L., McIndoo, J., Hiatt, A.C., and Lowe, B.,** Genetics of polyamine synthesis in tobacco-developmental switches in the flower, *Cold Spring Harbor Symp.* 50, 475, 1985.
13. **Bendeck de Cantu, L. and Kandeler, R.,** Significance of polyamines for flowering in *Spirodela punctata*, *Plant Cell Physiol.*, 30, 455, 1989.
14. **Martin-Tanguy, J., Perdrizet, E., Prevost, J., and Martin, C.,** Hydroxycinnamic acid amides in fertile and cytoplasmic male sterile lines of maize, *Phytochemistry*, 21, 1939, 1982.
15. **Ponchet, M., Martin-Tanguy, J., Marais, A., and Martin, C.,** Hydroxycinnamoyl acid amides and aromatic amines in the inflorescence of some *Araceae* species, *Phytochemistry*, 21, 2865, 1982.
16. **Tran Thanh Van, M., Thi Dien, M., and Chlyah, A.,** Regulation of organogenesis in small explants of superficial tissue of *Nicotiana tobacum* L., *Nature*, 314, 615, 1974.
17. **Kaur-Sawhney, R., Tiburcio, A.F., and Galston, A.W.,** Spermidine and flower bud differentiation in thin-layer explants of tobacco, *Planta*, 173, 615, 1988.
18. **Kaur-Sawhney, R., Kandpal, G., McGonigle, B., and Galston, A.W.,** Further experiments on spermidine-mediated floral bud formation in thin layer explants of Wisconsin 38 tobacco, *Planta*, 181, 212, 1990.
19. **Protacio, C.M. and Flores, H.E.,** Polyamine metabolism in *in vitro* flower system, *Plant Physiol.*, 89, S-586, 1989.
20. **Bernier, G.,** The control of floral evocation and morphogenesis, *Annu. Rev. Plant Physiol. Plant Mol. Biol.*, 39, 175, 1988.
21. **Tran Thanh Van, K., Toubart, P., Cousson, A., Darvill, A., Gollin, D.J., Chelf, P., and Albersheim, P.,** Manipulation of the morphogenetic pathway of tobacco explants by oligosaccharins, *Nature*, 314, 615, 1985.
22. **Smulders, M.J.M., Kemp, A., Barendse, G.W.M., Croes, A.F., and Wullems, G.J.,** Role of ethylene in auxin-induced flower bud formation in tobacco explants, *Physiol. Plant.*, 78, 167, 1990.
23. **Shih, L.M., Kaur-Sawhney, R., Fuhrer, J., Samanta, S., and Galston, A.W.,** Effects of exogenous 1,3-diaminopropane and spermidine on senescence of oat leaves. I. Inhibition of protease activity, ethylene production, and chlorophyll loss as related to polyamine content, *Plant Physiol.*, 70, 1592, 1982.
24. **Fuhrer, J., Kaur-Sawhney, R., Shih, L.M., and Galston, A.W.,** Effects of exogenous 1,3-diaminopropane and spermidine on senescence of oat leaves. II. Inhibition of ethylene biosynthesis and possible mode of action, *Plant Physiol.* , 70, 1597, 1982.
25. **Lee, S.H. and Park, K.Y.,** The mechanism of polyamines on ethylene biosynthesis in tobacco suspension cultures, *Korean J. Bot.*, 31, 267, 1988.
26. **Cohen, E., Malis-Arad, S., Heimer, Y., and Mizrahi, J.,** Participation of ornithine decarboxylase in early stage of tomato fruit development, *Plant Physiol.*, 70, 540, 1982.
27. **Slocum, R.D. and Galston, A.W.,** Changes in polyamine biosynthesis associated with postfertilization growth and development in tobacco ovary, *Plant Physiol.*, 79, 336, 1985.
28. **Furuya, M.,** Photomorphogenesis in ferns, in *Photomorphogenesis. Encyclopedia of Plant Physiology*, N.S., Vol. 16, Shropshire, W. Jr. and Mohr, H., Eds., Springer-Verlag, Berlin, 1983, 569.
29. **Slocum, R., Kaur-Sawhney, R., and Galston, A.W.,** The physiology and biochemistry of polyamines in plants, *Arch. Biochem. Biophys.*, 235, 283, 1984.
30. **Galston, A.W. and Kaur-Sawhney, R.,** Polyamines and senescence in plants, in *Plant Senescence: Its Biochemistry and Physiology*, Thomson, W.W., Nothnagel, E.A., and Huffaker, R.C., Eds., American Society of Plant Physiologists, Rockville, MD, 1987, 167.
31. **Cleland, R.E., Virk, S.S., Taylor, D., and Bjorkman, T.,** Calcium, cell walls and growth, in *Calcium in Plant Growth and Development,* Leonard, R.T. and Hepler, P.K., Eds., American Society of Plant Physiologists, Rockville, MD, 1990, 9.
32. **Goldberg, R. and Perdrizet, E.,** Ratio of free to bound polyamines during maturation in mung bean hypocotyl cells, *Planta* , 161, 531, 1984.
33. **Langebartels, C., Kerner, K., Schraudner, M., Trost, M., Heller, W., and Sandermann, H., Jr.,** Biochemical plant responses to ozone. I. Differential induction of polyamine and ethylene biosynthesis in tobacco, *Plant Physiol.*, 95, 882, 1991.
34. **Steward, F.C.,** Growth and development of cultivated cells. III. Interpretations of the growth from free cell to carrot plant, *Amer. J. Bot.*, 45, 709, 1958.
35. **Montague, M.J., Koppenbrink, J.W., and Jaworski, E.G.,** Polyamine metabolism in embryogenic cells of *Daucus carota*. I. Changes in intracellular content and rates of synthesis, *Plant Physiol.*, 62, 430, 1978.
36. **Montague, M.J., Koppenbrink, J.W., and Jaworski, E.G.,** Polyamine metabolism in embryogenic cells of *Daucus carota*. II. Changes in arginine decarboxylase activity, *Plant Physiol.*, 63, 341, 1979.
37. **Feirer, R.P., Mignon, G., and Litvay, J.D.,** Arginine decarboxylase and polyamines required for embryogenesis in wild carrot, *Science*, 223, 1433, 1984.

38. **Feirer, R.P., Wann, S.R., and Einspahr, D.W.,** The effect of spermidine synthesis inhibitors on *in vitro* plant development, *Plant Growth Regul.*, 3, 319, 1985.
39. **Fienberg, A.A., Choi, J.H., Lubich, W.P., and Sung, Z.R.,** Developmental regulation of polyamine metabolism in growth and differentiation of carrot culture, *Planta*, 162, 532, 1984.
40. **Sung, Z.R., Smith, R., and Horowitz, J.,** Quantitative studies of embryogenesis in normal and 5-methyltryptophan-resistant cell lines of wild carrot: the effects of growth regulators, *Planta*, 147, 236, 1979.
41. **Batchelor, K.W., Smith, R.A., and Watson, N.S.,** Dicyclohexylamine is not an inhibitor of spermidine synthase, *Biochem. J.*, 233, 307, 1986.
42. **Mengoli, M., Bagni, N., Luccarini, G., Nutti Ronchi, V., and Serafini-Fracassini, D.,** *Daucus carota* cell cultures: polyamines and effect of polyamine biosynthesis inhibitors in the preembryogenic phase and different embryo stages, *J. Plant Physiol.*, 134, 389, 1989.
43. **Fobert, P.R. and Webb, D.T.,** Effect of polyamines, polyamine precursors, and polyamine biosynthetic inhibitors on somatic embryogenesis from eggplant (*Solanum melongena*) cotyledons, *Can. J. Bot.*, 66, 1734, 1988.
44. **Bradley, P.M., El-Fiki, F., and Giles, K.L.,** Polyamines and arginine affect somatic embryogenesis of *Daucus carota*, *Plant Sci. Lett.*, 34, 397, 1984.
45. **Guha, S. and Maheshwari, S.C.,** *In vitro* production of embryos from anthers of *Datura*, *Nature*, 204, 497, 1964.
46. **Bourgin, J.P. and Nitsch, J.P.,** Obtention de *Nicotiana* haploides a partir d'étamines cultivées *in vitro*, *Ann. Physiol. Veg.*, 9, 377, 1967.
47. **Flores, H.E., Protacio, C.M., and Signs, M.W.,** Primary and secondary metabolism of polyamines in plants, *Rec. Adv. Phytochem.*, 23, 329, 1989.
48. **Higgins, P.J., Lyne, R., and Bowles, D.J.,** Polyamines and embryogenesis in barley anther culture, in *Amines in Plants*, European Phytochemical Society Meeting, England, December, 1987.
49. **Litz, R.E., and Schaffer, B.,** Polyamines in adventitious and somatic embryogenesis in mango (*Mangifera indica* L.), *J. Plant Physiol.*, 128, 251, 1987.
50. **Meijer, E.G.M. and Simmonds, J.,** Polyamine levels in relation to growth and somatic embryogenesis in tissue cultures of *Medicago sativa* L., *J. Expt. Bot.*, 39, 787, 1988.
51. **Flores, H.E.,** Polyamines and plant stress, in *Stress Responses in Plants: Adaptation and Acclimation Mechanisms*, Alscher, R G. and Cumming, J.R., Eds., Wiley-Liss, New York, 1990, 217.
52. **Chriqui, D., D'Orazi, D., and Bagni, N.,** Ornithine and arginine decarboxylases and polyamine involvement during *in vivo* differentiation and *in vitro* dedifferentiation of *Datura innoxia* leaf explants, *Physiol. Plant.*, 68, 589, 1986.
53. **Torrigiani, P., Altamura, M.M., Pasqua, G., Monacelli, B., Serafini-Fracassini, D., and Bagni, N.,** Free and conjugated polyamines during *de novo* floral and vegetative bud formation in thin cell layers of tobacco, *Physiol. Plant.*, 70, 453, 1987.
54. **Torrigiani, P., Altamura, M.M., Capitani, F., Serafini-Fracassini, D., and Bagni, N.,** *De novo* root formation in thin cell layers of tobacco: changes in free and bound polyamines, *Physiol. Plant.*, 77, 294, 1989.
55. **Choi, J.H. and Sung, Z.R.,** Induction, commitment, and progression of plant embryogenesis, in *Plant Biotechnology*, Kung, S.-D. and Arntzen, C.J., Eds., Butterworths, Boston, 1989, 141.
56. **Chailakhyan, M.K., Yanina, L.I., Devedzhian, A.G., and Lotova, G. N.,** Photoperiodism and tuber formation in grafting of tobacco onto potato, *Doklady Akad. Nauk. USSR.*, 257, 1276, 1981.
57. **Paynot, M., Perennec, P., Martin, C., Martin-Tanguy, J., Vernoy, R., and Javelle, F.,** Photoperiodisme, tuberisation, et phénolamides, *C.R. Acad. Sci. Paris*, 297, 87, 1983.
58. **Nitsch, J.P.,** Existence d'un stimulus photoperiodic non specifique capable de provoquer la tuberisation chez *Helianthus tuberosus* L., *Bull. Soc. Bot. Fr.*, 112, 333, 1965.
59. **Martin, C., Vernay, R., and Paynot, N.,** Physiologie végétale. Photoperiodisme, tuberisation, floraison et phénolamides, *C.R. Acad. Sci. Paris* , 295, 565, 1982.
60. **Palmer, C.E. and Smith, O E.,** Cytokinins and tuber initiation in the potato *Solanum tuberosum* L., *Nature*, 221, 279, 1969.
61. **Estrada, R., Tovar, P., and Dodds, J.H.,** Induction of *in vitro* tubers in a broad range of potato genotypes, *Plant Cell Tissue Organ Cult.*, 7, 3, 1986.
62. **Leubner, G. and Amrhein, N.,** Are hydroxycinnamoylputrescines causally involved in the tuberisation process of potato plants?, in *Proc. I Int. Symp. Molecular Biology of the Potato*, Bar Harbor, Maine, 1989.
63. **Protacio, C.M. and Flores, H.E.,** unpublished data, 1991.
64. **Flores, H.E.,** Studies on the physiology and biochemistry of polyamines in higher plants, *Ph.D. thesis*, Yale University, New Haven, 1983.
65. **Dumortier, F.M., Flores, H.E., Shekhawat, N.S., and Galston, A.W.,** Gradient of polyamines and their biosynthetic enzymes in coleoptiles and roots of corn, *Plant Physiol.*, 72, 915, 1983.
66. **Federico, R. and Angelini, R.,** Distribution of polyamines and their catabolic enzyme in etiolated and light-grown leguminosae seedlings, *Planta*, 173, 317, 1988.

67. **Friedman, R., Altman, A., and Bachrach, U.,** Polyamines and root formation in mung bean hypocotyl cuttings. I. Effects of exogenous compounds and changes in endogenous polyamines, *Plant Physiol.*, 70, 844, 1982.
68. **Jarvis, B.C., Shannon, P.R.M., and Jasmin, S.,** Involvement of polyamines with adventitious root development in stem cuttings of mung bean, *Plant Cell Physiol.*, 24, 677, 1983.
69. **Friedman, R., Altman, A., and Bachrach, U.,** Polyamines and root formation in mung bean hypocotyl cuttings, *Plant Physiol.*, 79, 80, 1985.
70. **Schwartz, M., Altman, A., Cohen, Y., and Arzee, T.,** Localization of ornithine decarboxylase and changes in polyamines content in root meristems of *Zea mays*, *Physiol. Plant.*, 67, 485, 1986.
71. **Altman, A., Schwartz, M., Cohen, Y., and Arzee, T.,** A role for polyamines in differentiation and growth of plant roots, in *The Biology and Chemistry of Polyamines*, Goldemberg, S. H. and Algranati, I. D., Eds., IRL Press, Oxford, 1990, 147.
72. **Biondi, S., Diaz, T., Iglesias, I., Gamberini, G., and Bagni, N.,** Polyamines and ethylene in relation to adventitious root formation in *Prunus avium* shoot cultures, *Physiol. Plant.*, 78, 474, 1990.
73. **Wang, S.Y. and Faust, M.,** Effect of growth retardants or root formation and polyamine content in apple seedlings, *J. Amer. Soc. Hort. Sci.*, 111, 912, 1986.

Chapter 13

EFFECTS OF LIGHT AND PLANT GROWTH REGULATORS ON POLYAMINE METABOLISM IN HIGHER PLANTS

Rajeev Rastogi and Peter J. Davies

TABLE OF CONTENTS

I. INTRODUCTION

The polyamines (PAs) are probably ubiquitously distributed in plant cells, and changes in PA levels and the activities of their biosynthetic enzymes have been implicated in various aspects of plant growth and development.[1-3]

Light[4] and plant hormones[5] clearly play important roles in the control of plant growth and development. Interestingly, striking changes in PA levels and their biosynthesis accompany many hormonal and light responses, and in a few cases PAs appear to mimic the effect of plant hormones. This has led some researchers to suggest that PAs may mediate the action of light and plant hormones or are part of their signal response pathways. However, unlike the animal systems where it has been documented that not only PA biosynthesis is considerably increased in response to hormones, but the PAs produced as a result are necessary for the hormone action,[6,7] in plants it is not clear whether changes in PA levels are essential for plant hormone action or they simply show correlations with plant responses to plant hormones.

The aim of this chapter is to review the effects of light and plant hormones on PA metabolism in higher plants. The effect of ethylene on PA metabolism is not discussed in this article, and readers' attention is directed to Chapter 5 in this volume, which explores, in detail, the metabolic and physiological interactions of ethylene and PAs.

II. CONTROL OF POLYAMINE METABOLISM BY LIGHT

Although light regulates a variety of developmental processes in plants, only a few studies have examined the relationship of light and PA metabolism. The effects of both the quality and duration of light have been investigated. When detached leaves of *Avena sativa* are incubated in darkness, their senescence, as measured by the loss of chlorophyll and protein, is accompanied by a decrease in arginine decarboxylase (ADC) activity and spermidine (Spd) levels.[8] White light, which retards senescence of such leaves, results in an increase in ADC activity and Spd levels, suggesting a correlation between delay of senescence and increased PA biosynthesis. Whether the increase in ADC activity and Spd content was specifically activated by light is, however, not clear.

In etiolated pea seedlings, red light exposure inhibits the growth of subapical epicotyl internodes but promotes the growth of buds. Red light treatment also decreased ADC activity in the epicotyls (Figure 1a), but increased it in the buds (Figure 1b) by threefold.[9] The red light effect on ADC activity in the two organs was seen within 2 h after red irradiation. This is prior to the maximum inhibition of the epicotyl growth rate by red light which occurs after 3 h.[10] Far-red light reversed the effect of red light on both growth and ADC activity in both the organs indicating phytochrome control of this response.[9] In the buds, the amounts of agmatine, putrescine (Put), and Spd were also increased 6 to 18 h after red light exposure.[11] A reverse effect of red light was noted on PA levels in the internodes. Whether the rise or fall of ADC activity in the pea seedlings is merely a consequence of changes in growth rate controlled by phytochrome was investigated by studying the effect of red light on ADC activity in the excised buds.[12] In buds excised with only a short piece of epicotyl, red light did not promote bud growth, but it still promoted ADC activity, and its effect on ADC activity was reversed by far-red light. These experiments indicate that in the dark-grown pea seedlings, changes in ADC activity in response to red light are not simply a consequence of altered growth, but phytochrome-regulated changes in ADC activity and PA levels may be causally linked to changes in growth.

Besides influencing PA biosynthesis, light also appears to regulate PA degradation. Diamine oxidase activity was significantly inhibited in etiolated lentil epicotyls after four days of exposure to red light (Table 1).[13] Phytochrome involvement in the regulation of diamine oxidase activity was shown by red/far-red reversibility (Table 1). In lentil epicotyls, therefore, red light

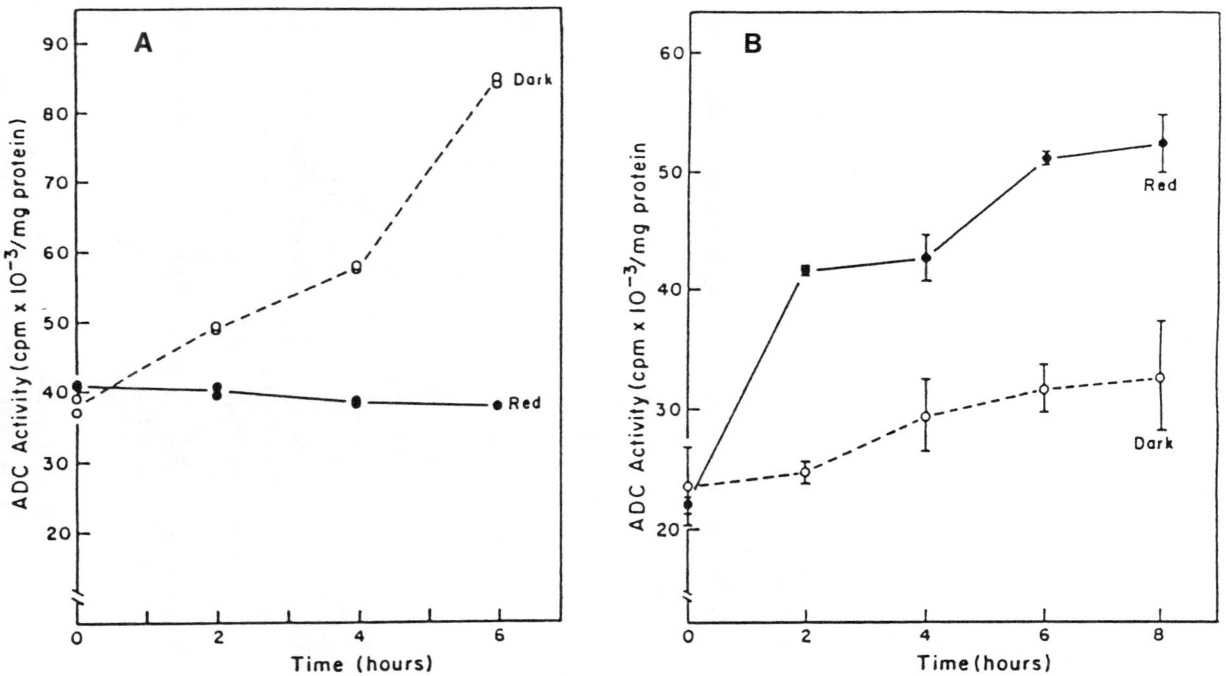

FIGURE 1. The kinetics of change in arginine decarboxylase activity (per mg protein) in epicotyls (A) and buds (B) of etiolated pea seedlings. Dark (empty circles); red light (solid circles).[9]

TABLE 1
Reversibility Experiment Showing Phytochrome Involvement in Light-mediated Control of Diamine Oxidase Activity in Lentil Epicotyls Exposed to Short Light Treatment

Lentil seedlings were grown in complete darkness for three days and then exposed daily to short light treatment (Red (R): 10 min.; Far-red (FR): 15 min.; R + FR: 10 + 15 min). Epicotyls were collected four days after the first light treatment, and the second internodes were excised and processed for diamine oxidase activity determination. Each value represents the mean ± SD of three separate experiments.[13]

	Diamine oxidase activity	
Light treatment	U/g fresh weight[b]	U/internode
DC[a]	7.14 ± 0.19	0.443 ± 0.012
R	5.44 ± 0.39	0.178 ± 0.013
FR	7.55 ± 0.25	0.330 ± 0.011
R + FR	8.39 ± 0.31	0.390 ± 0.014

[a] DC, dark control.
[b] One unit (u) of diamine oxidase activity = amount of enzyme catalyzing oxidation of 1 μmole substrate/min.

would be expected to result in increased PA levels, an effect opposite to that in peas. However, as the levels of PAs were not actually measured in lentil, a direct comparison is not possible.

Smith and Davies[14] investigated the effect of photoperiodism on PA metabolism in apical buds of G2 peas in relation to the induction of apical senescence. The photoperiodic control of senescence of the G2 peas is determined by dominant alleles at the *Sn* and *Hr* loci, and the continued meristematic activity under the short day conditions is associated with enhanced gibberellin activity.[15] Polyamine levels were high in apices of plants grown under a short-day photoperiod during which senescence is delayed, and as the apex senesced under long day treatment, PA levels declined. The decrease in PA concentration, mainly Spd, occurred at the time of reduction in bud size and growth, but not prior to the onset of senescence. Further, the transfer of plants from short-day to long-day regime also did not lead to a decrease in PA concentration before senescence symptoms became visible. Other experiments on the manipulation of the PA content indicated that the internal PA level correlated with apical bud size and vigor but did not show any relationship to bud senescence.[16] Thus, the photoperiodic mechanism which controls apical senescence in peas acts independently of PAs, though PAs appear to be required for continued bud growth.

III. MODULATION OF POLYAMINE METABOLISM BY PLANT HORMONES AND HERBICIDES

A. PLANT HORMONES

The effects of plant hormones on PA metabolism have been examined in many cases. Most studies have analyzed changes in PA levels and biosynthesis following hormone application, but one also employed mutants of differing hormonal status.

1. Auxins

The effects of auxins on PA metabolism were first reported in the tubers of *Helianthus tuberosus* by Bagni and co-workers. In the dormant tubers of this species which contain low PA titers, cell division could be stimulated by exogenous PAs[17] or by treatment with indoleacetic

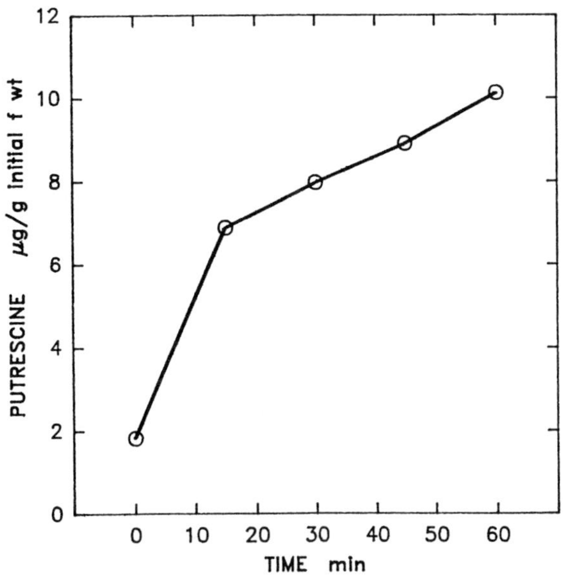

FIGURE 2. Putrescine levels in *Helianthus tuberosus* tuber slices during the first hour of activation by 2,4-dichlorophenoxyacetic acid.[19]

acid (IAA).[18] A large increase in PA concentration accompanied the IAA-induced growth. In another study, Bagni et al.[19] characterized the biochemical changes in the first hour after the break of dormancy in response to 2,4-D (2,4-dichlorophenoxyacetic acid) application. They found that the increase in PA concentration was not only dramatic but also was very quick. Putrescine levels increased 2 to 3 times in the first 15 min of activation and by the end of one hour they had increased by more than fivefold (Figure 2). The levels of Spd and spermine (Spm) showed a similar trend to that of Put. It was also shown that this increase in PA concentration, after auxin activation of the tubers, was concomitant with a decrease in the concentration of glutamine and arginine — the main precursor for Put synthesis in activated *H. tuberosus*.[20] Taken together the above findings suggest that PAs play an important role in the growth of *H. tuberosus* tubers and that the increase in PA levels at the initiation of the break of dormancy may be one of the primary events in the process.

Another auxin, naphthaleneacetic acid (NAA), also induced growth of *H. tuberosus* tubers, and resulted in the accumulation of Spd,[21] though not as dramatically as reported by Bagni et al.[18] Ornithine and arginine metabolism was examined to ascertain whether the auxin-stimulated growth of the tubers was dependent upon the auxin-induced activation of PA biosynthesis, but their incorporation into PAs was insignificant even in the presence of NAA. Since growth induced by NAA was paralleled by a consistent increase in Spd content, and agmatine was detected in the NAA-treated explants, it suggests that PA biosynthesis via the ADC pathway must take place but that it does not appear to be the controlling step. Although this latter study indicated a correlation between the elevated Spd concentration and stimulation of growth by NAA, a causal relationship between the two cannot be drawn because of the lack of evidence of incorporation of precursors into PAs. Thus, in *H. tuberosus* tubers it is not clear whether the increase in PA levels in response to auxins is a nonspecific event which accompanies growth or it is one of the triggers of that growth as suggested by Bagni et al.[19]

Indolebutyric acid (IBA)-induced root formation on mungbean hypocotyl cuttings is also accompanied by a 2- and 1.5-fold increase in Put and Spd levels, respectively.[22] The changes in PAs occurred 24 to 48 h after IBA treatment and, at that time, there was no visible growth of adventitious roots. The initiation of roots in these cuttings, however, occurs 24 h after auxin treatment. The application of PAs either alone or in combination with IBA did not promote the formation of adventitious roots. Inhibitors of PA biosynthetic enzymes inhibited the IBA-induced root formation, and the effects of these inhibitors were reversed by arginine, a precursor of Put, suggesting that the PAs may be necessary for the IBA-induced root formation.

The incorporation of arginine and ornithine into Put and Spd in the IBA-treated mung bean cuttings peaked with the periods of root primordia development and the subsequent root growth and elongation.[23] IBA considerably enhanced Put biosynthesis in both phases, but inhibited incorporation of Put into Spd, thereby resulting in an increased Put to Spd ratio. Thus the auxin-induced increase in PA levels during root formation appears to be due to an increase in PA biosynthesis.

Jarvis et al.[24] also found that the IBA-enhanced root formation was associated with an increase in the levels of Put, Spd and Spm which preceded the development of root primordia. Methylglyoxal-*bis*(guanylhydrazone) (MGBG), an inhibitor of *S*-adenosylmethionine decarboxylase (SAMDC), inhibited rooting and root development and reduced the IBA-induced increase in Spd and Spm levels. Unlike the above studies by Friedman et al.[22] where PAs with or without IBA failed to induce rooting, PAs induced rooting in this study, and this may be related to the different ages of cuttings used in the two studies.

Treatment with IAA was found to induce a fourfold increase in ornithine decarboxylase (ODC) activity during barley grain germination.[25] IAA also stimulated ADC activity in rice embryos.[26] In tomato ovaries, a threefold increase in ODC activity was associated with the naphthoxyacetic acid-induced parthenocarpy.[27] Since endogenous auxins are involved in normal post-fertilization growth in this organ, and this growth is specifically inhibited by DFMO (DL-α-difluoromethylornithine, an inhibitor of ODC) and Put reverses the DFMO effect,[28] it appears that auxin-mediated PA biosynthesis is required for tomato ovary development.

In a few cases, auxins have been shown to have an inhibitory effect on PA levels and their biosynthesis. In carrot, embryogenic cultures show a several-fold increase in PA levels over nondifferentiating controls as well as the non-embryogenic mutant lines. Auxins inhibited embryogenesis as well as suppressing the activities of ADC and SAMDC and reducing Spd and Spm levels.[29] IAA also inhibited the red light-promoted ADC activity in buds of pea seedlings.[30]

In another example where auxins appear to negatively regulate PA levels, Katoh et al.[31] showed that the transfer of barley callus growing on a 2,4-D containing medium to a medium free of 2,4-D resulted in a fivefold increase in the levels of Put and Spm. This increase in PA levels was also accompanied by elevated ODC activity. Apparently, the increase in PA levels was not related to the withdrawal of auxin from the medium; it probably was due to root formation as a result of auxin withdrawal because increase in PA levels did not take place until 60 to 90 days after the removal of auxin.

In callus cultures of *Nicotiana tabacum*, auxins exhibit a concentration-dependent effect on the titers of free Put and nicotine.[32] The callus grown in a low auxin-containing medium ($1 \mu M$) showed a very significant increase in the accumulation of Put and nicotine, as well as ADC and ODC activities, as compared to callus grown on a high auxin-containing medium ($11.5 \mu M$). The high Put levels and ADC and ODC activities at low auxin concentration are apparently related to high nicotine production rather than to the growth of callus, which, in fact, was reduced at low auxin concentration.

Auxins seem to regulate the activity of diamine oxidase, as well. In the cotyledons of pea seedlings, 2,4-D and IAA inhibited diamine oxidase activity.[33]

2. Cytokinins

Benzyladenine and kinetin promote the expansion of excised *Cucumis sativus* cotyledons. This cytokinin response was found to be accompanied by a significant increase in ADC activity and Put content (Figure 3).[34] The increases in ADC activity (Figure 3a) and Put levels (Figure 3b) were observed as early as 12 h after cytokinin treatment, with the maximum occurring at 2.5

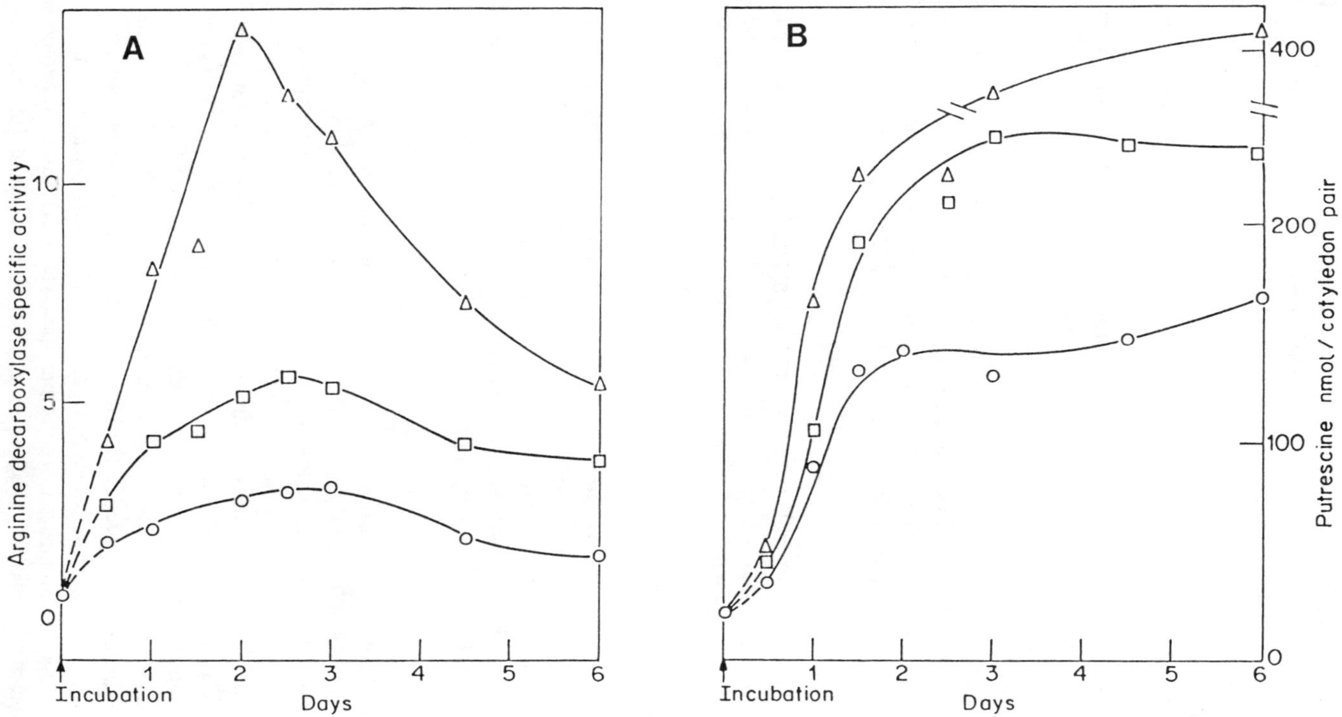

FIGURE 3. Changes in arginine decarboxylase (A), and putrescine levels (B) in *Cucumis sativus* cotyledons with duration of culture. Control (circles); 22 μ*M* benzyladenine (triangles); 46.5 μ*M* kinetin (squares).[34]

days. It was shown that the increase in ADC activity in response to cytokinin was due to a *de novo* synthesis of ADC protein as cycloheximide inhibited the increase in ADC activity. The cytokinins also reversed the inhibitory effects of abscisic acid and potassium on ADC activity and Put titers. Similarly, kinetin and benzyladenine enhanced Put levels by twofold in the cotyledons of lettuce seedlings.[35]

Walker et al.[36] used the cucumber cotyledon greening test to investigate the possible interaction of PAs and cytokinins in the greening process. Cytokinin application to excised dark grown cotyledons resulted in large increases in chlorophyll and Put levels. Application of PAs to cotyledons, however, decreased the synthesis of chlorophyll, whether or not cytokinins were present. The inhibitors of Put biosynthesis did not affect chlorophyll synthesis but reduced Put levels. These results indicate that the cytokinin-stimulated increase in Put levels may not be essential for the cytokinin-induced greening.

Choudhuri and Ghosh[26] observed that kinetin promoted ADC activity in rice embryos. Also, in high and intermediate vigor rice seeds, kinetin increased PA content and ADC activity, and decreased the activity of PA oxidase.[37]

Aging of pea leaves is accompanied by a decrease in PA levels and diamine oxidase activity. Kinetin delayed senescence and retarded the decrease in the activity of diamine oxidase and Spd and Spm levels.[38] Kinetin, however, had no effect on diamine oxidase activity in germinating pea seeds.[33] In the buds of etiolated pea seedlings where red light promotes growth and ADC activity, cytokinins further augmented the activity of ADC by threefold.[30]

3. Gibberellins

The effect of gibberellins on PA metabolism has been examined in a variety of systems. In many cases, gibberellin applications are followed by an increase in PA levels and the activities of their biosynthetic enzymes. In excised cotyledons of *Cucumis sativus*, gibberellic acid (GA_3) significantly stimulated the activity of ADC and also increased the levels of Put.[34] Further, GA_3 also restored the potassium- or abscisic acid-induced inhibition of ADC activity.[34] In germinating pea seedlings, GA_3 resulted in an increase in diamine oxidase activity.[33]

Gibberellic acid, ethylene, or submergence stimulate rapid internode growth in deepwater rice. The internodes show a six- to eightfold increase in the activity of SAMDC and a substantial increase in ADC activity.[39] The concentrations of Put and Spd also are significantly increased. Treatment of excised rice-stem sections with GA_3 or ethylene was found to enhance the activities of SAMDC and ADC, and also inhibit the decline in Put and Spd levels which occurred in the control.

In germinating barley grains, GA_3 enhanced ODC activity by almost four times over the control,[25] and alleviated the inhibitory effect of exogenous PAs on ODC activity. However, in the well known gibberellin-regulated system of α-amylase production in the barley aleurone layers, Lin[40] found that while active PA biosynthesis occurred in the aleurone layers, PA levels or the activities of PA biosynthetic enzymes were not significantly affected by added GA_3 during the eight-hour period of GA_3-induced α-amylase production. MGBG, however, substantially reduced the levels of Spd and α-amylase produced, and this inhibitory effect of MGBG was partially overcome by Spd. Taken together, these experiments show that the early action of GA_3 in the induction of α-amylase in barley aleurone layers is not via PA metabolism, but PAs, particularly Spd, may be involved in the GA_3-dependent amylase production.

During the GA_3-promoted internode elongation in light grown dwarf peas, threefold increase in ADC activity and 1.5-fold increase in Put and Spd levels were observed, which paralleled the promotive effect of GA_3 on internode elongation.[41] Furthermore, an inhibitor of GA_3 biosynthesis reduced stem elongation, ADC activity, and PA levels. Kaur-Sawhney et al.[42] showed that whereas the stimulation of ADC activity by GA_3 appeared three hours after its application and

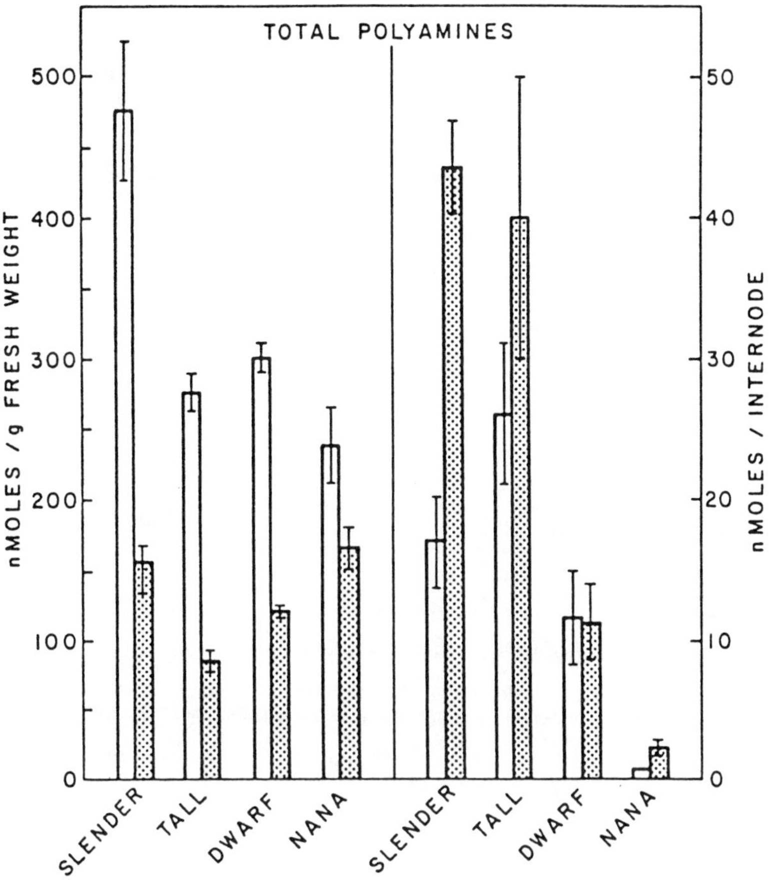

FIGURE 4. The concentrations (left) and amounts (right) of total polyamines (putrescine, spermidine and spermine) in young (clear) and mature (shaded) internodes of four genetic lines of peas with different internode length phenotypes three weeks after sowing. Bars indicate ± S.E.[43]

peaked after nine hours, the growth promotion of the internodes by GA_3 did not occur until 12 h with a distinct promotion occurring after 24 h. Since the increases in ADC activity and PA titers precede the GA_3 stimulation of internode growth, these results suggest that PAs may mediate the hormone-induced growth response.

A different approach to the question of whether gibberellins act via increasing PA levels was adopted by Smith et al.[43] They analyzed PA levels in pea-stem internodes from four different internode phenotypes that have been shown to be regulated by different levels of gibberellins, determined by mutations either in the gibberellin biosynthetic pathway or in the sensitivity to gibberellins. The four phenotypes used were tall, dwarf, nana, and slender. Tall plants contain a normal GA_1 complement; in the dwarf, the conversion of GA_{20} to GA_1 is blocked; the nana mutation blocks the production of all gibberellins, while slender plants grow as if they are saturated with active gibberellins, irrespective of the levels of extractable gibberellins. The quantities of PAs in the mature internodes were directly related to the amount of internode elongation, but PA concentrations (per gram fresh weight), except in slender, showed an inverse correlation with internode length (Figure 4), reflecting a decrease in cell number per unit internode length in the taller plants. In very young internodes with 14 to 20% of final internode length, PA concentrations were higher than that in mature internodes and very similar in all the

TABLE 2
The Effect of Polyamine Biosynthesis Inhibitors on the Levels of Total Polyamines (Putrescine, Spermidine and Spermine) in Internode 4-5 of Tall Peas, and Internode 8-9 of Nana Peas Treated With GA_3[43]

	Internode length	nmol/inter-node	nmol/g fresh wt.
		Tall	
Water	3.7 ± 0.2	23 ± 2	302 ± 13
DFMA	2.8 ± 0.1	10 ± 1	207 ± 12
DFMA + agmatine	3.1 ± 0.4	18 ± 2	289 ± 29
		Nana	
Water	0.2 ± 0.0	2.6 ± 0.3	734 ± 93
GA_3	3.2 ± 0.3	26.6 ± 4.0	533 ± 66
GA_3 + DFMO + DFMA	2.4 ± 0.1	13.8 ± 1.6	400 ± 30
GA_3 + DFMO + DFMA + Put	3.0 ± 0.2	22.9 ± 2.9	480 ± 111

different lines. Despite this lack of correlation between PAs and gibberellin-induced differences in internode lengths, DL-α-difluoromethylarginine (DFMA, an inhibitor of ADC) inhibited internode growth, ADC activity and PA levels in tall and GA_3-stimulated nana plants, while exogenous PAs or agmatine partially restored the growth of the DFMA-inhibited plants (Table 2), indicating that PAs may be required for the full expression of gibberellin-induced internode elongation. Based on their observations, Smith et al.[43] suggested that PAs probably play an important role in cell division rather than cell elongation in the gibberellin-induced internode growth.

The effects on PA metabolism of certain growth retardants which inhibit gibberellin biosynthesis have also been investigated. Ancymidol, chlormequat, and paclobutrazol promoted root initiation and growth in apple seedlings, and this was accompanied by a considerable increase in PA levels.[44] Similar results were obtained with triadimefon in bean hypocotyls,[45] but the changes in PA levels occurred several days after the increase in root primordia number was noted.

4. Abscisic acid (ABA)

ABA was found to further promote the red light-induced ADC activity in buds of etiolated pea seedlings.[30] In rice embryos, however, ABA markedly reduced the levels of Put and Spd, and the activity of ADC, but increased the activity of PA oxidase.[37] A decline in Put levels and ADC activity also accompanied the ABA-inhibited growth of excised *Cucumis sativus* cotyledons.[34]

Lin[46] examined the effect of ABA on PA metabolism in embryoless half-seeds of barley. ABA did not appreciably affect the incorporation of arginine or ornithine into Put, but inhibited their incorporation into Spd, and, as a result, Put accumulated. Apparently, ABA has an effect on Spd metabolism in this system.

B. HERBICIDES

Since herbicides inhibit growth, it is likely that they also affect PA metabolism. DiTomaso et al.[47] examined the effect of napropamide, a herbicide which inhibits DNA synthesis and cell division, on PA levels in pea root tips. Napropamide inhibited growth and development of roots (Figure 5a). Polyamine levels were not affected by the herbicide in the first 12 hours but after

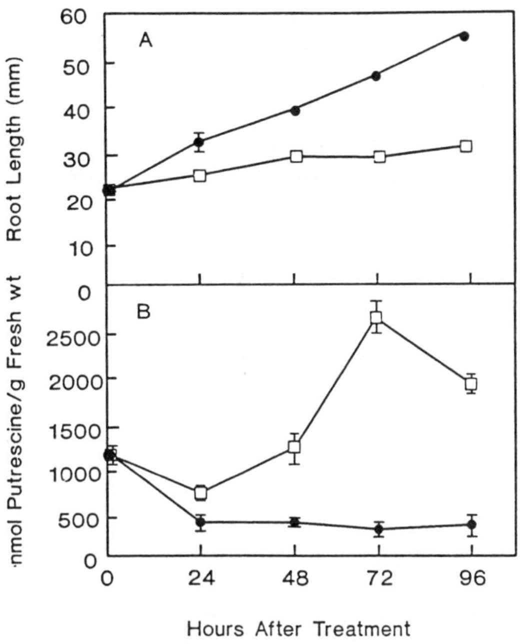

FIGURE 5. Time course of the effect of 10 μ*M* napropamide on root growth (A) of excised roots in White's medium, and putrescine accumulation (B) in 5 mm pea root sections grown in White's medium. Control (solid circle); napropamide (square). Values presented are means ±SE.[47]

48 and 72 h, Put levels were increased three- and eightfold, respectively (Figure 5b). Cadaverine levels also increased significantly after 48 h but the levels of Spd and Spm remained unaffected. The significance of increase of Put level in the herbicide-treated roots is not known but may be related to the herbicide-induced stress.

IV. CONCLUDING REMARKS

It is apparent from the information presented above that PA metabolism responds to light and plant hormones. In most cases, PA levels and/or the activities of their biosynthetic enzymes are increased in response to light and hormone treatments, but there are examples of inhibition of PA biosynthesis and activation of PA degradation as well. This correlation between changes in PA metabolism and altered growth patterns brought about by light and hormones may be especially significant in view of the fact that these two factors play important roles in plant growth and development. It is, however, not clear whether the changes in PA levels are simply secondary, unrelated effects of altered growth elicited by light and hormones, or whether they are mediators or are part of the signal response chain of light and hormone action, as has been suggested by some researchers. If PAs are to be regarded as second messengers, it is not unreasonable to expect that changes in PA titers and/or their biosynthesis would be quick and precede the physiological and morphological manifestation of the plant response to these factors. In some cases, PA metabolism does show changes before the growth effects of light and hormones become apparent[9,19,22,34,41] supporting the contention above. Conversely, in many instances, changes in PAs accompany or follow the plant responses, clearly indicating that the primary effect of light and plant hormones could not be on PA metabolism. In addition, where claims have been made that hormonal effects are mediated by PAs,[41] further investigation has sometimes shown that the increase in PA levels is not directly correlated with the effects produced by the hormone,[43] indicating that the action of the hormone cannot be entirely due to changes in PA levels. Some studies have shown that changes in PA metabolism are the result of morphological changes induced by plant hormones, whereas others have strongly indicated some kind of indirect but important role for PAs in light- and hormone-regulated plant responses. The shortage of experiments showing a reduction in PA levels and an inhibition of the hormonal response by PA inhibitors, and the reversal of the inhibitor effects by PAs, unfortunately precludes any definite conclusions as far as the role of PAs as second messengers is concerned.

Therefore, while it is clear that PAs are essential factors in many of the responses to plant hormones, it is, however, unlikely that the entire effects of plant hormones are mediated directly by changes in PA levels.

ACKNOWLEDGMENTS

We gratefully acknowledge financial support from the Herman Frasch Foundation and the Cornell University Biotechnology Program.

REFERENCES

1. **Slocum, R.D., Kaur-Sawhney, R., and Galston, A.W.,** The physiology and biochemistry of polyamines, *Arch. Biochem. Biophys.*, 235, 283, 1984.
2. **Smith, T.A.,** Polyamines, *Annu. Rev. Plant Physiol.*, 36, 117, 1985.
3. **Evans, P.T. and Malmberg, R.L.,** Do polyamines have roles in plant development?, *Annu. Rev. Plant Physiol. Plant Mol. Biol.*, 40, 235, 1989.
4. **Kendrick, R.E. and Kronenberg, G.H.M.,** Eds., *Photomorphogenesis in Plants*, Martinus Nijhoff Publishers, Dordrecht, 1986.
5. **Davies, P.J.,** Ed., *Plant Hormones and their Role in Plant Growth and Development*, Martinus Nijhoff Publishers, Dordrecht, 1987.
6. **Heby, O.,** Role of polyamines in the control of cell proliferation and differentiation, *Differentiation*, 19, 1, 1981.
7. **Tabor, C.W. and Tabor, H.,** Polyamines, *Annu. Rev. Biochem.*, 53, 749, 1984.
8. **Kaur-Sawhney, R., Shih, L.M., Flores, H.E., and Galston, A.W.,** Relation of polyamine synthesis and titer to aging and senescence in oat leaves, *Plant Physiol.*, 69, 405, 1982.
9. **Dai, Y.-R. and Galston, A.W.,** Simultaneous phytochrome-controlled promotion and inhibition of arginine decarboxylase activity in buds and epicotyls of etiolated peas, *Plant Physiol.*, 67, 266, 1981.
10. **Behringer, F.J., Davies, P.J., and Reid, J.B.,** Genetic analysis of the role of gibberellin in the red-light inhibition of stem elongation in etiolated pea seedlings, *Plant Physiol.*, 94, 432, 1990.
11. **Goren, R., Palavan, N., Flores, H., and Galston, A.W.,** Changes in polyamine titer in etiolated pea seedlings following red light treatment, *Plant Cell Physiol.*, 23, 19, 1982.
12. **Goren, R., Palavan, N., and Galston, A.W.,** Separating phytochrome effect on arginine decarboxylase activity from its effect on growth, *J. Plant Growth Reg.*, 1, 61, 1982.
13. **Angelini, R., Federico, R., and Mancinelli, A.,** Phytochrome-mediated control of diamine oxidase level in the epicotyl of etiolated lentil (*Lens culinaris medicus*) seedlings, *Plant Physiol.*, 88, 1207, 1988.
14. **Smith, M.A. and Davies, P.J.,** Effect of photoperiod on polyamine metabolism in apical buds of G2 peas in relation to the induction of apical senescence, *Plant Physiol.*, 79, 400, 1985.
15. **Davies, P.J., Birnberg, P.R., Maki, S.L., and Brenner, M.L.,** Photoperiod modification of [^{14}C] gibberellin A_{12} aldehyde metabolism in shoots of pea, line G2, *Plant Physiol.*, 81, 991, 1986.
16. **Smith, M.A. and Davies, P.J.,** Manipulation of the polyamine content and senescence of apical buds of G2 peas, *Plant Growth Regul.*, 3, 401, 1985.
17. **Bagni, N.,** Aliphatic amines as a growth-factor of coconut milk as stimulating cellular proliferation of *Helianthus tuberosus* (Jerusalem artichoke) *in vitro*, *Experientia*, 22, 732, 1966.
18. **Cocucci, S. and Bagni, N.,** Polyamine-induced activation of protein synthesis in ribosomal preparation from *Helianthus tuberosus* tissue, *Life Sci.*, 7, 113, 1968.

19. **Bagni, N., Malucelli, B., and Torrigiani, P.,** Polyamines, storage substances and abscisic acid-like inhibitors during dormancy and very early activation of *Helianthus tuberosus* tuber tissues, *Physiol. Plant.*, 49, 341, 1980.
20. **Bagni, N. and Speranza, A.,** Pathways of polyamine biosynthesis during the growth of *Helianthus tuberosus* parenchymatous tissue, in *Plant Growth Regulators*, Kudrev, T., Ivanova, I., and Karanov, E., Eds., Publishing House Bulgarian Academy of Science, Sofia, 1977, 75.
21. **Schuber, F. and Lambert, C.,** Metabolism of ornithine and arginine in Jerusalem Artichoke tuber tissue. Relationship with the biosynthesis of polyamines, *Physiol. Vég.*, 12, 571, 1974.
22. **Friedman, R., Altman, A., and Bachrach, U.,** Polyamines and root formation in mung bean hypocotyl cuttings I. Effects of exogenous compounds and changes in endogenous polyamine content, *Plant Physiol.*, 70, 844, 1982.
23. **Friedman, R., Altman, A., and Bachrach, U.,** Polyamines and root formation in mung bean hypocotyl cuttings II. Incorporation of precursors into polyamines, *Plant Physiol.*, 79, 80, 1985.
24. **Jarvis, B.C., Shannon, P.R.M., and Yasmin, S.,** Involvement of polyamines with adventitous root development in stem cuttings of mung bean, *Plant Cell Physiol.*, 24, 677, 1983.
25. **Kyriakidis, D.A.,** Effect of plant growth hormones and polyamines on ornithine decarboxylase activity during the germination of barley seeds, *Physiol. Plant.*, 57, 499, 1983.
26. **Choudhuri, M.M. and Ghosh, B.,** Purification and partial characterization of arginine decarboxylase from rice embryos (*Oryza sativa* L.), *Agric. Biol. Chem.*, 46, 739, 1982.
27. **Mizrahi, Y. and Heimer Y.M.,** Increased activity of ornithine decarboxylase in tomato ovaries induced by auxin, *Physiol. Plant.*, 54, 367, 1982.
28. **Cohen, E., Arad, S., Heimer, Y.M., and Mizrahi, Y.,** Participation of ornithine decarboxylase in early stages of tomato fruit development, *Plant Physiol.*, 70, 540, 1982.
29. **Fienberg, A.A., Choi, J.H., Lubich, J.H., and Sung, Z.R.,** Developmental regulation of polyamine metabolism in growth and differentiation of carrot culture, *Planta* 162, 532, 1984.
30. **Palavan, N., Goren, R., and Galston, A.W.,** Effects of some growth regulators on polyamine biosynthetic enzymes in etiolated pea seedlings, *Plant Cell Physiol.*, 25, 541, 1984.
31. **Katoh, Y., Hasegawa, T., Suzuki, T., and Fujii, T.,** Changes in the amounts of putrescine, spermidine and spermine in the hiproly barley callus after auxin withdrawal, *Agric. Biol. Chem.*, 49, 1027, 1985.
32. **Palazón, J., Pinol, M.T., Altabella, T., Cusidó, R., and Serrano, M.,** Auxin-induced regulation of amino acid and putrescine in the free state and nicotine content in cultured tobacco callus, *J. Plant Physiol.*, 128, 153, 1987.
33. **Srivastava, S.K., Prakash, V., and Naik, B.I.,** Regulation of diamine oxidase activity in germinating pea seeds, *Phytochemistry*, 16, 185, 1977.
34. **Suresh, M.R., Ramakrishna, S., and Adiga, P.R.,** Regulation of arginine decarboxylase and putrescine levels in *Cucumis sativus* cotyledons, *Phytochemistry*, 17, 57, 1978.
35. **Cho, S.-C.,** Effects of cytokinin and several inorganic cations on the polyamine content of lettuce cotyledons, *Plant Cell Physiol.*, 24, 27, 1983.
36. **Walker, M.A., Roberts, D.R., and Dumbroff, E.B.,** Effects of cytokinin and light on polyamines during the greening response of cucumber cotyledons, *Plant Cell Physiol.*, 29, 201, 1988.
37. **Mukhopadhyay, A., Choudhuri, M.M., and Ghosh, B.,** Changes in polyamines and related enzymes with loss of viability in rice seeds, *Phytochemistry*, 22, 1547, 1983.
38. **Srivastava, S.K., Raj, A.D.S., and Naik, B.I.,** Polyamine metabolism during aging and senescence of pea leaves, *Indian J. Exp. Biol.*, 19, 437, 1981.
39. **Cohen, E. and Kende, H.,** The effect of submergence, ethylene and gibberellin on polyamines and their biosynthetic enzymes in deepwater-rice internodes, *Planta*, 169, 498, 1986.
40. **Lin, P.P.C.,** Polyamine metabolism and its relation to response of the aleurone layers of barley seeds to gibberellic acid, *Plant Physiol.*, 74, 975, 1984.
41. **Dai, Y.-R., Kaur-Sawhney, R., and Galston, A.W.,** Promotion by gibberellic acid of polyamine biosynthesis in internodes of light-grown dwarf peas, *Plant Physiol.*, 69, 103, 1982.
42. **Kaur-Sawhney, R., Dai, Y.-R., and Galston, A.W.,** Effect of inhibitors of polyamine biosynthesis on gibberellin-induced internode growth in light-grown dwarf peas, *Plant Cell Physiol.*, 27, 253, 1986.
43. **Smith, M.A., Davies, P.J., and Reid, J.B.,** Role of polyamines in gibberellin-induced internode growth in peas, *Plant Physiol.*, 78, 92, 1985.
44. **Wang, S.Y. and Faust, M.,** Effect of growth retardants on root formation and polyamine content in apple seedlings, *J. Amer. Soc. Hort. Sci.*, 111, 912, 1986.
45. **Fletcher, R.A., Asare-Boamah, N.K., Kreig, L.C., Hofstra, G., and Dumbroff, E.B.,** Triadimefon stimulates rooting in bean hypocotyl, *Physiol. Plant.*, 73, 401, 1988.
46. **Lin, P.P.C.,** Effect of methylglyoxal-*bis*, (guanylhydrazone) and abscisic acid on polyamine metabolism in embryonectomized barley seeds, *Plant Growth Regulation*, 3, 257, 1985.
47. **DiTomaso, J.M., Rost, T.L., and Ashton, F.M.,** Herbicide-induced diamine accumulation in pea roots. The effect of napropamide on polyamine levels, *Plant Cell Physiol.*, 29, 1367, 1988.

Chapter 14

PHYSIOLOGICAL AND BIOCHEMICAL STUDIES ON THE ANTI-SENESCENCE PROPERTIES OF POLYAMINES IN PLANTS

Ravindar Kaur-Sawhney and Arthur W. Galston

TABLE OF CONTENTS

I. INTRODUCTION

Polyamines (PAs), among which the diamine putrescine (Put); triamine spermidine (Spd), and tetramine spermine (Spm) are the most common, appear to be ubiquitous in eukaryotic cells. They are known to occur in high levels in young, actively growing tissues and to decline with age in various organisms including plants. Their possible role in the control of senescence has been reviewed.[1-6]

The hypothesis that PA titer might be related to senescence of plant organs originated from attempts to regenerate cereal plants from mesophyll protoplasts.[7] Unlike solanaceous plants, which readily regenerate from protoplasts, cereals are recalcitrant, and it is only recently that cereal protoplast division and regeneration have become repeatable. We had observed that oat mesophyll protoplasts, which look healthy immediately upon isolation, soon deteriorate and become nonviable. We reasoned that if we could supplement the medium with substances that would retard leaf senescence, we might improve the viability of the protoplasts isolated from such leaves. Cytokinins, long known to repress the senescence of excised leaves,[8] produced no beneficial effect on protoplasts, but the amino acids arginine, ornithine, and lysine, reported to have antisenescence action on excised oat leaves[9,10] did retard senescence and increase the longevity of protoplasts in culture. Put and the diamine cadaverine (Cad), produced by decarboxylation of arginine and lysine, respectively, were even more active. Spd and Spm, which are synthesized from Put by the terminal addition of aminopropyl groups derived from methionine, were most effective in suppressing senescence when supplied at 0.05 to 10.0 mM. This is approximately the range of concentrations in which the PAs are known to exist in plant tissues,[11] thus supporting the hypothesis that PAs may function as senescence regulators *in vivo*. The morphological and biochemical effects of PAs on senescing tissues and the possible mechanisms involved in the antisenescence action of PAs will be discussed in this chapter. Part of this work has been reviewed in Galston and Kaur-Sawhney.[3]

II. ANTISENESCENCE EFFECTS OF POLYAMINES

Plant senescence is initiated and accompanied by a series of degradative events. In leaves, senescence is correlated with sharp increases in RNase and protease activities, which may lead to nucleic acid and protein breakdown, followed by disintegration of chloroplast structure and ultimately chlorophyll loss.

Polyamines, especially Spd, are generally abundant in young nonsenescent tissues, and decline as organs age and senesce.[3] The decline in PA titer seems to coincide with the appearance of senescence related symptoms. Since addition of exogenous PAs retards the appearance of these symptoms, it is possible that the onset of senescence may be caused by a decline in PA titer. Several reports have examined this hypothesis; most are supportive while some are not.

A. PROTOPLASTS

The antisenescence effects of exogenously applied PAs were first observed in freshly isolated mesophyll protoplasts of oat. These protoplasts senesce rapidly upon isolation, do not undergo sustained cell division, and lyse progressively with increasing time of culture. They show a progressive decline in their ability to incorporate ^{3}H - leucine into proteins, ^{3}H - uridine into RNA, and ^{3}H - thymidine into DNA.[12] Exogenous addition of PAs, especially Spd, to the isolation medium not only stabilizes the protoplasts against lysis but also promotes protein, RNA and DNA synthesis, with some inception of mitotic activity.[13,14] Pretreatment of leaves, either with senescence retardants such as kinetin and cycloheximide[15] or DL-α-difluoromethyl arginine (DFMA), a specific suicide inhibitor of arginine decarboxylase (ADC), an enzyme involved in the biosynthesis of Put, produced viable protoplasts which incorporated protein and nucleic acid precursors at a higher rate than protoplasts from nontreated leaves.[16] The action of cycloheximide, an inhibitor of protein synthesis, suggests that synthesis of protein is essential to the onset of senescence.

Another clear symptom of senescence in protoplasts is an increase in the activity of hydrolytic enzymes such as RNase and protease. These increases are drastically inhibited if protoplasts or the leaves from which they are isolated are incubated in the presence of L-arginine, L-lysine or PAs.[15] PAs also protect protoplasts against added pancreatic RNase, or the RNase released into the medium by lysing protoplasts.[17] Thus PAs are effective in inhibiting the senescence related symptons that occur progressively in freshly isolated cereal protoplasts.

B. LEAVES

Polyamine analysis in a variety of senescing plant organs and tissues reveals that the natural onset of senescence is correlated with lower levels of Put, Spd, and Spm. This is true both in intact leaves, which senesce sequentially on the plant, and in dark-induced senescence of excised leaves of several cereals, especially oats;[18] it is also true in dicots such as rape,[18] pea[19] and radish.[20] In general, the decline of Spd seems most significant, and in excised barley leaves this decline is repressed by kinetin, [21] which itself retards senescence in excised leaves.

The activity of ADC, which can produce Put through initial decarboxylation of arginine to agmatine, [2] decreases progressively during the aging of attached primary leaves, and during dark-induced senescence of excised leaves of oat. Conversely, it increases in excised leaves exposed to light, which retards senescence. By contrast, the activity of ornithine decarboxylase (ODC), the enzyme that catalyzes the biosynthesis of Put from ornithine, is high and constant in the attached leaf, irrespective of age, and decreases in excised leaves kept in the dark and in light, irrespective of senescence. The activity of *S*-adenosyl-L-methionine decarboxylase (SAMDC), an enzyme involved in the biosynthesis of Spd and Spm, also shows no correlation with senescence.[22] Thus, decreased ADC activity seems best correlated with both the decline of PAs and with leaf senescence.

Excision and dark incubation of oat leaves, which initiates senescence, causes an almost immediate rise in RNase activity.[18] The specific activity increases by about 50% after 1 h and by about 300% after 24 to 26 h. After about 6 h there is a sharp increase in proteolytic activity, and after about 24 h a decline in chlorophyll content. These senescence related processes are prevented by the exogenous addition of 1 m*M* Spm or other PAs.

A 30 to 60 min period of contact of the excised peeled oat leaf with 1 m*M* Spm is sufficient to prevent chlorophyll loss after 48 h of dark incubation, and the $t_{1/2}$ (time of contact required for 50% protection) is only about 5.5 min. PA determinations revealed insignificant disappearance of Spm from the bathing solution during the incubation time, indicating that only small quantities of Spm are required to protect the leaf against chlorophyll loss. The same short exposure of leaf to 1 m*M* Spm completely prevented the more than threefold rise in RNase and protease activites noted after 48 h of dark incubation.

Spermine is most effective when applied to the leaves immediately after excision and peeling and continues to be effective if added any time until about 48 h of dark incubation. The marked decrease in chlorophyll content, by more than 70% of the initial value in excised leaves floated on buffer in darkness, is partially prevented by a 10 min pulse of 1 m*M* Spm given imediately after excision of the leaf and is completely prevented by a 60 min treatment. Similar short pulse treatments with kinetin and cychloheximide were not as effective as Spm. Thus, Spm appears to be an effective antisenescence agent when applied to oat leaves during dark induced senescence. Although the exogenous concentrations required are perhaps 100 times those found in the cell, the permeability barrier to the entry of PAs is high, by virtue of their catonic nature at physiological pH. Spermidine, Put and Cad are also effective in reducing chlorophyll breakdown in the dark in many cereal and dicot leaves, the order of activity being Spm> Spd>

Put = Cad.[18] A tenfold higher concentration of L-arginine, L-ornithine, and L-lysine, are also effective, per se or after conversion to PAs, in decreasing chlorophyll breakdown in leaf segments of oat.[10,15] Again this is correlated with inhibition of the dark-induced rise in protease activity.[23] These effects may be due to electrovalent attachment of protonated amino groups in PAs to anionic sites on macromolecules.[23]

Although chlorophyll loss is one of the relatively early events in senescence and PAs are known to retard chlorophyll breakdown in many plant leaves, chlorophyll loss in dark- or light-incubated suspensions of either intact or disrupted chloroplasts from radish leaves is not affected by PAs.[20] The author thus believes that PAs act through indirect mechanisms rather than by direct inhibition of chlorophyll degradation. 1,3-Diaminopropane (Dap), an oxidation product of Spd and Spm, occurs naturally in many cereal plants.[2] The oxidation is catalyzed by PA oxidase, a cell wall-localized enzyme in oat leaves.[24] We observed[25] that endogenous levels of Dap, like those of Put and Spd, decrease in attached oat leaves with increasing age of seedlings and in excised leaves with increasing time of dark incubation. Diaminopropane accumulated following exogenous application of Spd to excised dark-incubated oat leaves. Exogenous Dap, like Spd, inhibits senescence, possibly by preventing the increase in protease activity and ethylene production that precedes chlorophyll breakdown.[26] Thus Dap, like other PAs, may act as an antisenescence agent in cereals, especially oats. A similar role has been suggested for Dap arising from exogenous application of Spd and Spm to dark incubated excised rice leaves.[27]

The effect of PAs in retarding chlorophyll breakdown is localized to peeled areas of excised oat leaves,[18] and around cut edges and near large veins of soybean leaves.[28] These results suggest that PAs are not well transported into and through these leaf cells.

The senescence of excised leaves in the light proceeds at a much slower rate than in the dark, and the mechanism of control of senescence by PAs in light also appears different. ADC activity falls in excised dark incubated leaves, but rises if the leaves are incubated in light. Since PA administration in the dark prevents senescence, it is possible that the retardation of senescence by light might be mediated by light activation of the biosynthesis of PAs via ADC. Paradoxically, in the light, all exogenously applied PAs cause a photobleaching, thus promoting, rather than retarding, the breakdown of chlorophyll. However, as in the dark, they inhibit the rise in RNase and apparent protease activity. This photobleaching of chlorophyll in the presence of PAs is proportional to the duration of exposure to high irradiance (16.5 Wm^{-2}) fluorescent light.[18] Others have reported similar bleaching of excised barley[21] and *Hydrilla* leaves.[29]

C. ORGANS AND TISSUES

PAs also seem to be involved in senescence of many plant organs and storage tissues. The endogenous levels of PAs decreased with increasing age of *Helianthus tuberosus* tubers[30] and were also low in dormant potato tubers.[31] The senescence of nonphotosynthetic tissue cultures of Paul's Scarlet rose was also reduced by addition of micro quantities of PAs to the media of stationary phase cultures.[32] The deleterious effects of wounding and aging in potato tubers were inhibited by Spd.[33] While there are as yet no detailed studies on the relation of PA titers and whole-plant senescence, the data obtained with seedling leaves *in situ* suggest a causal connection between the two.

III. CORRELATION BETWEEN ENDOGENOUS TITERS AND EFFECTS OF EXOGENOUS APPLICATION OF POLYAMINES ON SENESCENCE

The effects of exogenous PAs do not always correlate well with the trends of endogenous PA titers. Birecka et al,[34] while comparing chlorophyll and protein degradation in detached dark-exposed leaves of *Avena*, *Nicotiana*, and *Heliotropium*, found that older *Heliotropium* leaves had very low PA titer and exhibited a weak senescence syndrome. In contrast, leaves of *Avena* and *Nicotiana* had higher levels of PAs and showed marked senescence despite insignificant PA decline. The authors thus suggested that endogenous PAs may have effects different than those reported for exogenous PAs. Likewise, no correlation between PA titers and senescence was observed in tissues high in PAs (tobacco leaves and potato tubers).[33] In some instances, exogenous PAs actually promoted senescence.

In contrast, *Alcobaça* and other varieties of tomato with extended shelf-lives and delayed senescence have a higher than usual titer of Put, which increases rather than showing the usual decline during ripening.[35] In Liberty, another variety of tomato, an unusual increase in Put during ripening is again correlated with increased storage life.[36] In several varieties of tomatoes, showing usual patterns of PA decline and senescence, exogenous PAs supplied through the peduncle retard senescence. These data provide additional evidence that endogenous PA levels are inversely correlated with senescence and that endogenous PAs may have effects similar to those of exogenous PA applications.

There are other instances in which the correlation between PA decline and onset of senescence does not hold. For example, in a line of peas whose senescence is regulated by photoperiod, senescence begins before PA concentration declines in apical buds, thus ruling out PA control of senescence.[37] However, in the same line of peas, defruiting, which prevents senescence in apical buds, increases PA titer and bud size.[37] Likewise, application of GA_{20} increased internode growth, bud size, and Spd titer, but not Put or Spm titer,[38] while the application of exogenous Spd through injection into the hollow stem did not delay senescence nor increase PA levels significantly. This probably indicates the operation of mechanisms regulating PA levels in pea tissue, since exogenous application of Spd does raise PA levels in other plants, such as oats. By contrast, injection of 1-amino-1-cyclopropylcarboxylic acid (ACC) into the stem accelerated senescence of the apical bud and lowered PA levels concommitently. Aminoethoxyvinylglycine (AVG), an inhibitor of ethylene biosynthesis, raised Put levels, but also induced yellowing or bleaching at some nodes above the point of treatment. The joint application of DFMA and DL-α-difluoromethylornithine (DFMO), an irreversible inhibitor of ODC, decreased PA concentration somewhat, but failed to reduce senescence. The authors conclude that there is no convincing evidence for the role of PAs in the control of senescence in this system.

No significant changes in PAs, except a decline in Spd, which was suppressed by exogenous kinetin, was reported during senescence of intact or excised barley leaves exposed to darkness.[21] Furthermore, in contrast with the previous work of Kaur-Sawhney and Galston,[18] Srivastava[21] found that exogenous PAs did not retard senescence of barley leaf pieces in the dark; however Spd and Spm did increase it in the light, probably because of photobleaching.

No correlation between the levels of endogenous PAs and senescence was found in intact leaves and dark incubated excised leaves obtained from oat and barley seedlings sprayed with DFMA alone or together with DFMO, although both ADC activity and PA levels were markedly inhibited.[39] In contrast we have reported earlier that similar treatment of oat seedlings with DFMA retarded senescence.[16] Protoplasts obtained from such treated leaves were more viable with higher PA titers and ability to incorporate protein and nucleic acid precursors. It is difficult at present to rationalize these differences of experience.

IV. POSSIBLE MECHANISMS INVOLVED IN ANTISENESCENCE ACTION OF POLYAMINES

A. ACTION ON MEMBRANES

There are several reports indicating that PAs may retard senescence via their interaction with membranes.[40] PAs are known to bind to the negatively charged phospholipid components or other anionic sites on membranes, thus altering the stability and permeability characteristics of such membranes. Physiological concentrations of Spd or Spm have reportedly stabilized protoplasts of oat mesophyll cells,[13] *Chlamydomonas reinhardii*[41] and bacteria.[40] PAs are also known to stabilize thylakoid membranes of barley chloroplasts and prevent chlorophyll loss from senescing barley leaf discs.[42] Reduction by PAs of betacyanin efflux from wounded beet root discs, probably by the stabilization of membranes against senescence-related permeability changes, has also been reported.[43] This observation, however, is probably due to a chemical reaction between betacyanin and PAs in the external medium resulting in the decolorization of the red pigment,[44] and thus do not indicate PA-induced reduction of leakage. Applied, PAs also failed to prevent anthocyanin leakage from senescing petals of *Tradescantia*.[45]

Polyamine binding is also known to affect membrane fluidity and thus PAs may modulate the activities of membrane-associated enzymes indirectly.[1] In addition, Roberts et al.,[46] using microsomal membranes from *Phaseolus*, reported that exogenously applied PAs associate with membrane lipids, thereby substantially reducing membrane fluidity. The authors caution that some physiological effects attributed to exogenous application of PAs may thus reflect artificially induced membrane rigidification. However, increased membrane fluidity is known to be associated with senescence[47] and PAs may thus retard senescence by reducing this tendency.

B. INTERACTION WITH Ca^{2+}

Since PAs, at cellular pH, are positively charged and are known to bind strongly to the acidic phosphate groups of nucleic acids,[48,49] they could possibly affect senescence by virtue of their cationic property. This appears to be true since treatments with Ca^{2+}, a divalent cation, interfered with the antisenescence effects of spermine, during dark-induced senescence of oat leaves.[18] Exogenous $CaCl_2$ applied alone was without effect.

Further evidence for a competitive electronegative binding on membranes between PAs and Ca^{2+} is based on studies using peeled oat leaves. Spermidine and DAP were shown to reduce ethylene synthesis by inhibition of ACC synthase and of the conversion of ACC to ethylene.[26] Addition of Ca^{2+} could competitively reduce these effects, indicating that the sites of action on the membrane of the PAs and Ca^{2+} are the same. Similar results were obtained by Apelbaum et al.[50] using apple tissue. Correlating microsomal membrane microviscosity with ethylene production in apple disks, Ben-Arie et al.[51] showed the effects of Spm and Ca^{2+} to be temperature dependent. At temperatures below 12°C, Ca^{2+} inhibited more than Spm whereas above 12°C the reverse was true. They concluded that the effects of Spm and Ca^{2+} on ethylene biosynthesis, which may be involved in senescence, are probably mediated via changes in the fluidity of the microsomal membrane with different modes or sites of action.

In senescing orange peel discs,[52] inhibition of ethylene production or action by various inhibitors stimulated incorporation of [3,4-^{14}C] methionine into Spd. Unlabeled Spd inhibited the transfer of label to ACC but did not stimulate label incorporation into Spd. This inhibition was partially overcome by the addition of Ca^{2+}, a result consistent with earlier senescence studies.[18,26] Unlike the situation in leaves (see above), Ca^{2+}, like PAs, is known to have beneficial effects in delaying senescence in leaves[53] and in maintaining membrane integrity.[54] Since both are cations the prevention of the PA effects by Ca^{2+} could be due to competitive charge neutralization of some specific sites on the cellular membrane.

C. ANTI-ETHYLENE ACTION

Since ethylene is a known senescence promoter and PAs are senescence inhibitors, the fact that PAs and ethylene share a common precursor, *S*-adenosylmethionine (SAM), suggests a mechanism for their interaction. Even-Chen et al.[52] used [3,4-^{14}C] methionine to trace carbon flow into either PAs or ethylene in aged orange peel discs. Inhibition of ethylene synthesis by a variety of treatments resulted in a two- to fourfold increase in the specific labeling of Spd, suggesting utilization of a common precursor. Unlabeled Spd did not promote transfer of label to Spd but did inhibit the transfer to ACC.

The most common hypothesis tested is that PAs and ethylene regulate each other's synthesis, either directly or by metabolic competition for SAM. It is now well documented that PAs inhibit ethylene formation in a number of plant tissues, including oat leaves,[26] apple fruit, bean and tobacco leaf explants,[50] *Tradescantia* petals, and mung bean hypocotyls.[45] Putrescine, Spd, and Spm were reported to inhibit auxin-induced ethylene production and the conversion of methionine and ACC to ethylene in apple tissue. In peeled, dark-incubated oat leaves, Spd and Dap were shown to reduce ethylene levels both by inhibition of ACC synthase and by conversion of ACC to ethylene.[26]

However, these interactions are still far from being clearly understood. In dark incubated oat leaves, Spd and Dap inhibited the rise in ethylene levels and also retarded the senescence related chlorophyll loss.[26] Exogenous application of PAs reduced ethylene titer in senescencing *Trandescantia* petals, but leakage of anthocyanin from these petals was not prevented.[45] Inhibitors of PA synthesis, when added to cut carnations, increased both ethylene production and senescence, while addition of an ACC synthase inhibitor decreased ethylene production and increased PA titer, causing a marked delay in the onset and progress of senescence.[55] Recently, Downs and Lovell[56] found that exogenous application of Put or Spd (0.1 to 10 m*M*) neither inhibited ethylene production nor protected cut carnation flowers from senescence. In contrast, Wang[57] found that the inhibition of ethylene production by PAs probably explains the ability of PAs to extend the vase-life of carnations. Some reports have suggested that PA effects on senescence are artifacts; exogenously applied PAs are nonspecifically toxic and may interfere with ethylene synthesis, thus resulting in inhibition of senescence.[46]

D. FREE RADICAL SCAVENGING

As mentioned earlier, PAs at physiological pH are mainly protonated and thus polycationic, and it has been suggested that they may achieve many of their physiological effects by associating with negative charges on nucleic acids and phospholipids.[1] Kitada et al.[58] reported that Spm inhibits NADPH- and ascorbic acid-dependent lipid peroxidation in liver microsomes and suggested that Spm may associate with phospholipids and thus stabilize membranes. Lipid peroxidation is an inherent feature of senescence, and any tendency of PAs to inhibit peroxidation reactions should retard senescence. Recently, Drolet et al.[59] have provided evidence indicating that PAs are effective scavengers of superoxide free radicals generated in a number of chemical and *in vitro* enzyme systems. The inhibition was concentration-dependent and related to the number of amino groups on the scavengers. For example, Spd and Spm, which have three and four amino groups respectively, were more effective scavengers of superoxide than the diamines, Put and Cad, suggesting the involvement of amino groups. In addition, the production of superoxide radicals by senescing microsomal membranes was inhibited by PAs, as was the superoxide-dependent conversion of ACC to ethylene. It is therefore conceivable that the effect of PAs in retarding senescence could occur via their free radical-scavenging capability.

E. CONTROL OF MACROMOLECULAR SYNTHESIS AND BREAKDOWN

Increased PA biosynthesis and titers are correlated with enhanced protein and nucleic acid synthesis during cellular proliferation in a variety of procaryotic and eucaryotic organisms.[60,61] PAs are also known to stimulate protein synthesis when added to cell-free translation systems,[62] and to enhance yields of full-length translational products in lactating mouse mammary glands.[63] *In vitro*, stimulation of protein synthesis is correlated with increased PA synthesis following auxin-induced activation of dormant *Helianthus* tissue.[64] Also, exogenous PAs promote apparent protein and nucleic acid synthesis in oat protoplasts by retarding lysis.[13]

Increased PA biosynthesis during the transition from the G_1 to the S phase of the cell cycle, preceding the onset of DNA synthesis in dividing cells, appears to be a universal phenomenon in animals[65] and plants.[66] DFMO-mediated depletion of PAs causes a rapid decrease in the frequency of polysome complexes, which can be reversed by addition of PAs.[67] Also, the activities of several enzymes regulating nucleic acid synthesis and repair are affected by PAs.[1] Thus, it has been suggested that PAs achieve many of their physiological and biochemical effects through indirect control of the synthesis of macromolecules.

The senescence-retarding effect of PAs in oat protoplasts and in dark-induced excised leaves is correlated with increases in the incorporation of labeled precursors into proteins, RNA and DNA.[13,15] Such treatments also result in limited mitotic activity[14] and the inhibition of senescence-induced rise in ribonuclease and protease activities.[18] With regard to both of these enzymes, PAs appear to exert a dual effect, inhibiting both *de novo* synthesis and enzyme activity.[17,23,68]

V. CONCLUSIONS

The evidence supporting a role for PAs in the control of senescence is twofold: (1) in a number of plants, the titer of PAs, especially Spd and Dap, declines as leaves senesce, and (2) the application of exogenous PAs delays senescence in excised leaves or leaf segments of several monocots and dicots.

The mechanism of senescence inhibition by PAs seems related to their cationic nature and to effects on membrane stability. PAs can inhibit ethylene formation and action. They can also function as free radical scavengers, thus serving to protect membranes from oxidative damage, and events initiated by ethylene formation.

There are some examples in which the titers of PAs do not decline with increased senescence, and in which exogenous PAs do not retard senescence. In such cases, it appears that senescence is controlled by other factors or by more than one pathway.

ACKNOWLEDGMENTS

Our research on polyamines over the years has received generous support, especially from the National Science Foundation.

REFERENCES

1. **Slocum, R.D., Kaur-Sawhney, R., and Galston, A.W.,** The physiology and biochemistry of polyamines in plants, *Arch. Biochem. Biophys.*, 235, 283, 1984.
2. **Smith, T.A.,** Polyamines, *Annu. Rev. Plant Physiol.*, 36, 117, 1985.
3. **Galston, A.W. and Kaur-Sawhney, R.,** Polyamines and senescence in plants, in *Plant Senescence: Its Biochemistry and Physiology,* Thompson, W.W., Nothnagel, E.A., and Huffaker, R.C., Eds., American Society of Plant Physiologists, Rockville, MD, 1987, 167.
4. **Evans, P.T. and Malmberg, R.L.,** Do polyamines have roles in plant development?, *Annu. Rev. Plant Physiol. Plant Mol. Biol.*, 40, 235, 1989.
5. **Flores, H.E., Protacio, C.M., and Signs, M.W.,** Primary and secondary metabolism of polyamines in plants, *Recent Adv. Phytochem.*, 23,329, 1989.
6. **Galston, A.W. and Kaur-Sawhney, R.,** Polyamines in plant physiology, *Plant Physiol.*, 94, 406, 1990.

7. **Brenneman, F. and Galston, A.W.,** Experiments on the cultivation of protoplasts and calli of agriculturally important plants 1. Oat (*Avena sativa* L.), *Biochem. Physiol. Pflanzen.*, 168, 453, 1975.
8. **Richmond, A.E. and Lang, A.,** Effect of kinetin on protein content and survival of detached *Xanthium* leaves, *Science*, 125, 650, 1957.
9. **Martin, C. and Thimann, K.V.,** Role of protein synthesis in the senescence of leaves. II. The influence of amino acids on senescence, *Plant Physiol.*, 50, 432, 1972.
10. **Von Abrams, G.J.,** An effect of ornithine on degradation of chlorophyll and protein in excised leaf tissue, *Z. Pflanzenphysiol.*, 72, 410, 1974.
11. **Slocum, R.D. and Galston, A.W.,** Changes in polyamine biosynthesis associated with postfertilization growth and development in tobacco ovary tissues, *Plant Physiol.*, 79, 336, 1985.
12. **Fuchs, Y. and Galston, A.W.,** Macromolecular synthesis in oat leaf protoplasts, *Plant Cell Physiol.*, 17, 475, 1976.
13. **Altman, A., Kaur-Sawhney, R., and Galston, A.W.,** Stabilization of oat leaf protoplasts through polyamine-mediated inhibition of senescence, *Plant Physiol.*, 60, 570, 1977.
14. **Kaur-Sawhney, R., Flores, H.E., and Galston, A.W.,** Polyamine-induced DNA synthesis and mitosis in oat leaf protoplasts, *Plant. Physiol.*, 65, 368, 1980.
15. **Kaur-Sawhney, R., Adams, W.R., Jr., Tsang, J., and Galston, A.W.,** Leaf pretreatment with senescence retardants as a basis for oat protoplast improvement, *Plant Cell Physiol.*, 18, 1309, 1977.
16. **Tiburcio, A.F., Kaur-Sawhney, R., and Galston, A.W.,** Polyamine metabolism and osmotic stress. II. Improvement of oat protoplasts by an inhibitor of arginine decarboxylase, *Plant Physiol.*, 82, 375, 1986.
17. **Galston, A.W., Altman, A., and Kaur-Sawhney, R.,** Polyamines, ribonuclease and the improvement of oat leaf protoplasts, *Plant Sci. Lett.*, 11, 69, 1978.
18. **Kaur-Sawhney, R. and Galston, A.W.,** Interaction of polyamines and light on biochemical processes involved in leaf senescence, *Plant Cell Environ.*, 2, 189, 1979.
19. **Srivastava, S.K., Raj, A.D.S., and Naik, B.I.,** Polyamine metabolism during ageing and senescence of pea leaves, *Indian J. Exp. Bot.*, 19, 437, 1981.
20. **Altman, A.,** Retardation of radish leaf senescence by polyamines, *Physiol. Plant.*, 54, 189, 1982.
21. **Srivastava, B.I.S.,** Polyamine changes during senescence and tumorogenesis in plants, *Mech. Ageing Develop.*, 40, 17, 1987.
22. **Kaur-Sawhney, R., Shih, L.M., Flores, H.E., and Galston, A.W.,** The relation of polyamine synthesis and titer to ageing and senescence in oat leaves, *Plant Physiol.*, 69, 405, 1982.
23. **Kaur-Sawhney, R., Shih, L.M., Cegielska, T., and Galston, A.W.,** Inhibition of protease activity by polyamines. Relevance for control of leaf senescence, *FEBS Lett.*, 145, 345, 1982.
24. **Kaur-Sawhney, R., Flores, H.E., and Galston, A.W.,** Polyamine oxidase in oat leaves: a cell wall-localized enzyme, *Plant Physiol.*, 68, 494, 1981.
25. **Shih, L.M., Kaur-Sawhney, R., Fuhrer, J., Samanta, S., and Galston, A.W.,** Effects of exogenous 1,3-diaminopropane and spermidine on senescence of oat leaves. I. Inhibition of protease activity, ethylene production, and chlorophyll loss as related to polyamine content, *Plant Physiol.*, 70, 1592, 1982.
26. **Fuhrer, J., Kaur-Sawhney, R., Shih, L.M., and Galston, A.W.,** Effects of exogenous 1,3-diaminopropane and spermidine on senescence of oat leaves. II. Inhibiton of ethylene biosynthesis and possible mode of action, *Plant Physiol.*, 70, 1597, 1982.
27. **Cheng, S.H., Shyr, Y.Y., and Kao, C.H.,** Senescence of rice leaves XII. Effects of 1,3-diaminopropane, spermidine and spermine, *Bot. Bull. Acad. Sin.*, 25, 191, 1984.
28. **Cheng, S.H. and Kao, C.H.,** Localized effect of polyamines on chlorophyll loss, *Plant Cell Physiol.*, 24, 1463, 1983.
29. **Kar, R.K. and Choudhuri, M.A.,** Effects of light and spermine on senescence of *Hydrilla* and spinach leaves, *Plant Physiol.*, 80, 1030, 1986.
30. **Bagni, N., Serafini-Francassini, D., and Torrigiani, P.,** Polyamines and cellular growth processes in higher plants, in *Plant Growth Substances*, Wareing, P.F., Ed., Academic Press, London, 1982, 473.
31. **Kaur-Sawhney, R., Shih, L.M., and Galston, A.W.,** Relation of polyamine biosynthesis to the initiation of sprouting in potato tubers, *Plant Physiol.*, 69, 411, 1982.
32. **Muhitch, M.M., Edwards, L.A., and Fletcher, J.S.,** Influence of diamines and polyamines on the senescence of plant suspension cultures, *Plant Cell Rept.*, 2, 82, 1983.
33. **Altman, A. and Bachrach, U.,** Involvement of polyamines in plant growth and senescence, in *Advances in Polyamine Research*, Caldarera, C.M., Zappia, V., and Bachrach, U., Eds., Vol. 3, Raven Press, New York, 1981, 365.
34. **Birecka, H., DiNolfo, T.E., Martin, W.B., and Frohlick, M.W.,** Polyamines and leaf senescence in pyrrolizidine alkaloid bearing *Heliotropium* plants, *Phytochemistry*, 23, 991, 1984.

35. **Dibble, A.R.G., Davies, P.J., and Mutschler, M.A.,** Polyamine content of long keeping alcobaca tomato fruit, *Plant Physiol.*, 86, 338, 1988.
36. **Saftner, R.A. and Baldi, B.G.,** Polyamine levels and tomato fruit development: possible interaction with ethylene, *Plant Physiol.*, 92, 547, 1990.
37. **Smith, M.A. and Davies, P.J.,** Effect of photoperiod on polyamine metabolism in apical buds of G–2 peas in relation to the induction of apical senescence, *Plant Physiol.*, 79, 400, 1985.
38. **Smith, M.A., Davies, P.J., and Reid, J.B.,** Role of polyamines in gibberellin-induced internode growth in peas, *Plant Physiol.*, 78, 92, 1985.
39. **Birecka, H., Birecka, M., and Treton, K.,** Endogenous polyamine levels and dark-induced leaf senescence, *Plant Physiol. Suppl.*, 93, 129, 1990.
40. **Mager, J.,** The stabilizing effect of spermine and related polyamines on bacterial protoplasts, *Biochem. Biophys. Acta*, 36, 529, 1959.
41. **Hasnaim, S.E., Khan, M.A., and Upadhyaya, K.C.,** The effect of polycations on thermal damage of *Chlamydomonas reinhardii* protoplasts, *Indian J. Exp. Biol.*, 18, 1037, 1980.
42. **Popovic, R.B., Kyle, D.J., Cohen, A.S., and Zalik, S.,** Stabilization of thylakoid membranes by spermine during stress-induced senescence of barley leaf discs, *Plant Physiol.*, 64, 721, 1979.
43. **Naik, B.I. and Srivastava, S.K.,** Effect of polyamines on tissue permeability, *Biochemistry*, 17, 1885, 1978.
44. **Smith, T.A. and Crocker, S.J.,** The decolorization of betacyanin by polyamines, *Phytochemistry*, 24, 2436, 1985.
45. **Suttle, J.C.,** Effect of polyamines on ethylene production, *Phytochemistry*, 20, 1477, 1981.
46. **Roberts, D.R., Dumbroff, E.B., and Thompson, J.E.,** Exogenous polyamines alter membrane fluidity in bean leaves — a basis for potential misinterpretation of their true physiological role, *Planta*, 167, 395, 1986.
47. **Borochov, A., Halevy, A., Borochov, H., and Shinitsky, M.,** Microviscosity of plasmalemmas in rose petals as affected by age and environmental factors, *Plant Physiol.*, 61, 812, 1978.
48. **Cohen, S.S.,** *Introduction to the Polyamines*, Prentice-Hall, Englewood Cliffs, N.J., 1971.
49. **Bachrach, U.,** *Function of Naturally Occurring Polyamines*, Academic Press. N.Y., 1973.
50. **Apelbaum, A., Burgoon, A.C., Anderson, J.D., Lieberman, M., Ben-Arie, R., and Mattoo, A.K.,** Polyamines inhibit biosynthesis of ethylene in higher plant tissue and fruit protoplasts, *Plant Physiol.*, 68, 453, 1981.
51. **Ben-Arie, R., Lurie, S., and Mattoo, A.K.,** Temperature-dependent inhibitory effects of calcium and spermine on ethylene biosynthesis in apple discs correlate with changes in microsomal membrane microviscosity, *Plant Sci. Lett.*, 24, 239, 1982.
52. **Even-Chen, Z., Mattoo, A.K., and Goren, R.,** Inhibition of ethylene biosynthesis by aminoethoxyvinylglycine and by polyamines shunts label from 3,4 (^{14}C) methionine into spermidine in aged orange peel discs, *Plant Physiol.*, 69, 385–388, 1982.
53. **Poovaiah, B.W. and Leopold, A.C.,** Deferral of leaf senescence with calcium, *Plant Physiol.*, 52, 236, 1973.
54. **Williams, R.J.P.,** Calcium chemistry and its relation to biological function, *Symp. Soc. Exp. Biol.*, 30, 1, 1976.
55. **Roberts, D.R., Walker, M.A., Thompson, J.E., and Dumbroff, E.B.,** The effects of inhibitors of polyamine and ethylene biosynthesis on senescence, ethylene production and polyamine levels in cut carnation flowers, *Plant Cell Physiol.*, 25, 315, 1984.
56. **Downs, C.G. and Lovell, P.H.,** The effect of spermidine and putrescine on the senescence of cut carnations, *Physiol. Plant.*, 66, 679, 1986.
57. **Wang, C.Y. and Baker, J.E.,** Extending vase life of carnations with aminooxyacetic acid, polyamines, EDU and CCCP, *Hort. Sci.*, 15, 805, 1980.
58. **Kidata, M., Igarashi, K., Hirose, S., and Kitagawa, H.,** Inhibition by polyamines of lipid peroxide formation in rat liver microsomes, *Biochem. Biophys. Res. Commun.*, 87, 388, 1979.
59. **Drolet, G., Dumbroff, E.B., Legge, R.L., and Thompson, J.E.,** Radical scavenging properties of polyamines, *Phytochemistry*, 25, 367, 1986.
60. **Abraham, A.K. and Pihl, A.,** Role of polyamines in macromolecular synthesis, *Trends Biochem. Sci.*, 6, 106, 1981.
61. **Heby, O.,** Role of polyamines in the control of cell proliferation and differentiation, *Differentiation*, 19, 1, 1981.
62. **Tabor, C.W. and Tabor, H.,** Polyamines, *Annu. Rev. Biochem.*, 53, 749, 1984.
63. **Takemoto, T., Nagamatsu, Y., and Oka, T.,** The study of spermidine-stimulated polypeptides in cell-free translation of mRNA from lactating mouse mammary gland, *Biochem. Biophys. Acta.*, 740, 73, 1983.
64. **Cocucci, S. and Bagni, N.,** Polyamine-induced activation of protein synthesis in ribosomal preparation from *Helianthus tuberosus* tissue, *Life Sci.*, 7, 113, 1968.
65. **Fuller, D.J.M., Gerner, E.W., and Russell, D.H.,** Polyamine biosynthesis and accumulation during the G_1 to S phase transition, *J. Cell Physiol.*, 93, 81, 1977.

66. **Cohen, E., Arad, S., Heimer, Y.H., and Mizrahi, Y.,** Polyamine biosynthetic enzymes in the cell cycle of *Chlorella*: Correlation between ornithine decarboxylase and DNA synthesis at different light intensities, *Plant Physiol.*, 74, 385, 1984.
67. **Rudkin, B.B., Mamont, P.S., and Seiler, N.,** Decreased protein-synthetic activity is an early consequence of spermidine depletion in rat hepatoma tissue-culture cells, *Biochem. J.*, 217, 731, 1984.
68. **Kaur-Sawhney, R., Altman, A., and Galston, A.W.,** Dual mechanisms in polyamine-mediated control of ribonuclease activity in oat leaf protoplases, *Plant Physiol.*, 62, 158, 1978.

Chapter 15

CHANGES IN POLYAMINE METABOLISM IN RESPONSE TO ABIOTIC STRESS

Hector E. Flores

TABLE OF CONTENTS

I. INTRODUCTION

Polyamine metabolism and function in plant cells have many elements in common with the processes studied in microbial and animal systems. For example, plant and bacterial cells share the existence of alternate pathways for putrescine synthesis, mediated by arginine (ADC) and ornithine decarboxylase (ODC), respectively. Many of the roles ascribed to plant polyamines in plants are also essentially those proposed for other systems, such as control of the cell cycle, cell division and morphogenesis. Polyamines in plant systems, however, have several unique features; the sharing of a common precursor for polyamine and ethylene biosynthesis, the widespread occurrence of hydroxycinnamic acid conjugates, and the existence of over 1000 polyamine-derived alkaloids are some of the metabolic features found only in plant species (see Chapters 4 and 5). Polyamines have been involved in functions which only occur in plant systems, such as phytochrome and plant hormone-mediated processes, and control of plant senescence; these are reviewed in Chapters 13 and 14, respectively. The oldest known and perhaps the most dramatic changes in polyamine metabolism which characterize plant cells are those obtained when plants are challenged with abiotic stresses. Although the functional significance of these responses is not fully understood, it is clear that major shifts in nitrogen and polyamine metabolism can occur when plant cells are starved for nutrients, or exposed to low pH, osmotic shock or environmental pollutants. In this chapter, we will review the physiology and biochemistry of these and other changes. Several recent reviews have covered some of these aspects.[1-3] Polyamine metabolism can also be affected when plants are exposed to biotic stress, involving mainly exposure to plant pathogenic fungi and viruses. Because these alterations involve mostly polyamine-derived secondary metabolites, they will not be covered in this review (see Chapter 4). Discussions of polyamines and stress responses have also been related to their metabolic/functional antagonism with ethylene and their possible role as antisenescence compounds; these topics are covered in Chapters 5 and 15, respectively.

II. POLYAMINE METABOLISM AND IONIC STRESS

A. THE "CLASSIC" RESPONSE: PUTRESCINE AND POTASSIUM DEFICIENCY

The work on the remarkable changes in polyamine, and especially putrescine (Put) levels which plant cells undergo as the result of nutritional deficiencies dates back to the pioneering studies of Richards and Coleman, and of Terence Smith at the Long Ashton Research Station (Bristol, U.K.). The work of Smith in particular not only developed the phenomenology of polyamine responses to stress, but over the course of over two decades established the general features of polyamine biosynthesis and degradation in plants.

Since the beginning of the century, it has been well established that plant nitrogen metabolism can undergo dramatic alterations as a result of environmental changes (see Chibnall).[4] For example, when seedlings of leguminous species or detached cereal leaves are grown in darkness without an external supply of nitrogen, they accumulate asparagine and glutamine. The levels of amides accumulating under such conditions are much higher than those obtained as a result of proteolysis, and a role in the detoxification of ammonia released by amino acid oxidation has been proposed for these compounds.[4] Several mineral nutrient deficiencies have also been known to cause increases in soluble nitrogen; for example, phosphorus-deficient barley plants accumulate nitrate and amides.[5] Potassium deficiency in the same plants also leads to a rapid disappearance of protein, a marked increase in amino and amide nitrogen, and accumulation of nitrate in the advanced stages of deficiency.[6] The major free amino acids accumulating in significant levels during potassium shortage were lysine, arginine, glycine, serine, leucine, and tyrosine. During the course of these studies, an unidentified, ninhydrin-positive compound accumulated as the external symptoms of deficiency became acute, and finally became the major organic soluble nitrogen compound present in the plant. This compound was present only in trace amounts under normal conditions; its accumulation under K^+ deficiency was suppressed if alkali metals such as rubidium or sodium were supplied. Chromatographically, the compound behaved exactly like the diamine putrescine. After isolation and crystallization, the unknown compound was shown to be indistinguishable from putrescine in any of its various salt forms (picrate, hydrochloride, or platinichloride).[7] In addition to barley, this early work showed that wheat and red clover leaves of K^+-deficient plants also accumulated Put.[8] As shown for barley, maximum Put accumulation coincided with the appearance of severe nutrient deficiency symptoms. Putrescine accumulated in the stem and leaves rather than in the roots, and none of it was found associated with the protein fraction. The work of Smith and others later showed that Put accumulates in K^+-deficient leaves of tomato, cabbage, radish, lettuce, pea, beet, oat, rye, tobacco, banana, maize, *Ribes nigrum,* broad bean, and spinach.[9-14] Thus, diamine accumulation as a result of potassium starvation is widespread among mono- and dicotyledonous species, and may well be a universal response.

The above studies also gave the first insights into the pathways of polyamine biosynthesis in higher plants. When K^+-deficient barley and clover leaves were incubated with DL-2-^{14}C-ornithine, Put quickly became labeled, thus suggesting that a direct decarboxylation mechanism, such as that catalyzed by ornithine decarboxylase (ODC) was involved in the Put increase.[15] Further work, however, showed that agmatine (Agm), the direct decarboxylation product of arginine, was present in K^+-starved plants and accumulated with a similar time course as that for Put.[16] In fact, when L-^{14}C-arginine was fed to K^+-deficient shoots, a more rapid label of Put was obtained than that resulting from feeding labeled ornithine. Feeding labeled arginine to barley shoots yielded Agm, which in turn was converted to labeled Put, thus confirming the close metabolic relationships between these two compounds. Furthermore, Agm feeding caused the appearance of *N*-carbamoylputrescine (NCP), the decarboxylation production of citrulline;[17] synthetic NCP fed to barley shoots was hydrolyzed readily to Put. Thus, it appeared as if NCP was a metabolic intermediate between Agm and Put, even though this compound was not always detectable in K^+-deficient plants.[17] In *Phaseolus* and *Sesamum*, however, NCP clearly accumulates when K^+ is limiting.[18-19] These studies strongly suggested the existence of an alternative pathway for Put biosynthesis, which had previously been known only in bacteria.[20] This was confirmed by the detection and characterization of the enzymes involved in the pathway, arginine decarboxylase (ADC), agmatine iminohydrolase, and NCP amidohydrolase.[9,12,21] The first and third enzymes were found to increase in K^+-deficient plants, while no difference was detected in agmatine iminohydrolase. These results thus established that the first rate-limiting step in Put formation under mineral deficiency is that catalyzed by ADC.

The studies discussed above make it quite clear that Put accumulation in K^+-deficient plants can be quite considerable; this diamine reaches levels at least one order of magnitude higher than in control plants.[22] Interestingly, although spermidine (Spd) and spermine (Spm) levels also increase under these conditions, the magnitude of the change is always significantly less than that observed for Put. In some cases, Put accumulation can be quite remarkable. For example, Put in K^+-deficient barley and clover can reach levels of 23 μmol/g dry weigh. In dying tissue,[9] Put titer can reach up to 140 μmol/g dry weight.[9] Thus, Put accounts for at least 20% of the total plant nitrogen.[23] As discussed below, although Put can accumulate under other mineral nutrient deficiencies, the magnitude and specificity of the increase appears to be most marked in the case of potassium. This has led several researchers to propose use of the endogenous Put levels as a measure of the potassium status of the plant.[13,24-25] In a study with apple seedlings, Hoffman and Samish[24] noted that both potassium excess and deficiency caused Put accumulation, and proposed that the concentration of K^+ in the growth medium be adjusted as to provide a minimum level of Put in the leaves. A study with lucerne (*Medicago sativa*) also showed Put to be a sensitive indicator of K^+ status. Putrescine levels were lowest only in plants which were producing over 90% of the maximum dry matter yield.[25] Although magnesium deficiency also resulted in Put increase, K^+-deficiency was the most important nutritional factor regulating the plant levels of Put. Most notably, Put accumulation was quite dramatic before visual symptoms of K^+ deficiency became apparent; a drop from 3% to 2.5% of K^+ in the dry matter was sufficient. In black currant leaves,[13] the Put concentration changed from virtually negligible levels to 1 to 5 μmol/g fresh weight when the potassium concentration in the medium was dropped below 1 meq/liter. Taking into account the two cationic sites present in Put, it was estimated that this diamine could compensate between 30 to 70% of K^+ at the lowest levels of this nutrient.

B. OTHER MINERAL NUTRIENT DEFICIENCIES

Although the Put levels of up to 0.15 to 0.20% dry weight (1 to 5 m*M*) attained during potassium shortage are very striking, it is now clear that deficiencies of other mineral elements can also cause significant accumulation of Put and other organic nitrogen constituents. In barley, agmatine concentration was increased by deficiencies of potassium, phosphorus, calcium, sulfur, magnesium, and manganese, but only K^+ deficiency appeared to increase Put levels significantly.[26] A more systematic study found that Mg^{2+} deficiency did cause a considerable increase in Put.[27] As in the case of Ca^{2+} deficiency, these rises in Put caused by starvation for nutrients other than K^+ are smaller and take longer to develop.[7] Magnesium deficiency also caused Put accumulation in barley, pea, bean, and radish leaves, although it had no effect on spinach leaves.[14] In this study, it was apparent that spermidine and spermine increased significantly in radish, pea, and bean leaves deficient in Mg^{2+} or K^+, but not in barley seedlings. Ornithine, arginine, and citrulline increased in sulfur deficient flax *(Linum usitatissimum)*.[28] Yoshida later reported[29] that sulfur deficiency also led to Put increase, although this report has not been confirmed. Interactions between various nutrient deficiencies have also been reported. In barley, the severity of K^+ deficiency symptoms is strongly influenced by the external supply of phosphorus.[23] As expected, Put content rose when potassium was reduced. However, for large amounts of Put to be accumulated, phosphorus (P) supply had to be in excess of K^+. When P was low relative to K^+, a significant increase in Agm level developed. In tobacco, Put increases under deficiency of K, P, Ca, Mg, Fe, Mn, S, and B,[10] but to a larger extent in the K^+ and P-deficient plants. Calcium deficiency and high salt (KCl) increased putrescine 1.5-fold in barley leaves.[9]

C. AMMONIUM NUTRITION AND SALT STRESS

The source of inorganic nitrogen is well known to affect the way in which plants assimilate this nutrient,[30] as well as the pool of organic nitrogen compounds. In relation to polyamine metabolism, the work of Le Rudulier and Goas has clearly shown the dramatic accumulation of Put which ensues when plants are grown in ammonium instead of nitrate as sole nitrogen source.[31-33] Soybean seedlings, grown both in the light and dark, showed a 100-fold Put increase when fed with ammonium.[31] The K^+ concentration was twice as great in the nitrate- vs. the ammonium-fed plants, leading to the hypothesis that the observed Put accumulation was simply the result of K^+ deficiency caused by the different nitrogen (N) regime. However, it later became clear that when experimental conditions were controlled so as to result in an identical internal K^+ content in all treatments, ammonium nutrition still led to a dramatic accumulation in Put.[32] Thus, K^+ deficiency does not appear to mediate the alterations in polyamine metabolism caused by ammonium nutrition. When 5-^{14}C-ornithine was fed to soybean seedlings, the conversion into Put was more rapid in ammonium- than in nitrate-fed plants;[34] ornithine incorporation into arginine was about the same on either medium. As shown previously with K^+ deficient plants, the increase in Put in ammonium-grown plants was correlated with ADC activity; ODC or citrulline decarboxylase were not detected.[35] Agmatine was metabolized into Put via NCP, and the rate of conversion was accelerated in the presence of ammonium in the growth medium.[36] The activity of Agm iminohydrolase was only 20% higher in ammonium- and in nitrate-grown plants. Finally, Put and Agm appeared to be oxidized more slowly when grown on ammonium medium.[37] In addition to soybean, ammonium nutrition has also been shown to result in Put accumulation in barley, corn, wheat, tomato, tobacco, pea, pepper, petunia, and eggplant.[27,38-39] As in the case of K^+ deficiency, this response also seems to be universal and to involve the ADC-mediated pathway.[40] The interactions between K^+ and ammonium nutrition have been studied in tobacco and tomato leaves.[38] Putrescine and cadaverine (Cad) levels increased in tobacco leaves with additional ammonium in the nutrient solution, in the absence of additional K^+. However, when 10 m*M* K^+ was added, Put became undetectable although the Cad level was almost doubled. The concentration of Cad reached the remarkable value of 140 mol/g fresh weight. Similar changes in Cad were observed in tomato leaves.

Extensive literature in Russian has accumulated concerning the effects of salinity stress on polyamine metabolism. The comprehensive monograph by Strogonov summarizes the earlier findings,[41] and more recent work is covered by Shevyakova.[42] Most of this work has been done on horse bean (*Vicia faba*) and cotton. A considerable accumulation of Put and Cad was found when peas and horse bean plants were grown in NaCl above 50 m*M*. Cadaverine was increased in pea seedlings grown in saline medium.[43] Under these conditions, Cad-derived alkaloids (lupinine, anabasine, and aphylline) were also detected. It is interesting that some of these alkaloids are also found in another family, Chenopodiaceae, in which halophytic species are common. The above increases in Cad were transient, as the diamine decreased back to control levels in 10-day-old plants. Later studies, however, have failed to detect increases in free Cad or in alkaloids in salt-treated pea seedlings.[44] In cotton, plants treated with Na_2SO_4 showed an increase in Put which was paralleled by a decrease in arginine.[45] Putrescine was metabolized faster in the salt-stressed plants, although this was not due to increased conversion into Spd. This increased Put metabolism apparently cannot be attributed to faster oxidation by an amine oxidase, since this enzyme activity actually declines with salinity.[41] In *Oenothera biennis*, salt treatments harsh enough to cause severe necrosis of leaf tissues led to a 6- to 10-fold increase in Put, and increases of similar magnitude in Spd and Spm.[46]

As indicated by the above studies, the effect of saline conditions on diamine accumulation is not always clear cut, and the mechanism involved is generally less understood than the changes caused by mineral deficiencies or ammonium nutrition. For example, salt stress caused Put accumulation in the leaves of mung bean seedlings, but no changes in ADC or ODC were detected in the aerial organs; however, a two- to eightfold increase in ODC was observed in the roots.[47] In pea and maize seedlings grown in the presence of various salts, there was only a 10 to 30% increase in Put on a dry weight basis.[48] Later work, however, could not show any significant changes in Put, Spd, or Spm when plants of *Vicia faba*, *Atriplex* spp., or *Salicornia europea* were treated with up to 750 m*M* NaCl.[49] In barley, horse bean, cotton, and peas, feeding Put and Cad in excess of 1 m*M* leads to loss of turgor and causes formation of necrotic spots, which mimic the symptoms of extreme salinization.[50] In contrast, Ashour and Thalouth[51] found that when *Vicia faba* plants that had been grown at a high salt concentration were sprayed with 100 ppm Put, the growth of the plants was improved. Strogonov and Shevyakova have suggested[42,52] that Put formation during moderate salinization is of adaptive significance. They propose that when fed in low amounts (10 to 100 μ*M*) Put can be a growth stimulant, but at higher concentration it may have toxic effects. This and related findings have led these workers to propose that accumulation of diamines in salt-stressed plants is a direct cause of salt toxicity. As we shall discuss below, however, experiments with exogenous polyamines are difficult to interpret, and may give results which have little or nothing to do with the function of endogenous polyamines.

D. ACID STRESS

The effect of low pH on polyamine metabolism was first reported in bacteria. It is well known that *Escherichia coli* possesses both a constitutive, biosynthetic ADC pathway, and a biodegradative form of the enzyme that is inducible by arginine. When the bacteria are transferred from pH 7 to pH 4, ADC activity (presumably the biodegradative form) increases from virtually undetectable levels to 270 μl CO_2 per mg dry weight per hour.[20] The amines produced by this and other amino acid decarboxylases may help raise the medium pH to values more favorable to growth.

Smith and Sinclair reported the effect of acid stress on barley.[53] When seven day old seedlings were switched from control medium (water) to a dilute (25 m*M*) solution of HCl, the activity of ADC and NCP amidohydrolase increased 2-fold within 3 days. Similar increases in enzyme activity were obtained with H_2SO_4 feeding. When excised *Cucumis sativus* cotyledons were fed with 10 m*M* HCl, a doubling of ADC activity was observed.[54] The effect of external pH on polyamine metabolism has been reported under closely controlled conditions using peeled oat leaf segments.[55] Putrescine levels are low and stable when the leaf segments are floated on buffers in the range of pH 6 to 8. As the pH is decreased below 6, and independently of the buffer system employed, there is a progressive increase in Put titer, at least down to pH 3.5. The acid-induced rise in Put is detectable within 3 h and peaks at about 9 h; it is also readily reversed when the leaf segments are transferred back to pH 6.0 or above. In this system, the kinetics and magnitude of Put increase are paralleled by an increase in ADC activity, and by incorporation of label from *U*-^{14}C-arginine into Put.

E. OSMOTIC STRESS

The effect of osmotic stress on polyamine metabolism was discovered during our attempts to culture cereal mesophyll protoplasts and stabilize them by exogenous polyamine application.[56] To our surprise, we found that the polyamine pattern of freshly isolated oat mesophyll protoplasts was strikingly different from that of the oat mesophyll cells from which they are derived.[57] The Put content in protoplasts was at least 5 to10 times higher, while Agm, Spd, and Spm remained essentially unchanged on a per cell basis. This dramatic change occurred in the 2- to3-h period during which excised peeled oat leaf segments were treated with cellulolytic enzymes. We further found that the increase in Put was not due to the toxic, low molecular weight components which are commonly present in crude commercial cellulase preparations, because similar increases in Put were obtained with desalted/dialyzed enzymes. Since it was well known that the osmoticum present in the protoplast isolation medium can frequently cause changes in membrane potential and inhibit photosynthesis, we tested the hypothesis that the increased Put content of protoplasts is a direct result of osmotic stress. Freshly peeled oat leaf segments were incubated in a solution containing only phosphate buffer and 0.4 M sorbitol or mannitol. Within a few hours, the Put content of these osmotically stressed leaf segments increased to about the same level as that found in protoplasts, and reached about 60 times the initial value after 6 h. Incubation in control buffer resulted in no significant change in Put level. In contrast, Spd levels increased only by 30 to 50%, and Spm titer showed no change. We were also able to show that other osmotica, both permeating and nonpermeating (sorbitol, mannitol, proline, betaine, and sucrose) could induce similar increases in Put.[58] The response was also found in other cereals (wild oat, barley, corn, and wheat). Although some increase in Put occurred as a result of the injury caused by leaf peeling, the accumulation caused by osmotic shock was at least one order of magnitude higher, and the response was concentration dependent. A sorbitol concentration of 0.4 *M* produced optimal Put accumulation. A half-maximal effect was obtained with 0.2 *M*, and 0.6 *M* sorbitol gave a supraoptimal depression (80% of maximum accumulation). In the cereal leaf system (both stressed and unstressed), most if not all of the Put pool could be accounted for by "free" Put, readily extractable in perchloric acid. No significant extra Put was released by acid hydrolysis of the leaf extracts. We therefore concluded that osmotic treatment did not cause the release of Put from a "bound" form, but actually induced *de novo* Put biosynthesis.

As discussed in Chapter 2, several irreversible specific inhibitors of polyamine biosynthesis have become available in the last decade, and have proved very useful in the analysis of polyamine metabolism and function.Their application in plant systems has been reviewed recently.[59-60] The inhibitors with the most potential as physiological tools are the enzyme-activated (suicide type) irreversible inhibitors of ODC and ADC, α–difluoromethylornithine (DFMO), and α-difluoromethylarginine (DFMA). We tested these inhibitors in order to elucidate the mechanism of Put increase during osmotic stress, and we tested the effect of specific inhibitors of polyamine biosynthesis.

When oat leaf segments were stressed with sorbitol, the increase in Put was paralleled by a 50 to 70% rise in ADC activity; little or no change was observed in ODC activity. DFMA added at 1m*M* completely prevented the osmoticum-induced increase in Put, and concomitantly inhibited ADC activity *in vivo* in a concentration-dependent manner.[56-57] Over the time span of the experiment, DFMA did not affect the Spd and Spm titers. In contrast, DFMO had no effect on ODC activity *in vivo* or *in vitro*, but potentiated the Put and ADC response to sorbitol and increased Spd levels. This regulatory phenomenon remains unexplained, although it is clear that DFMO facilitates Spd formation from Put under osmotic shock. *In vitro* assays confirmed the specificity of DFMA, and showed that there was no cross inhibition between DFMO and ADC, or DFMA and ODC, respectively. Thus, our experiments showed that under osmotic shock, oat leaf cells were induced to synthesize Put through an activation of the pathway mediated by ADC. That ADC was indeed preferentially activated by osmotic shock was suggested by two additional observations: (1) D-arginine and L-canavanine, both competitive inhibitors of ADC, also prevented the increase in Put; and (2) α-methylornithine, a competitive inhibitor of ODC, had the same effect as DFMO on stressed leaves.

Most of the reported increases in Put in stressed plants occur over a period of several days. The response to osmotic shock, however, was massive and rapid; the increase in Put under acid stress also occurs within a few hours.[55] In the osmotic shock system, we found that the increase in Put and ADC had a lag time of 1 to 2 h.[58] We hypothesized that protein synthesis may be involved in the response, and tested it using cycloheximide (CH). We showed that CH (10 to 50 μg/ml) prevented the rise in Put and Spd, but it did so only if it was present at time zero. If the addition of CH was delayed for 1 h, 60% of its ability to inhibit the Put rise was lost. After 2 h of exposure to the osmoticum, CH addition was ineffective. These results suggest that the primary event in this stress-induced phenomenon occur very rapidly, and require *de novo* protein synthesis. Whether the requirement is for new synthesis of ADC vs. a regulatory molecule or post-translational modification is not known. Interestingly, the tissue needs to be stressed for several hours in order for the full rise in Put to occur. If leaf segments are floated on sorbitol for 1 h and then transferred to buffer, and putrescine titer determined after a total of 4 h, the rise is significant, but much smaller than if the tissue has been stressed for the full 4 h period.[56,58] Consistent with our early results,[56] The incorporation of label from ^{14}Carginine into Put in the osmotically stressed cereal leaves was not accompanied by a comparable increase in the label of Spd or Spm.[61] This is in agreement with the finding that under other types of ionic stress, there is little if any change in the polyamines proper.[1] It further suggests that the stress response involves not only a rise in Put biosynthesis, but also a block in polyamine synthesis.

It is not clear if the results obtained with excised, osmotically shocked cereal leaves are relevant to a real drought stress situation. We did find that when oat seedlings grown in pots were allowed to wilt by withholding water over a period of a few days, Put titer and ADC activity rose significantly.[58] These changes were detectable only after the second day of treatment and were coincident with measurable signs of stress, such as wilting and protein loss. However, the potassium status of these plants was not monitored during the experiment, and it is therefore not possible to conclude that these results are solely the result of wilting and not due to nutrient starvation. We also attempted to correlate the response to osmoticum with the differential sensitivities of barley cultivars to drought. When peeled leaf segments of Excelsior (drought-resistant variety), Proctor (drought susceptible) and Prior (intermediate susceptibility) were osmotically shocked, all three cultivars showed approximately the same response (Put increase and ADC activity). Along the same line, abscisic acid (ABA), which has been linked to responses to drought stress, could induce significant proline accumulation in excised cereal leaves, but only a slight increase in Put. Polyethyleneglycol (PEG) treatment did not significantly increase Put levels, although it is very effective in reproducing the symptoms of drought stress.[56-58]

F. PHYSIOLOGICAL IMPACT AND SIGNIFICANCE

It is clear from the above discussion that polyamine metabolism is extremely sensitive to a variety of challenges which can be grouped under the label of "ionic stress." In particular, K^+ deficiency, ammonium nutrition, acid stress, and osmotic shock can lead to rapid and massive increases in diamine levels, most commonly Put. The response is mediated via the ADC pathway, as shown by rapid labeling from arginine via Agm and NCP, correlated increases in ADC activity but not in ODC, and inhibition of the response (in the case of osmotic shock and acid stress)[55,58] by application of DFMA. Because of the remarkable similarities in the physiology and biochemistry of Put accumulation under various stress conditions, it is tempting to conclude that a common mechanism underlies these responses. What the nature of the mechanism is and what is the physiological significance of this massive shift in the pools of organic nitrogen is at present unclear. In the case of K^+ deficiency, Put accumulation can result in titers as high as 1.2% dry weight,[1] representing at least 20% of the total nitrogen. If only because of this massive recruitment of available nitrogen into a single compound, we must inquire about the consequences of these dramatic changes for the plant cell.

In their early work with K^+-deficient barley, Coleman and Richards[8] suggested that deficiency in this nutrient can result in an imbalance between inorganic anions and cations, the final consequence of which is acidification of the cellular pH. In this context, the accumulation of Put may be part of an homeostatic mechanism that prevents excessive fluctuation of the cellular pH. Data from several systems are consistent with this view. For example, it has been estimated that at least 30% of the cation deficit in K^+-deficient black currant can be balanced by an increase in Put.[13] If the level of Agm is taken into account, as suggested by Smith,[1] then both amines could compensate between 45 and 100% of the deficiency in K^+. There are, however, few if any systems in which the full balance of anions and cations is known, and thus these estimates can be subject to a very wide error margin. As discussed above, the conditions of ion imbalance which occur under K^+ deficiency can be mimicked by an external supply of excess hydrogen ions,[53-55] or by growing the plant with ammonium as the nitrogen source,[31-37] which also leads to acidification of the cytoplasm. Sulfur dioxide fumigation of pea seedlings causes a significant increase in free and bound putrescine.[62] Since SO_2 absorption into the cells results in higher acidity as a result of the formation of HSO_3^- and SO_3^{2-} ions, and eventually SO_4^{2-}, amine accumulation may also, in this case, compensate for the relative deficit of cations. All these results are consistent with, but do not prove, the pH-stat hypothesis advanced by Smith,[1] that Put accumulation can compensate for the lowered pH in the cytoplasm of the stressed cells. The definitive test of his hypothesis will have to involve direct measurements of cytoplasmic pH, which are now possible with NMR techniques, as well as detailed kinetic analysis of the changes in pH and Put, and probing with biosynthesis inhibitors. This should be possible in any of the stress systems which show a fast response, such as osmotic shock and acid stress. It should also be feasible to induce relatively fast K^+ depletion by using specific ionophores applied to peeled leaves or cell cultures, and by following the parameters outlined above.

Several experiments involving exogenous polyamine application have been used to argue that stress-induced Put accumulation is actually a deleterious response. For example, Coleman and Richards[8] fed Put through cut barley leaves, and found necrotic areas characteristic of K^+ deficiency. Agmatine and Put concentrated at the tip of the leaf, where degeneration typically began.[23] A similar time course has been reported in black currant leaves.[13] In barley, horse bean, cotton, and peas, feeding Put and Cad in excess of 1 m*M* leads to loss of turgor and induces necrotic symptoms similar to those caused by salt treatment.[42] While striking, these results may be difficult to relate to the effects of increases of endogenous polyamines. A physiological explanation for the effects of exogenous diamine feeding has recently been provided by DiTomaso et al.[63] Treatment of excised corn roots with exogenous Put resulted in a wound response characterized by membrane depolarization and increase in leakage of potassium. In contrast to previous work, it was shown that Put did not replace calcium in maintaining membrane stability, but instead caused significant membrane damage. It was suggested that the basis for exogenous Put toxicity was the presence of an apoplastic diamine oxidase, and that damage to the plasma membrane was caused by the oxidation products. Di- and polyamine oxidases have been reported present in the cell wall (see Chapter 3). The results of DiTomaso et al.[63] provide an explanation for the toxic effects observed by Coleman and Richards and Shevyakova in their feeding experiments, but also raise the question of how physiologically meaningful are results obtained with exogenous Put application. It is unlikely that endogenously produced Put will be oxidized to the same extent as that which is applied externally. We suggest that manipulation of internal polyamine concentrations with specific inhibitors may be a more appropriate way of probing the potentially deleterious effect of stress-induced diamine accumulation. A recent report indicates that this is in fact a viable approach. As shown above,[56-57] oat leaf protoplasts are labile cells and accumulate high levels of Put. When freshly isolated oat mesophyll protoplasts are isolated and grown in the presence of DFMA, the resulting levels of Put are very low compared to the untreated controls.[64] Protoplasts treated with DFMA show significantly enhanced ability to synthesize protein and RNA, and have higher titers of Spd and Spm, although they still fail to divide consistently.

Most studies concerning polyamines and stress responses have considered these metabolites in isolation from the rest of intermediary nitrogen metabolism. Recent studies, however, suggest that it may be useful to integrate polyamine responses with nitrogen fluxes in the plant cell. During our studies on osmotic shock,[56,61] we observed that Put increase is strongly dependent on precursor availability. When leaf segments were osmotically shocked in the dark, the levels of Put were only about one third of those reached when leaves were stressed in the light.[61] If the dark-stressed leaves were fed with arginine, the level of stress-induced Put then became comparable to that of light-incubated stressed leaves. Thus, even though Put biosynthesis is induced by stress via the ADC pathway, the size of the arginine precursor pool appears to be limiting, and is actually a major determinant of the magnitude of the response. Other studies support this view and suggest an interesting hypothesis. It is well known that mineral deficiencies and salt stress can lead to large increases in ammonia. Lovatt has shown that ammonia can be effectively removed by increased activity of *de novo* arginine biosynthetic pathway.[65] Arginine can, in turn, be converted to Put. Most interestingly, this increased arginine and Put biosynthesis occurs in stress-tolerant cultivars, but is inhibited in stress-sensitive plants; as a result, ammonia accumulates and toxicity results. For example, the salt-tolerant muskmelon (*Cucumis melo*) can grow relatively well in solutions containing 30 to 60 m*M* NaCl, and makes arginine *de novo* under these conditions.[65] In contrast, the salt-sensitive summer squash (*Cucurbita pepo*) shows a significant reduction in growth under salt stress, correlated with increased ammonia accumulation and decreased arginine biosynthesis. Similarly, cell suspension cultures of the drought-sensitive alfalfa cultivar Mesilla accumulate high levels of ammonia when exposed to water stress caused by PEG.[66] A drought-tolerant population (Mes-3) derived from cv. Mesilla shows decreased ammonia accumulation, and increased Put levels. Slocum and Weinstein[67] have proposed that stress-induced Put accumulation is part of a mechanism for ammonia detoxification. In support of this proposal, they have shown that when oat leaf segments are osmotically stressed, the levels of Put and ammonia rise in parallel. When the leaf segments are stressed in the presence of DFMA, Put accumulation is inhibited as expected, and ammonia rises even further. This suggests at least a partial role for Put in sequestering NH_3.

III. POLYAMINE METABOLISM AND ENVIRONMENTAL CHALLENGES

In the last five years, several other abiotic challenges have been reported to cause significant changes in polyamine metabolism. Based on the available information, these responses do not appear to be the result of changes in ionic balance and are therefore treated separately in this section. If anything, these findings make for an even more complicated picture than we had until now. These studies also suggest that polyamines may have other functions during stress responses, such as the ability to serve as free radical scavengers (ozone stress) or to stabilize cellular membranes (chilling, heat, and drought injury).

A. ENVIRONMENTAL POLLUTANTS

Ozone (O_3) is a major component of air pollution, and is considered to have serious effects on the vegetation in Europe and North America. Even when applied at low levels, O_3 stress can lead to a significant decline in net photosynthesis, cause leaf injury, and accelerate senescence.[68] Subacute levels of O_3 can induce a number of biochemical changes, some of which may form part of the plant's defense mechanism(s). Reactions to this oxidative stress include increases in ascorbic acid, peroxidases, phenolic compounds, and polyamines. When exogenous polyamines were fed to tomato and tobacco plants,[69-70] there was a significant suppression of O_3-induced leaf injury. In O_3-treated barley leaves, ADC activity increased before the injury was apparent.[71] Spermidine content increased significantly during O_3 treatment, both in young and old leaves. When DFMA was applied to the leaves, the rise in ADC activity was prevented and injury to ozone exposure was considerably enhanced; little visual injury was seen in the leaves of O_3-free controls treated with DFMA. These results suggest that polyamines may have a protective role against O_3 damage. Polyamines have also been observed to accumulate in O_3-treated Norway spruce and wheat.[72,73] The mechanism involved is not clear, but because O_3 effects may be largely due to oxidative damage, the possibility that polyamines may be involved in free radical scavenging has been investigated. The levels of superoxide radical formed enzymatically with xanthine oxidase, or chemically from riboflavin or pyrogallol were inhibited *in vitro* by Put, Spd, or Spm at 10 to 50 m*M*.[74] Superoxide radical production and superoxide-dependent conversion of 1-amino-cyclopropane-1-carboxylic acid were also inhibited by polyamines added to microsomal membrane preparations. These findings, together with *in vivo* studies showing that polyamines can retard senescence and inhibit lipid peroxidation (see Chapters 5 and 14), were attributed to the radical-scavenging ability of these compounds. A later study, however, has disputed these conclusions.[70] Leaf injury of the O_3 sensitivity of the tobacco cultivar Bel W 3 could be prevented by feeding Put, Spd, or Spm through the root, these exogenous treatments were correlated with a two- to threefold increase in soluble free and conjugated Put and Spd; in particular, Put and Spd associated with cell wall and membrane fractions were elevated four to sixfold. However, when free polyamines were assayed *in vitro* for their radical-scavenging properties, very low rate constants were found. On the other hand, Put conjugates had relatively high rate constants. It was thus concluded that free polyamines could not account for the protection against O_3 damage, assuming a role as free radical scavengers. It was suggested that the protective effect of exogenous free polyamines was mediated by their prior conversion to conjugated forms. Consistent with this hypothesis, it was recently found[75] that monocaffeoyl-Put, an effective scavenger of oxyradicals, was present in the apoplastic fluid of tobacco leaves exposed to O_3. This extracellular location could allow free radical scavenging at the site where they are first generated.

B. CHILLING INJURY

Recent work has explored the interactions between polyamines and fruit quality during low temperature/low oxygen storage.[76-81] During cold storage (0°C in air, 1% O_2) of Chinese cabbage (*Brassica campestris*), all polyamines declined linearly with storage time.[76] The decline was slower in low O_2-stored samples than in air-stored fruits. In the meristematic tissues, polyamines increased in both conditions during the first six weeks of storage. The chilling injury of zucchini squash (*Cucurbito pepo*) can be reduced by storage in a low O_2 atmosphere.[77] The decrease in surface pitting under these conditions was paralleled by higher levels of Spd and Spm. Putrescine and Spd also increased in the skin tissue during exposure to chilling temperatures. No significant increase in polyamines, however, was observed when the fruits were stored at nonchilling temperatures. The symptoms of chilling injury in squash could also be reduced by preconditioning the fruit for 2 days at 10°C.[78] Preconditioning led to a significant increase in Spd and Spm, but not in Put levels. The increases in Spd and Spm were correlated with elevated activity of *S*-adenosylmethionine decarboxylase.[79] In the controls, chilling injury was correlated with increases in Put, but decreases in Spd and Spm. In this case, then, it is the polyamines proper and not the diamines that appear correlated with a protective mechanism underlying preconditioning. The levels of chloroform-soluble fluorescent products, presumably a result of lipid peroxidation, were reduced in the preconditioned zucchini fruits.[78] Based on these results, the authors suggested that polyamines may prevent chilling injury in squash by a mechanism involving protection of membrane lipids. Similar correlations have been found during low O_2 storage of Macintosh apples.[80-81] In this case, both Put and Spd increased by two- to sixfold in apples stored under a controlled atmosphere.[80] Polyamines also inhibited the *in vitro* activity of polygalacturonase. In support of these correlations, pressure infiltration of apples with polyamines resulted in a significant increase in fruit firmness.[81] This treatment also inhibited the development of chilling injury (brown core).

C. HEAT AND DROUGHT STRESS

Most studies relating polyamines to stress responses have been done with the common, widespread compounds, Put, Spd, and Spm. The usual response to stress is an increase in one or more of the polyamines already present at the start of the stress. A recent finding suggests that, at least under heat stress, plant cells can be induced to synthesize a novel array of polyamines. Thermophilic bacteria usually contain only Spd when grown under "normal" conditions (25 to 35°C). When the temperature is increased above 50°C, a whole new series of polyamines appears, mostly symmetrical analogs of Spd and Spm.[82] The most common of these "thermopolyamines" are norspermidine (caldine) and norspermine (thermine). These uncommon, long-chained polyamines are apparently essential for continued protein synthesis at high temperatures, both *in vitro* and *in vivo*.[82] The pattern of uncommon polyamines made in bacterial thermophiles depends on the growth temperature. In extreme conditions (80°C) *Thermus thermophilus* produces a new polyamine composed of five aminopropyl residues, which has been named caldopentamine.

Until recently, the presence of "thermopolyamines" was assumed to be restricted to thermophilic prokaryotes. Studies with drought-tolerant and heat-tolerant cultivars of alfalfa and cotton, respectively, have given surprising results. When cell cultures of drought-tolerant alfalfa lines are exposed to water deficit caused by PEG, significant amounts of caldine and thermine accumulate.[83,84] Pollen and cell cultures of heat-tolerant cotton genotypes also accumulate these uncommon polyamines when exposed to high temperature.[84] A compound tentatively identified as caldopentamine also appears under these conditions. Little is known as to how "thermopolyamines" are synthesized. The building block is presumably 1,3-diaminopropane (DAP), which is derived from the action of polyamine oxidase (PAO) on Spd. Addition of aminopropyl units to this precursor by an aminopropyltransferase (APT) will yield the final product. PAO and APT activities have in fact been detected in shoot meristems and cell cultures of alfalfa, and exhibit a one- to fivefold stimulation when the cells are exposed to drought stress.[84,85] In summary, this exciting work has shown that plant cells have the ability to synthesize uncommon polyamines, previously reported only in thermophilic bacteria, when challenged with drought or heat stress. Although the evidence is so far correlative, the biosynthesis of these uncommon polyamines may be related to stress responses in drought/heat-tolerant plants.[86] The physiology and biochemistry of uncommon polyamines is discussed in detail in Chapter 8.

IV. CONCLUDING REMARKS

Work done by several groups during the last three decades has established that various environmental challenges which can be classified as "ionic stresses" can lead to rapid and massive alterations in polyamine metabolism. Specifically, the ADC-mediated pathway of Put biosynthesis appears to be "activated" under K^+ deficiency, by low pH, ammonium nutrition, and osmotic stress. It is only in the case of the first three stresses that the response appears to be equally applicable to mono- and dicotyledonous plants. The changes caused by osmotic stress, so far, seem to be more restricted to species of the Gramineae. While the physiology and biochemistry of the response to ionic stress is well delineated, several important questions remain unanswered. For example, it is not yet clear if *de novo* synthesis of ADC is actually induced by any of the above stresses. Increases in ADC activity are usually 50 to 100%, while Put accumulation can reach, in many cases, over one order of magnitude higher. Whatever the regulatory mechanism(s) involved, their study should be a priority in future work. The availability of ADC cDNA clones and of antibodies against ADC should greatly facilitate future studies (See Chapter 9). The genetic engineering of plants in which we can precisely control polyamine levels is now a real possibility. We will also have to develop tools to look at the latter steps in polyamine biosynthesis, namely Spd and Spm, which seem to be impaired during ionic stress. As discussed above, the physiological significance of these responses remains obscure in spite of recent progress. Recent studies have made clear that experiments involving application of exogenous polyamines may lead to equivocal interpretations. However, the availability and judicious use of enzyme-activated irreversible inhibitors should prove powerful probes for polyamine function in stressed plants.

The results described in section III are a useful reminder of the further surprises that polyamines may hold for us in the future. The effects of ozone pollution, heat, and cold stress, discovered only in the last few years, have opened several new dimensions in the studies of polyamines and abiotic challenges. It is very difficult at the present time to visualize an underlying common mechanism to all these responses. My educated guess is that there is probably a common metabolic denominator for the responses to ionic stress, but at least another distinct mechanism responsible for the latter responses. In spite of all these unknowns, it is my opinion that polyamine changes in relation to abiotic stress deserve a continuing research effort, and that such studies may help develop crops which are better able to withstand environmental insult. For all the excitement that awaits us, it would be wise to constantly remind ourselves of the need to integrate polyamine physiology within the context of intermediary nitrogen metabolism, and of their metabolic and functional interactions with plant growth regulators.[60,87]

ACKNOWLEDGMENTS

The author would like to thank Ms. Paula Sgrignoli for assistance with typing and editing this manuscript, and Dr. Yao-ren Dai for useful discussion and suggestions. This work was funded in part by a grant from the Electric Power Research Institute (Palo Alto, CA.).

REFERENCES

1. **Smith, T.A.,** Putrescine and inorganic ions, *Recent Adv. Phytochem.,* 18, 7, 1984.
2. **Flores, H.E., Young, N.D., and Galston, A.W.,** Polyamine metabolism and plant stress, in *Cellular and Molecular Biology of Plant Stress*, Key, J.L. and Kosuge, Y., Eds., Alan R. Liss, New York, 1985, 93.
3. **Flores, H.E.,** Polyamines and plant stress, in *Stress Responses in Plants: Adaptation and Acclimation Mechanisms*, Alscher, R.G. and Cumming, J.R., Eds., Wiley-Liss, New York, 1990, 217.
4. **Chibnall, A.C.,** *Protein Metabolism in the Plant,* Yale University Press, New Haven, 1939.
5. **Richards, F.J. and Templeman, W.G.,** Physiological studies in plant nutrition: IV. Nitrogen metabolism in relation to nutrient deficiency and age in leaves of barley, *Ann. Bot.*, 50, 367, 1936.
6. **Richards, F.J. and Berner, J.R.,** Physiological studies in plant nutrition: XVII. A general survey of the free amino acids of barley as affected by mineral nutrition with special reference to potassium supply, *Ann. Bot.*, 18, 15, 1954.
7. **Richards, F.J. and Coleman, R.G.,** Occurrence of putrescine in potassium-deficient barley, *Nature,* 170, 460, 1952.
8. **Coleman, R.G. and Richards, F.J.,** Physiological studies in plant nutrition: XVIII. Some aspects of nitrogen metabolism in barley and other plants in relation to potassium deficiency, *Ann. Bot.*, 20, 393, 1956.
9. **Smith, T.A.,** L-Arginine carboxylase of higher plants and its relation to potassium nutrition, *Phytochemistry*, 2, 241, 1963.
10. **Takashashi, T. and Yoshida, D.,** Relationship between the accumulation of putrescine and the nutrition of tobacco plant, *J. Soil Sci. Manure (Jpn.)* , 31, 39, 1960.
11. Steward, F.C., Hulme, A.C., Freiberg, S.R., Hegarty, M.P., Pollard, J.K., Rabson, R., and Barr, R.A., Physiological investigations on the banana plant, *Ann. Botany,* 24, 83, 1960.
12. **Smith, T.A.,** Agmatine iminohydrolase in maize, *Phytochemistry*, 8, 2111, 1969.
13. **Murty, K.S., Smith, T.A., and Bould, C.,** The relation between the putrescine content and potassium status of black currant leaves, *Ann. Bot.*, 356, 687, 1971.
14. **Basso, L.C. and Smith, T.A.,** Effect of mineral deficiency on amine formation in higher plants, *Phytochemistry*, 13, 875, 1974.
15. **Coleman, R.G. and Hegarty, M.P.,** Metabolism of *dl*-ornithine-2-^{14}C in normal and potassium deficient barley, *Nature*, 172, 376, 1957.
16. **Smith, T.A. and Richards, F.J.,** The biosynthesis of putrescine in higher plants and its relation to potassium nutrition, *Biochem. J.*, 84, 292, 1962.
17. **Smith, T.A. and Garraway, J.L.,** *N*-carbamoylputrescine — an intermediate in the formation of putrescine in barley, *Phytochemistry*, 3,23, 1964.
18. **Crocomo, O.J., Cattini, M.A., and Zago, E.A.,** Accumulation of amines and amino acids in relation to the potassium level in leaves of beans (*Phaseolus vulgaris*), *Arq. Biol. Technol.*, 17, 93, 1974.
19. **Crocomo, O.J., Basso, L.C., and Brasil, O.G.,** Formation of *N*-carbamylputrescine from citrulline in *Sesamum*, *Phytochemistry*, 9, 1487, 1970.
20. **Gale, E.F.,** The production of amines by bacteria: 1. The decarboxylation of aminoacids by strains of *Bacterium coli, Biochem.*, 34, 392, 1940.
21. **Smith, T.A.,** *N*-carbamylputrescine amidohydrolase of higher plants and its relation to potassium nutrition, *Phytochemistry*, 4, 599, 1965.
22. **Smith, T.A.,** Putrescine, spermidine and spermine in higher plants, *Phytochemistry*, 9, 1479, 1970.
23. **Hackett, C., Sinclair, C., and Richards, F.J.,** Balance between potassium and phosphorus in the nutrition of barley. 1. The influence on amine content, *Ann. Bot.*, 20, 393, 1965.
24. **Hoffman, M. and Samish, R.M.,** Free amine content in fruit tree organs as an indicator of the nutritional status with respect to potassium, *Rec. Adv. Plant Nutr.*, 1, 189, 1971.

25. **Smith, G. S., Lauren, D. R., Cornforth, I. S., and Agnew, M. P.,** Evaluation of putrescine as a biochemical indicator of the potassium requirements of lucerne, *New Phytol.*, 419, 1982.
26. **Sinclair, C.,** Relation between mineral deficiency and amine synthesis in barley, *Nature*, 213, 214, 1967.
27. **Smith, T.A.,** Amine levels in mineral deficient *Hordeum vulgare* leaves, *Phytochemistry*, 12, 2093 1973.
28. **Coleman R.G.,** Occurrence of ornithine in sulphur-deficient flax and the possible place of ornithine and citrulline in the arginine metabolism of some higher plants, *Nature*,181, 776, 1958.
29. **Yoshida, D.,** Formation of putrescine from ornithine and arginine in tobacco plants, *Plant Cell Physiol.*, 10, 393, 1969.
30. **Poulton, J.E., Romeo, J.T., and Conn, E.E.,** *Plant Nitrogen Metabolism*, Plenum Press, New York, 1989, 474.
31. **Le Rudulier, D. and Goas, G.,** Mise en evidence et dosage de quelques amines dans les plantules de *Soja hispida* Moench privees de leurs cotyledons et cultivees en presence de nitrates, d'uree et de chlorure d' ammonium, *Compt. Rend. Acad. Sci. (Paris) Ser. D*, 279, 161, 1971.
32. **Le Rudulier, D. and Goas, G.,** Influence des ions ammonium et potassium sur l'accumulation de la putrescine chez les jeunes plantes de *Soja hispida* Moench privees de leurs cotyledons, *Physiol. Veg.*, 13, 125, 1975.
33. **Le Rudulier, D. and Goas, G.,** Contribution a l'etude de l'accumulation de putrescine et de la putrescine dans les plantules de *Glycine max* (L.), *Merr. Physiol. Veg.*, IX:, 609, 1979.
34. **Le Rudulier, D. and Goas, G.,** Devenir de l'ornithine ^{14}C-5 dans les jeunes plantes de *Soja hispida* Moench. privees de leurs cotyledons et cultivees en presence de chlorure d'ammonium et de nitrates, *C. R. Hebd. Seances Acad. Sci. Ser. D*, 278, 1039, 1974.
35. **Le Rudulier, D. and Goas, G.,** Etude de l'activite de l'arginine decarboxylase dans les jeunes plantes de *Glycine max*, privees de leurs cotyledons, *Phytochemistry*, 14, 1723, 1975.
36. **Le Rudulier, D. and Goas, G.,** Metabolisme azote des junes plantes de *Soja hispida* Moench. privees de leurs cotyledons. Recherche des voies de formation de la putrescine, *C. R. Hebd. Seances Acad. Sci. Ser. D,* 279, 161, 1974.
37. **Le Rudulier, D. and Goas, G.,** Devenir de la putrescine-1,4-^{14}C chez *Glycine max*, *Physiol. Plant.*, 40, 87, 1977.
38. **Hohlt, D.E., Maynard, D.N., and Barker, A.V.,** Studies on the ammonium tolerance of some cultivated Solanaceae, *J. Amer. Soc. Hort. Sci.*, 95, 345, 1970.
39. **Klein, H., Priebe, A., and Jager, H.J.,** Putrescine and spermidine in peas; efects of nitrogen source and potassium supply, *Physiol. Plant.*, 45, 497, 1979.
40. **Le Rudulier, D. and Goas, G.,** Biogenese de la *N*-carbamylputrescine et de la putrescine dans les plantules de *Glycine max* (L.), *Merr. Physiol. Veg.*, 18, 609, 1980.
41. **Strogonov, B.P.,** *Physiological Basis of Salt Tolerance in Plants*, Israel Program for Scientific Translations, Jerusalem, 193, 284.
42. **Shevyakova, N.I.,** Metabolism and the physiological role of diamines and polyamines in plants, *Soviet Plant Physiol.*, 28, 1052, 1981.
43. **Prikhod'ko, L.S. and Klyshev, L.K.,** Nitrogen metabolism of pea shoots at diferent substrate salinity, *Tr. Inst. Botan. Akad. Naus. SSR*, 20, 166, 1964.
44. **Anderson, J.N. and Martin, R.O.,** Identification of cadaverine in *Pisum sativum*, *Phytochemistry*, 12, 443, 1973.
45. **Shevyakova, N. I., Arutyunova, N. V., and Strogonov, B. P.,** Disturbance in arginine and putrescine metabolism in cotton leaves in the presence of excessive Na_2SO_4, *Fiziol. Rast.*, 19, 594, 1981.
46. **Bagni, C. and Cremonini, M.,** Induction and repression of polyamine synthesis in *Oenothera biennis* by diferent salts, *Inf. Bot. Ital.*, 5, 103, 1973.
47. **Friedman, R., Altman, A., and Bachrach, U.,** Polyamine involvement in plant response to salt stress, *Plant Physiol.*, 72S, 137, 1983.
48. **Jager, H. J. and Priebe, A.,** Zum problem der durch salinitat induzierten putrescinbildung in pflanzen, *Oecol. Plant.*, 10, 267, 1975.
49. **Priebe, A. and Jager, H.J.,** Effect of NaCl on the levels of putrescine and related polyamines of plants differing in salt tolerance, *Plant Sci. Lett.*, 12, 365, 1978.
50. **Shevyakova, N.I.,** On the stimulating and toxic effects of diamines on plants, *Soviet Plant Physiol.*, 13, 472, 1966.
51. **Ashour, N.I., and Thalouth, A.T.,** Effect of putrescine on growth and photosynthetic pigments of broad bean plants grown under chloride salinization conditions, *Biochem. Physiol. Pflanzen*, 162, 203, 1971.
52. **Strogonov, B.P., Shevyakova, N.I., and Kabanov, V.V.,** Diamines in plant metabolism under conditions of salinization, *Fiziol. Rast.*, 19, 1098, 1972.
53. **Smith, T.A. and Sinclair, C.,** The effect of acid feeding on amine formation in barley, *Ann. Bot.*, 31, 103, 1967.
54. **Suresh, M.R., Ramakrishna, S., and Adiga, P.R.,** Regulation of arginine decarboxylase and putrescine levels in *Cucumis sativus* cotyledons, *Phytochemistry*, 17, 57, 1978.
55. **Young, N.D. and Galston, A.W.,** Putrescine and acid stress. Induction of arginine decarboxylase activity and putrescine accumulation by low pH, *Plant Physiol.*, 71, 767, 1983.

56. **Flores, H.E.,** *Studies on the Physiology and Biochemistry of Polyamines in Higher Plants*, Ph. D. thesis.,Yale University, New Haven, CT., 1983.
57. **Flores, H.E. and Galston, A.W.,** Polyamine and plant stress: activation of putrescine biosynthesis by osmotic shock, *Science*, 217, 1259, 1982.
58. **Flores H.E. and Galston, A.W.,** Osmotic stress-induced polyamine accumulation in cereal leaves. I. Physiological parameters of the response, *Plant Physiol.*, 75, 102, 1984.
59. **Slocum, R.D. and Galston, A.W.,** Inhibition of polyamine biosynthesis in plants and plant pathogenic fungi, in *Inhibition of Polyamine Metabolism*, McCann, P.P, Pegg, A.E. and Sjoerdsma, A., Eds., Academic Press, San Diego, 305, 1987.
60. **Flores, H.E., Protacio, C.M., and Signs, M.W.,** Primary and secondary metabolism of polyamines in plants, *Rec. Adv. Phytochem.*, 23, 329, 1989.
61. **Flores, H.E. and Galston, A.W.,** Osmotic stress-induced polyamine accumulation in cereal leaves. II. Relation to amino acid pools, *Plant Physiol.*, 75, 110, 1984.
62. **Priebe, A., Klein, H., and Jager, H.J.,** Role of polyamines in SO_2 polluted pea plants, *J. Expt. Bot.*, 29, 1045, 1978.
63. **DiTomaso J.M., Shaff,J.E., and Kochian, L.V.,** Membrane-mediated putrescine transport and its role in stress-induced phytotoxicity, *Plant Physiol.*, 89, 147, 1989.
64. **Tiburcio, A.F., Kaur-Sawhney, R., and Galston, A.W.,** Polyamine metabolism and osmotic stress: 11. Improvement of oat protoplasts by an inhibitor of arginine decarboxylase, *Plant Physiol.*, 82, 375, 1986.
65. **Lovatt, C.J.,** Stress alters ammonia and arginine metabolism, in *Polyamines and Ethylene: Biochemistry, Physiology, and Interactions*, Flores, H.E., Arteca, R.N., and Shannon, J.C., Eds., Amer. Soc. Plant Physiol., Rockville, MD., 1990, 166.
66. **Phillips, G.C., Rodriguez-Garay, B., Kuehn, G.D., and Lovatt, C.J.,** Patterns of ammonia accumulation in cell cultures of water-deficit-stressed alfalfas and heat-stressed cottons, *Agron. Soc. 79th Ann. Meeting*, 1987, 152.
67. **Slocum, R.D. and Weinstein, L.H.,** Stress-induced putrescine accumulation as a mechanism of ammonia detoxification in cereal leaves, in *Polyamines and Ethylene: Biochemistry, Physiology, and Interactions*, Flores, H.E., Arteca, R.N., and Shannon, J.C., Eds., Amer. Soc. Plant Physiol., Rockville, MD., 1990, 157.
68. **Heagle, A.S.,** Ozone and crop yield, *Annu. Rev. Phytopathol.*, 27, 397, 1989
69. **Ormrod, D.P. and Beckerson, D.W.,** Polyamines as antiozonants for tomato, *Hort. Sci.*, 21, 1070, 1986.
70. **Bors, W., Langebartels, C., Michel, C., and Sandermann, Jr., H.,** Polyamines as radical scavengers and protectants against ozone damage, *Phytochemistry*, 28, 1589, 1989.
71. **Rowland-Bamford, A.J., Borland, A.M., Lea, P.J., and Mansfield, T.A.,** The role of arginine decarboxylase in modulating the sensitivity of barley to ozone, *Env. Pollution*, 61, 95, 1989.
72. **Dohmen, G.P., Koppers, A., and Langebartels, C.,** Biochemical response of Norway spruce *Picea abies* (L.), towards 14-month exposure to ozone and acid mist: effects on amino acid, glutathione and polyamine titers, *Environ. Pollut.*, 64, 375, 1990.
73. **Raab, M.M. and Weinstein, L.H.,** Polyamine and ethylene metabolism in *Triticum aestivum* (var. Vona), in *Polyamines and Ethylene: Biochemistry, Physiology, and Interactions*, Flores, H.E., Arteca, R.N., and Shannon, J.C., Eds., Amer. Soc. Plant Physiol., Rockville, MD., 1990, 408.
74. **Drolet, G., Dumbroff, E.B., Legge, R.L., and Thompson, J.E.,** Radical scavenging properties of polyamines, *Phytochemistry*, 25, 367, 1986.
75. **Langerbartels, C., Kerner, K.J., Leonardi, S., Schraudner, M., Trost, M., Heller, W., and Sandermann, Jr., H.,** Biochemical plant responses to ozone. I. Differential induction of polyamine and ethylene biosynthesis in Tobacco, *Plant Physiol.*, 95, 882, 1991.
76. **Wang, C.Y.,** Changes of polyamine content in chinese cabbage during storage in air or low oxygen atmosphere, *J. Food Quality*, 11, 289, 1988.
77. **Wang, C.Y. and Zuo, L.J.,** Effect of low-oxygen storage on chilling injury and polyamines in zucchini squash, *Sci. Hort.*, 39, 1, 1989.
78. **Kramer, G.F. and Wang, C.Y.,** Correlation of reduced chilling injury with increased spermine and spermidine levels in zucchini squash, *Physiol. Plant.*, 76, 479, 1989.
79. **Kramer, G.F. and Wang, C.Y.,** Effects of chilling and temperature preconditioning on the activity of polyamine biosynthetic enzymes in zucchini squash, *J. Plant Physiol.*, 136, 115, 1990.
80. **Kramer, G.F., Wang, C.Y., and Conway, W.S.,** Correlation of reduced softening and increased polyamine levels during low-oxygen storage of 'Mcintosh' apples, *J. Amer. Soc. Hort. Sci.*, 114, 942, 1989.
81. **Kramer, G.F., Wang, C.Y., and Conway, W.S.,** Inhibition of softening by polyamine application in 'Golden Delicious' and 'McIntosh' apples, *J. Amer. Soc. Hort. Sci.*, (in press), 1991.
82. **Oshima, T.,** Unusual polyamines in an extreme thermophile, *Thermus thermophilus*, *Adv. Polyamine Res.*, 4, 479, 1983.

83. **Rodriguez-Garay, B., Phillips, G.C., and Keuhn, G.D.,** Detection of norspermidine and norspermine in *Medicago sativa* L. (alfalfa), *Plant Physiol.*, 89, 525, 1989.
84. **Kuehn, G.D., Bagga, S., Rodriguez-Garay, B., and Phillips, G.C.,** Biosynthesis of uncommon polyamines in higher plants and their relation to abiotic stress responses, in *Polyamines and Ethylene: Biosynthesis, Physiology, and Interactions,* Flores, H.E. and Arteca, R.N., Eds., Amer. Soc.Plant Physiol., 1990, 190.
85. **Bagga, S., Huehn, G.D., and Phillips, G.C.,** Polyamine oxidase and aminopropyltransferase ezymes involved in biosynthesis of uncommon polyamines in alfalfa, *Agron. Abstr.,* 1989, 173, 1989.
86. **Kuehn, G.D., Rodriguez-Garay, B., Bagga, S., and Phillips, G.C.,** Novel occurrence of uncommon polyamines in higher plants, *Plant Physiol.*, 94, 855, 1990.
87. **Flores, H.E., Arteca, R.N., and Shannon, J.C.,** Eds., *Polyamines and Ethylene: Biochemistry, Physiology, and Interactions*, Amer. Soc. Plant Physiol., Rockville, MD., 1990, 425.

Analytical Methods for Polyamines and Enzymes of Polyamine Metabolism

Chapter 16

CHROMATOGRAPHIC METHODS FOR THE IDENTIFICATION AND QUANTITATION OF POLYAMINES

Mary A. Smith

TABLE OF CONTENTS

I. INTRODUCTION

The commonly studied aliphatic amines, cadaverine, putrescine, spermidine, and spermine have been associated with many physiological processes in plants. However, the precise role of these substances remains unknown. To investigate the physiological function of polyamines, several chromatographic methods have been developed to isolate them and to quantitate their levels in biological tissues. Among the methods used are thin-layer chromatography (TLC), reversed phase high performance liquid chromatography (HPLC), ion-pair reversed phase HPLC, and ion exchange chromatography. Some of these methods have been extended to assay tissues for polyamine precursors, metabolites, conjugates, and homologs, as well as aromatic and heterocyclic amines.

Although several methods are available, investigators continue to improve existing methods and to develop simpler, more rapid, and sensitive chromatographic assays. This chapter describes procedures for some of the more frequently used techniques and comments on the usefulness of a few of the alternative methods that are available. Several of the HPLC methods have been subjects of previous reviews.[1,2]

The most widely used procedures for measuring polyamines in plants employ HPLC of precolumn derivatives such as 5-dimethylaminonaphthalene-1-sulfonyl chloride (dansyl chloride) or benzoyl chloride. In the absence of HPLC availability, TLC of dansyl derivatives is most often a method of choice. However, TLC does not provide the sensitivity or the reliability of separation of HPLC.

II. THIN-LAYER CHROMATOGRAPHY

The analysis of polyamines by TLC is a simple and rapid procedure, considering that several samples can be run in parallel on the same plate. This procedure can be used to separate amines, amine conjugates, certain amino acids, and metabolic intermediates that may be present in a single sample. It is also useful for the identification of radiolabeled polyamines, their precursors and metabolites.[3] Two methods for visualizing amines include prechromatographic derivatization with dansyl chloride with fluorimetric detection of resolved components and postchromatographic staining with ninhydrin, followed by colorimetric monitoring. In depth reviews of TLC methods have been published previously.[2]

A. EXTRACTION OF POLYAMINES

1. Free Polyamines

Polyamines can be extracted from plant tissue with trichloroacetic acid (TCA), perchloric acid (PCA), or a mixture of hydrochloric acid and TCA. A general scheme for the extraction of polyamines is shown in Figure 1. Most often the tissue is homogenized in cold 5 to 10% PCA (100 mg ml^{-1}) and held on ice for 30 min to 1 h. For quantitative purposes, an internal standard such as 1,6-diaminohexane (hexanediamine) may be added to the medium. The homogenate is centrifuged at 4°C for 10 to 30 min at 3,000 to 26,000 × g depending upon the sample. The supernatant is analyzed for free polyamines, but a portion of the supernatant and the pellet may be treated further to release soluble and insoluble bound amines.

2. Hydrolysis of Bound Polyamines

The pellet is washed in PCA and then resuspended in the original volume of PCA by vortexing. Bound amines in an aliquot of the PCA suspension are hydrolyzed in a final concentration of 6 N HCl at 110°C for 18 h in a sealed ampule. After heating, the sample is filtered through glass wool, dried under a stream of air at 80°C, and resuspended in PCA. An aliquot of the supernatant containing the soluble bound polyamines is treated with HCl in same way.[4]

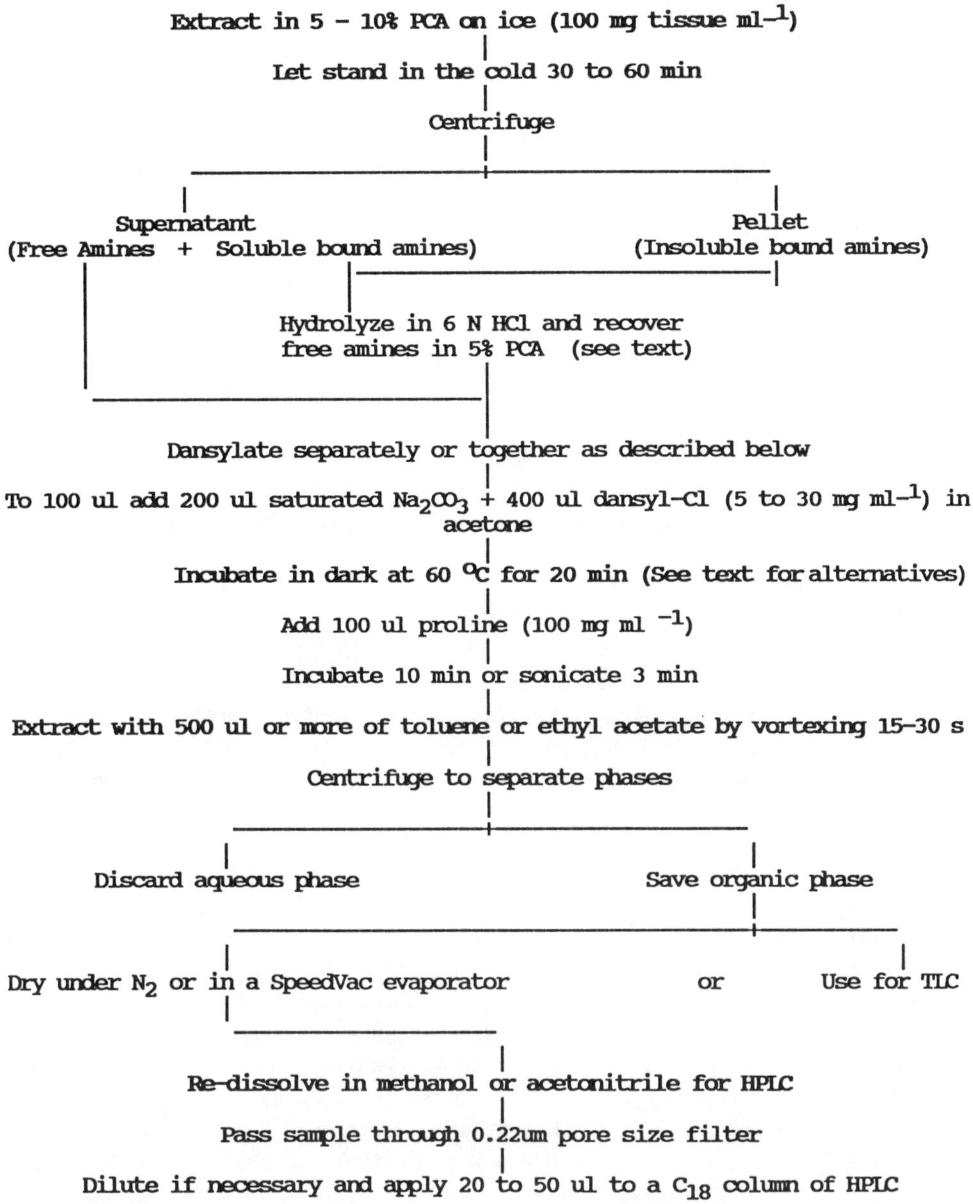

FIGURE 1. Scheme for the extraction and dansylation of polyamines.

3. Purification of Extracts

Some extracts may require purification before separation by chromatography, e.g., to remove interfering carbohydrates.[5-6] The supernatant is adjusted to pH 9.0 with NaOH and applied to a small column containing 0.5 g of silica gel (60 to 200 mesh, J. T. Baker, Phillipsburg, NJ, U.S.). The column is washed with distilled water (adjusted to pH 8.0 with NaOH) and then eluted with 0.03 N HCl. The eluate is dried at 45°C and redissolved in 0.03 N HCl. In a very similar procedure, silica gel Sep Paks (Waters Associates, Milford, MA, U.S.) have been used to purify polyamine extracts.[2,7]

4. Preparation of Standards

Polyamine standards are obtained as hydrochloride salts. Solutions of the aliphatic di- and polyamines are made up in 1 to 10 m*M* concentrations in water, or in low concentrations of acids (e.g., 0.5% PCA or 0.1 N HCl). Standards can be stored in plastic tubes at –20°C for at least 2 months.[8-10] Plastic is used because polyamines bind to the surface of glass. For derivatization, standards are diluted to picomole levels in 50 to 100 μl of the extraction medium.

B. DANSYLATION OF POLYAMINES

Dansyl chloride is a widely used fluorogenic agent in chromatographic procedures for polyamine analysis. It reacts stoichiometrically with primary and secondary amines. However, it is a nonspecific reagent that will react with aliphatic amines, aromatic and heterocyclic amines, alkaloids, amino acids, phenols, aliphatic alcohols, purine bases and some sugars.[11] Polyamines are detected and quantified easily because they occur in much higher concentrations then most of the competing substances in an extract. Interference from amino acids is avoided by partitioning them into an aqueous phase while extracting the polyamines into an organic solvent after the dansylation reaction is completed. In depth reviews of the dansylation of polyamines, which describe conditions for the reaction and the products of the reaction, have been published elsewhere.[2,11]

1. Derivatization Procedure

For the derivatization reaction, dansyl chloride is made up in acetone at a concentration of 10 mg ml^{-1} (Figure 1). This mixture can be stored for 24 h at 4°C in the dark.[9] Since dansyl derivatives are light sensitive, the reaction is performed in amber colored or aluminum foil covered reactivials (Pierce, Rockford, IL, U.S.). An aliquot of the PCA extract (100 to 200 μl) is added to 400 μl of dansyl chloride in acetone and 200 μl of saturated Na_2CO_3. The mixture is incubated at 60°C for 1 hour in a heating block equipped with a magnetic stirrer (Pierce, Rockford IL, U.S.).[1,12] Magnetic spin bars are placed in each vial for constant mixing to avoid phase separation at this temperature.[2] Kabra et al.[9] reported that incubation at 70°C is optimal, but in a recent systematic study of temperature and reaction time, Gennero et al.[13] found optimal conditions to be 60°C for 20 min. Others sonicate the reaction mixture at room temperature for 2 to 3 h or incubate at room temperature in the dark for 16 h or overnight.[2,14]

After incubation, excess dansyl chloride is removed by treating the reaction with 100 μl of proline (100 mg ml^{-1}) at room temperature for 30 min.[12] Kabra et al.[9] report 10 min is adequate, but Heby and Andersson[14] found the reaction to be quantitative after 3 min of sonication. Dansylamines are extracted from the mixture with 500 μl of toluene or ethyl acetate by vortexing 15 to 30 s. The organic layer containing the polyamines is separated from the aqueous phase by low speed centrifugation, removed and dried under nitrogen. A good procedure for quantitative purposes may be to transfer a known portion of the organic layer (50%) to an Eppendorf tube and evaporate under nitrogen or in a SpeedVac Evaporator (Savant, Farmingdale, NJ). The dry residue is dissolved in a suitable solvent for chromatography.

A blank sample and authentic polyamines are treated separately with dansyl chloride as described to permit quantitative analysis. Derivatives can be stored in methanol at 4°C in the dark for up to 4 months without substantial decomposition (< 10%) or in benzene up to one month at –20°C.[9,15]

Dansylation should be performed with pure dansyl chloride to achieve good analytical results. Purification of a poor sample of dansyl chloride can be accomplished on a silica gel G-60 column, 70 to 230 mesh (Merck, Darmstadt, Federal Republic of Germany).[2] Ten grams of dansyl chloride are dissolved in 20 ml of toluene and then applied to a 12 × 100 mm Silica gel column. The dansyl chloride is eluted with toluene and then evaporated to dryness *in vacuo*.

2. Post-Derivative Cleanup

Escribano et al.[16] found that lichens have high concentrations of salts and phenols which may interfere with the separation of the polyamines. To purify the sample after derivatizing, 150 to 600 µl of 5 *M* KOH in methanol is added to the toluene extract. The mixture is incubated for 30 to 45 min at 50°C. Then 1.5 ml of an aqueous solution of 100 to 200 mg each of KH_2PO_4 and Na_2HP0_4 is added to the mixture and the polyamines are extracted with 3 to 4 ml of toluene.[2,16]

C. SEPARATION OF DANSYLAMINES BY TLC

Dansylamines can be separated on precoated thin-layer silica gel plates (200 to 250 µm) such as silica gel G-60 (Merck) or LK6D plates (Whatman, Clifton NJ U.S.).[8,11,17,18] Samples in a nonpolar solvent such as benzene or toluene are loaded onto a plate that has been activated at 110°C for 1 h. The plate is developed in a solvent system of cyclohexane:ethyl acetate (5:4 v/v) or chloroform:triethylamine (25:2 v/v).[8,18]

After chromatography, the dansyl derivatives are visualized quickly under a UV light source (360 nm). Unknown fluorescent spots are identified by comparing their R_f values with those of polyamine standards run on the same plate. Evaluation of the plate is done rapidly before the fluorescent spots fade. The plate should be protected from the light to prevent light-destruction of the derivatives on the silica surface. Once the fluorescent areas have been marked and identified, they are scraped off the plate into centrifuge tubes containing 2 to 3 ml of ethyl acetate. The tubes are vortexed vigorously to extract the dansylamines from the gel. After centrifugation, the fluorescence of the supernatant is monitored in a spectrofluorometer using an excitation wavelength in the range of 350 to 365 nm and emission within 495 to 510 nm. For quantitative purposes, a blank sample is prepared by scraping a blank area off the plate into ethyl acetate and subjecting the supernatant to fluorescence monitoring.

Direct scanning of TLC plates has also been used for polyamine determination using a chromatogram analyzer. Quantitation is achieved by measuring peak heights or peak areas. Scanning can be done immediately or later on plates that have been properly stored.[2] Oil-treated plates maintain fluorescence intensity for at least a week when stored in the dark. For oil treatment, air-dried plates are dipped into a TLC tank filled with 10% paraffin or silicon oil in cyclohexane. The plate is dried in the hood for 5 min and then dipped into the oil again. Before scanning, the plate is equilibrated for 30 min in the room atmosphere that will be used for scanning.

D. TLC OF RADIOLABELED POLYAMINES

TLC is useful for the separation of radiolabeled polyamines, labelled precursors and metabolites.[3,19] After pulse labeling, tissue is washed with excess distilled water and then homogenized in methanol:chloroform:water (12:5:3 v/v), 50 mg ml^{-1} on ice. After 30 min, the homogenate is centrifuged. An aliquot of the supernatant is subjected to liquid scintillation counting to determine the total label taken up by the tissue. Twenty-five to 50 µl of the supernatant is applied to a silica gel plate which is developed in a solvent system of butanol:acetic acid:water (4:1:1 V/V); the amines are detected by staining the plates with ninhydrin.

In preparation for ninhydrin staining, the plates are dried at 90°C for 30 min. Then they are sprayed with 0.1% (w/v) ninhydrin in n-butanol saturated with water. To develop the ninhydrin stain, plates are heated at 100°C for 10 min. Amines spots, identified by R_f values, are scraped off the plate into 200 µl methanol. This mixture is added to 4 ml of scintillation cocktail and monitored for radioactivity in a liquid scintillation counter.

III. SEPARATION OF POLYAMINES BY REVERSED PHASE HPLC

A plethora of reversed phase HPLC methods have been developed for separating and identifying polyamines and related compounds in biological tissues. Precolumn and postcolumn derivatization with a variety of agents have been used to enhance sensitivity of detection of polyamines. Although dansyl chloride and benzoyl chloride are widely used derivatizing agents for precolumn modification of polyamines, several other chromophoric agents have been employed including dabsyl chloride (dimethylaminoazobenzene sulfonyl chloride), *o*-Phthalaldehyde (OPA), fluorescamine (4-phenylspiro [furan-2-(3H),1′-phthalan]-3,3′-dione),and 9-fluorenylmethyl chloroformate (FMOC). For postcolumn labeling of polyamines, OPA is frequently used.

Separations of derivatized and underivatized polyamines are achieved on high resolution reversed phase C_{18} columns (15 to 30 cm × 2.2 to 5 mm i.d.) packed with 5 to 10 µm size octadecyl silica particles. In all of the HPLC procedures, the life of the C_{18} column is maintained by using a few protective measures. A guard column packed with C_{18} material is inserted in front of the main column. HPLC grade solvents for the elution systems are ultra-filtered through < 0.50 µm pore size membranes and degassed *in vacuo* before use. Samples are filtered as well. For convenience, small samples can be filtered rapidly through ultra-filtration centrifuge tubes (Rainin, Woburn, MA, U.S.).[1]

A. REVERSED PHASE HPLC OF PRECOLUMN DERIVATIVES

1. HPLC of Dansylamines

This is a favored method for assaying polyamines because preparation of the derivatives is fast and simple as previously described for TLC. Fluorescence monitoring is very sensitive with limits of detection at the femtomolar level.[9,10,12,20]

a. Separation of Dansylamines

Dansylamines taken up in methanol or acetonitrile are applied to a reversed phase column in 20 to 50 µl aliquots by using a fixed volume injector loop. Separation is achieved with either an isocratic or a gradient elution solvent system of water:methanol, water:acetonitrile, or acetonitrile:phosphate buffer, pH 4.4.[1,9,12,21] The eluent is monitored with a fluorescence detector, using an excitation wavelength in the range of 333 to 365 nm and emission within 485 and 522 nm. Optimal wavelengths for excitation and emission depend on the instrument used and can be determined by scanning a dansylated sample.

Sample chromatograms of polyamine standards and derivatives prepared from plant extracts are shown in Figures 2A and 2B.[1,12] Polyamines were separated within 27 min by using an elution gradient system of methanol:water. Although several side products appear on the chromatogram, they elute well before the dansylamines. The elution system allowed excellent resolution of putrescine, cadaverine, spermidine, spermine, and the internal standard, hexanediamine. After eluting spermine the gradient was increased to 100% methanol and gradually lowered to starting conditons for re-equilibration for 5 min before a new sample application.

The speed of separation of dansylamines can be increased by using a different mobile phase, steepening the elution gradient, and/or by using a higher resolution reverse phase column (e.g., 3 µm particle size).[9,10,21-24] In some instances, ion-pairing agents such as octanesulfonate or heptanesulfonate have been included in the mobile phase to enhance separation.[7,10,20,22-24]

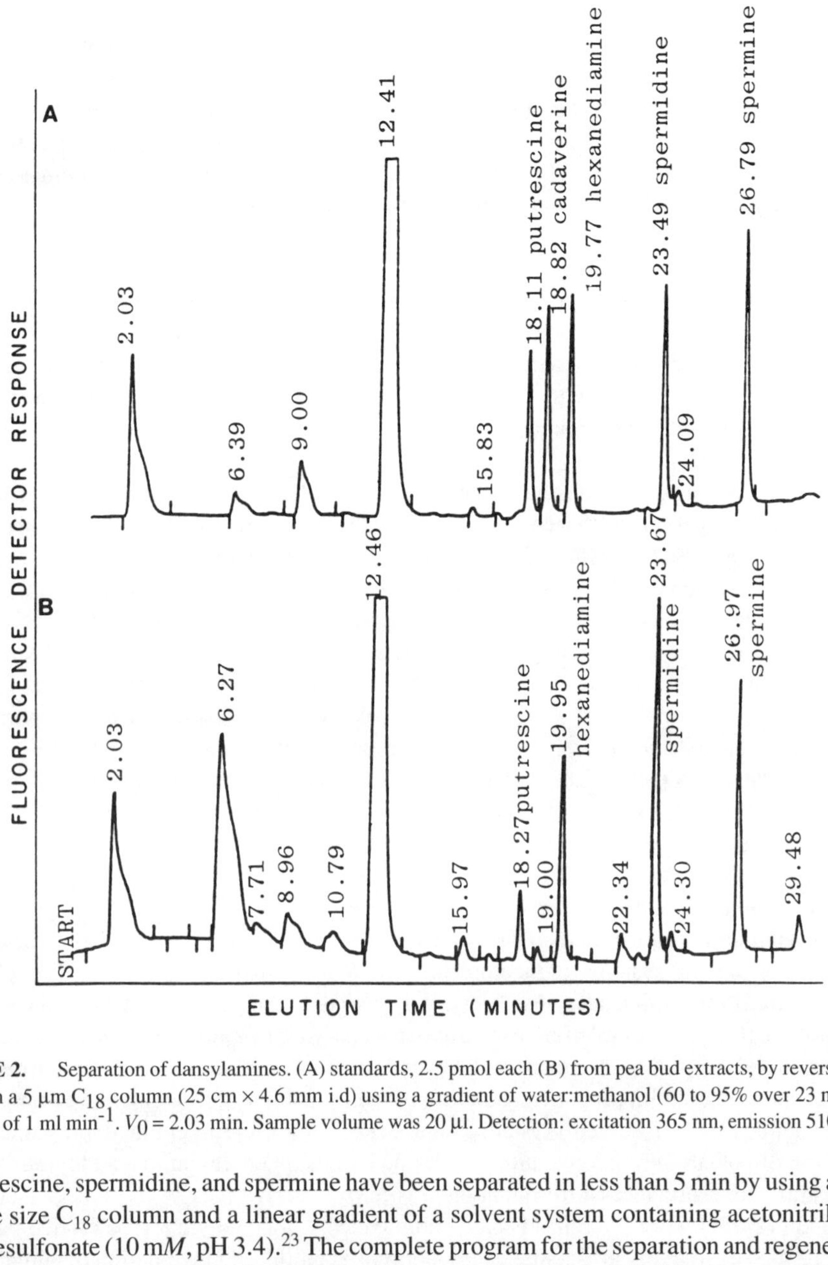

FIGURE 2. Separation of dansylamines. (A) standards, 2.5 pmol each (B) from pea bud extracts, by reverse phase HPLC on a 5 μm C_{18} column (25 cm × 4.6 mm i.d) using a gradient of water:methanol (60 to 95% over 23 min) at a flow rate of 1 ml min^{-1}. V_0 = 2.03 min. Sample volume was 20 μl. Detection: excitation 365 nm, emission 510 nm.[12]

Putrescine, spermidine, and spermine have been separated in less than 5 min by using a 3 μm particle size C_{18} column and a linear gradient of a solvent system containing acetonitrile, and heptanesulfonate (10 m*M*, pH 3.4).[23] The complete program for the separation and regeneration of the column was less than 8 min at room temperature. Using a modified gradient, rapid elution of polyamines, acetylated polyamines, and an internal standard, heptanediamine was also achieved with good resolution. Although chromatograms of derivatives from tissue extracts are not shown, the method has been applied to biological samples.[23,25] It is reported that excellent separation and quantification was obtained.

Analysis of extracts for agmatine has not been successfully achieved by HPLC of dansyl derivatives. The reaction of agmatine with dansyl chloride results in the formation of three fragmentation products including mono-dansyl agmatine, *bis*-dansyl agmatine, and *bis*-dansyl putrescine.[11,26,31]

2. HPLC of Benzoylamines

The principles for benzoylation are the same as for dansylation. However, not as many side products are formed and the derivatives are not light sensitive.[2] The method is widely used because it involves simple and rapid sample preparation. The procedure is inexpensive since only 5 to 10 μl of benzoyl chloride is needed per sample. It is also a convenient method, considering the availability of UV detectors. Detection of benzoylated polyamines is not as sensitive as for dansyl derivatives, but the limits of detection are between 0.01 and 1 nmol.[8]

a. Preparation of Benzoylamines

Extracts are prepared in PCA as described earlier. An aliquot of the supernatant, 0.1 to 1 ml is added to 1 ml of 2 N NaOH with 5 to 10 μl of benzoyl chloride. After incubation at room temperature for 20 min, the reaction is terminated with 2 ml of aqueous saturated NaCl. Benzoylamines are extracted with 2 ml of diethyl ether. The mixture is centrifuged and then the ether layer is removed and evaporated under nitrogen. The dry residue is dissolved in 100 to 200 μl of methanol or acetonitrile.[8,27]

Recently, several suggestions have been reported for the improvement of the benzoylation process. It has been observed that compounds in crude extracts inhibit the benzoylation reaction. Thus, it is suggested that derivatization be preceded by ion exchange chromatography.[28] Asotra et al. [29] found that derivatization is improved by dissolving the benzoyl chloride in methanol to a final concentration of 1.5%. They also reported that an incubation temperature of 37°C gives the best peak slope and maximum peak areas, and that the optimum reaction time is 18 to 20 min. They found that an incomplete reaction occurs below 37°C, and that increased duration of the reaction leads to extra peaks and decreased peak areas. Verkaelen et al.[30] reported that chloroform enhances the efficiency of extraction over diethyl ether by about 20%.[30] Finally, another recent recommendation is that HPLC analysis be performed on fresh derivatives. Storage of the derivatives in methanol at -20°C for a few days results in extra peaks on the chromatogram.[31]

b. Separation of Benzoylamines

Benzoylamines are separated on a reversed phase column using either isocratic or gradient elution with solvent systems of acetonitrile:water or methanol:water.[8,31,32] The eluate is monitored by UV detection at an absorbance wavelength of 254 nm. Isocratic elution with 64% methanol at a flow rate of 1 ml min^{-1} can be used to separate a mixture of putrescine, cadaverine, norspermidine, spermidine, homospermidine, and spermine in 20 min.[8] However, the resolution of putrescine and diaminopropane can be improved by using a step gradient elution program with a methanol:water solvent system (Figure 3).[31] A different gradient program allows separation of polyamines and two aromatic monoamines, phenethylamine and tyramine.[31]

Agmatine has been successfully quantitated as benzoyl derivatives, but it has been incorrectly identified in earlier work.[8,31] It has been recommended that benzoylation be carried out at pH 10 to 11, in order to avoid artifactual degradation of agmatine.[31] Degradation of agmatine has been observed when extracts are benzoylated at pH 9 or below, the main products being benzoyl-putrescine and benzoyl-urea.

3. HPLC of *o*-Phthalaldehyde Derivatives

o-Phthalaldehyde (OPA), in the presence of reducing agents such as 2-mercaptoethanol or ethanethiol, reacts rapidly with primary amines to form strong fluorescent derivatives, 1-alkylthio-2-substituted isoindoles.[33] Since it reacts with only primary amines, it is more selective than other fluorogenic agents. It is used for pre- and postcolumn derivatization, but postcolumn modification of polyamines is more often preferred because OPA derivatives are not very stable.[34]

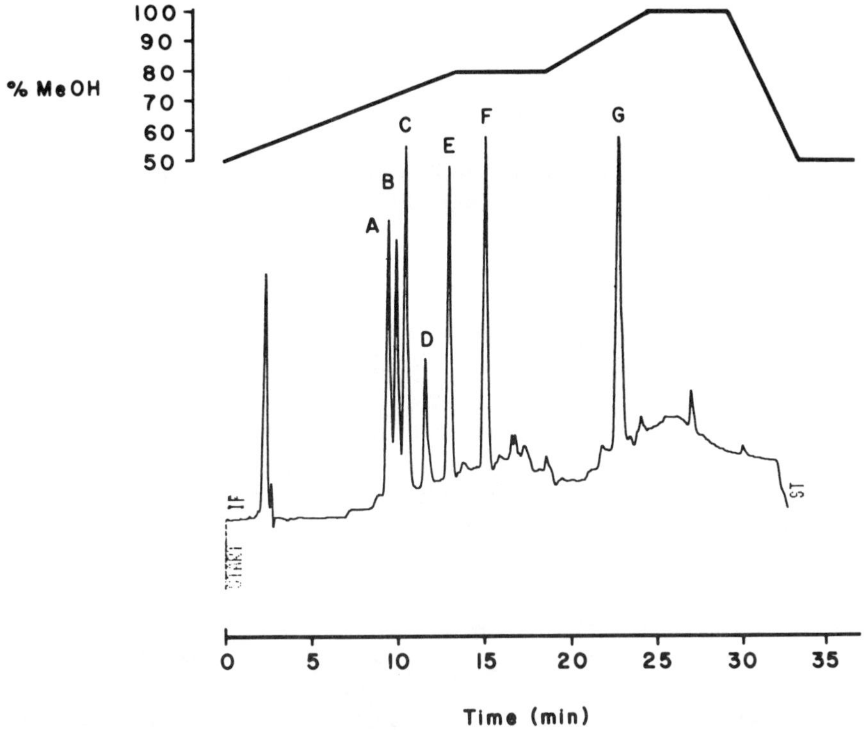

FIGURE 3. Separation of 2 nmoles each of diamine, polyamine, agmatine, and urea standards by reversed phase HPLC using gradient elution Program I as described by Slocum et al.[31] (A) putrescine, (B) diaminopropane, (C) cadaverine, (D) urea, (E) spermidine, (F) spermine, (G) agmatine.

HPLC of precolumn OPA derivatives has been recently reported for rapid and sensitive detection of polyamines from plant tissue.[28] Separation of the aliphatic polyamines and related substances took 12 to 14 min with sensitivity of detection as low as 1 to 2 pmol. The method involves a lengthy procedure for the preparation of the DCMA (N,N-dimethylcyclohexylamine) buffer for the solvent elution system. The DCMA buffer is stirred and fresh OPA reagent is stored, both overnight before use. In addition, prepurification of the extract by ion exchange is recommended. However, the authors state that sample preparation and analysis time is comparable to other current methods. One disadvantage is that the procedure does not lend itself to automation due to the instability of the derivatives.

a. Preparation of OPA Derivatives

PCA extracts are treated first by ion exchange chromatography.[28] One to 2 ml of the extract are applied to a 0.7 μm × 2.3 cm cation exchange column (H^+ form, 200 to 400 mesh, Bio-Rad AG 50 W - X 4 resin). The column is eluted with 2.8 ml of 0.7 M NaCl in 0.1 M sodium phosphate buffer (pH 8.0), followed by 1 to 2 ml of water, then 3.0 ml of 1 N HCl and finally with 2.4 ml of 6 N HCl. The 6 N HCl eluate is dried *in vacuo* at 40°C and the residue is taken up in 200 μl of water. An aliquot of the sample (e.g., 50 μl) is mixed with an equal volume of the derivatizing reagent for 90 s at room temperature.

The derivatizing reagent is prepared by dissolving 43 mg of OPA and 32 μl of 2-mercaptoethanol in 0.8 ml of methanol. This OPA / thiol mixture (OPT) is diluted to 8 ml with 0.4 N sodium borate buffer (pH 10.8). Storage of the OPT reagent overnight in a crimp-cap vial at room temperature reduces excess background fluorescence during the assay. The reagent can be stored under these conditions for months.[28]

b. Separation of OPA Derivatives

A 20 μl sample of OPA derivatives is applied to a reversed phase column. The solutes are eluted with a gradient solvent system of (A) methanol and (B) 0.1 *M* phosphate (pH 4.0) with 0.05 *M N,N*-dimethylcyclohexylamine (DCMA) in 50% (v/v) methanol:water. Details for making up the DCMA buffer for solvent B is described in the original paper. With this elution program the authors were able to resolve agmatine, spermine, norspermidine, spermidine, 3,5-homospermidine, 4,4-homospermidine, 1,3-diaminopropane, putrescine, and cadaverine in a single sample within 14 min. The sensitivity of detection of the derivatives was as low as 1 to 2 pmol.

4. HPLC of Other Precolumn Derivatives

Several reagents that are suitable for modifying polyamines for HPLC have been applied to biological extracts, but not to plant tissue. For example, dabsyl chloride derivatives of aliphatic amines and acetyl conjugates can be separated in 22 min with detection limits less than 2 pmol, and with the benefit of detection in the visible light range (436 nm).[35] Secondly, HPLC determination of polyamines as fluorescamine derivatives has been reported to be a simple and highly sensitive fluorimetric method with detection limits as low as 5 to 10 pmol. However, a major disadvantage in its use is that the derivatives have a short life of several hours.[36] The reagent has been used more for postcolumn derivatization.[2]

Very recently polyamine determination by precolumn derivatization with 9-fluorenylmethyl chloroformate (FMOC) has been reported.[37] The derivatization procedure is rapid, and the products of the reaction are very stable and highly fluorescent. Putrescine, spermidine, and spermine were separated by HPLC in 12 min with sensitivity of detection at the picomole level. The method is reported to be highly specific and selectively extracts polyamines without interference with tissue amino acids and other biogenic amines.

B. REVERSED PHASE HPLC ION-PAIRING AND POSTCOLUMN DERIVATIZATION

Reversed phase chromatography with ion-pairing can also be used to resolve a large number of polyamines and related compounds.[38-41] Ion-pairing agents such as octanesulfonate or heptanesulfonate included in the solvent elution system facilitate separation of components in a mixture. Since compounds such as the polyamines are neutralized in the presence of ion-pairing agents their retention on the column is increased. Retention and selectivity of the amines and other compounds of interest can be controlled by the concentration and hydrophobicity of the ion-pairing agents, the nature of the mobile phase, pH and temperature.[38,42] Thus several approaches are possible for improving separation of a variety of neutral and charged solutes in a sample.

Greenberg and Cohen[41] developed an ion-pairing reversed phase HPLC procedure for analyzing samples for *S*-adenosyl methionine (SAM) metabolites and polyamines. Prior to HPLC, the PCA extract was fractionated on a 1 × 5 SP-Sephadex column (Pharmacia Fine Chemicals, Piscataway, NJ, U.S.) prepared in 10 m*M* HCl. The sample was adjusted to pH 3 to 4 with KOH, chilled on ice for 30 min, and then centrifuged. The pellet was treated with cold 10 m*M* HCl and centrifuged again. The supernatant was applied to the Sephadex column. Methionine and 1-aminocyclopropane-carboxylate (ACC) were eluted with 35 ml of 50 m*M* HCl, putrescine and S-adenoyslhomocysteine (SAH) were eluted with 35 ml of 150 m*M* HCl, and SAM, decarboxylated SAM (dSAM), and polyamines were eluted with 40 ml of 0.5 *M* HCL. The fractions were dried *in vacuo* at 32°C and taken up in appropriate volumes of 10 m*M* HCl. Each fraction was analyzed separately by HPLC under different chromatographic conditions.

The amine containing fractions were applied to a C_{18} column and eluted with a gradient solvent system:buffer (A) 0.1 *M* NaH_2PO_4, 8 m*M* heptanesulfonic acid, 0.1 m*M* Na_2EDTA (pH 2.5), and acetonitrile (132:1, v/v) and buffer (B) 0.2 *M* NaH_2PO_4, 8 m*M* heptanesulfonic acid (pH 3.1, v/v) and acetonitrile (7.3, v/v). *S*-adenosyl-containing compounds were detected at 254 nm and the amines were derivatized with OPA on line after the eluent passed through a UV monitor. For postcolumn derivatization, the OPA reagent was pumped into the column effluent through a T-tube at a flow rate equal to that of the mobile phase. The derivatives were then passed through a flow cell for fluorogenic monitoring using excitation at 334 to 346 nm and emission at 425 nm. Detection of the underivatized SAM and metabolites by UV adsorption and polyamines as OPA derivatives by fluorometry was achieved at the pmol level.[41]

Ion-pair reversed phase HPLC procedures have also been developed to separate basic hydroxycinnamic amides in tissue extracts. *N-p*-coumaryl-, *N*-caffeoyl and *N*-feruloylputrescine, *N-p*-coumaroyl, *N*-caffeoyl and *N*-feruloylspermidine can be resolved on a reversed phase column with a methanol-water gradient system including tetrabutylammonium phosphate as an ion-pairing agent. The amides were detected by a spectrophotometer at 310 nm and the minimum detectable amount was 10 ng at the detector sensitivity level used.[43]

IV. ION EXCHANGE CHROMATOGRAPHY

Ion exchange chromatography followed by postcolumn derivatization with OPA has been employed to analyze samples for polyamines, polyamine analogs, aromatic amines, and heterocyclic amines.[44-51] However, the technique is not widely used since automation of the procedure requires an amino acid analyzer which is expensive and not readily available to most laboratories. The method is time consuming also , requiring as long as 2 to 3 h for separation and additional time for column regeneration and re-equilibration. Elution programs are complicated, involving two- and three-step solvent systems and two-stage temperature regimes. However, automated ion exchange chromatography can be a powerful tool for analysis of large number of interrelated amino compounds.

In an improved ion exchange method, Villanueva et al.[50,51] used a high performance amino acid analyzer to separate components in a mixture of amino standards that contained ornithine, lysine, arginine, putrescine, cadaverine, agmatine, ammonia, ethanolamine, histamine, aminopropane, diaminopropane, acetyl- and carbamyl- putrescine, norspermidine, spermidine, homospermidine, thermine, spermine, and 1,7-diaminoheptane, which served as an internal standard. They were able to perform the analysis in 72 min using two buffers, one temperature, and fluorimetric detection of OPA derivatives. The method was sensitive at the picomole level.

V. CONCLUSIONS

There are many chromatographic methods available for polyamine separation and identification with HPLC being the most widely used. Most HPLC methods allow good separations and involve simple sample preparations. The various methods differ in sensitivity , selectivity, ease of sample preparation, and speed of separation. Thus it is difficult to recommend one method over another. The most sensitive HPLC approach is fluorimetric detection of dansyl derivatives. Recent studies show dansyl derivatives can be prepared more rapidly than previously reported and separations can be done in less than 10 min. HPLC of benzoyl chloride derivatives is not as sensitive, but sample preparation is simple and inexpensive, and detection is achieved with a UV detector. A more selective agent for polyamine determination than dansyl or benzoyl chloride is OPA. Although OPA can be used for precolumn modification of polyamines, automation is limited by the instability of the derivatives. A new technique, HPLC with precolumn derivatization with FMOC is reported to offer the advantages of selectivity, sensitivity, stability, ease of preparation, and rapid separation, but it has not been applied to plant extracts. Rapidity of separations by most of the HPLC methods described has been improved by ion pairing. This technique offers the potential for further improvements of current methods and extending analysis to many more compounds.

ACKNOWLEDGMENTS

I thank Dr. C. B. Smith for computer assistance and Ms. T. A. Gaither for help with the literature search.

REFERENCES

1. **Smith, M.A. and Davies, P.J.,** *Modern Methods of Plant Analysis*, New Series, 5, Springer-Verlag, Berlin, 1987, 209.
2. **Seiler, N.,** Liquid chromatographic methods for assaying polyamines using prechromatographic derivatization, *Meth. Enz.*, 94, 10, 1983
3. **Flores, H.E. and Filner, P.,** Polyamine catabolism in higher plants. Characterization of pyrroline dehydrogenase, *Plant Growth Regul.*, 3, 277, 1985
4. **Slocum R. and Galston, A.W.,** Changes in polyamine biosynthesis associated with postfertilization growth and development in tobacco ovary tissues, *Plant Physiol.*, 79, 336.
5. **Cohen, S.S., Balint, R., and Sindhu, R.K.,** The synthesis of polyamines from methionine in intact and disrupted leaf protoplasts of virus- infected chinese cabbage, *Plant Physiol.*, 68, 1150, 1981.
6. **Grettie, D.P., Bartos, D., Bartos, F., Smith, R.G., and Campbell, R.A.,** *Advances In Polyamine Research*, Vol. 2, Raven Press, New York, 1978, 13.
7. **Brossat, B., Straczek, J., Belleville, F., Nabet, P., and Metz, R.,** Determination of free and total polyamines in human serum and urine by ion-pairing high performance liquid chromatography using a radial compression module. Application to blood polyamine determination in cancer patients treated or not treated with an ornithine decarboxylase inhibitor, *J. Chromatogr.*, 277, 87, 1983.
8. **Flores, H.E. and Galston, A.W.,** Analysis of polyamines in higher plants by high performance liquid chromatography, *Plant Physiol.*, 69, 701, 1982.
9. **Kabra, P.M., Lee, H.K., Lubich, W.P., and Marton, L.J.,** Solid-phase extraction and determination of dansyl derivatives of unconjugated and acetylated polyamines by reversed-phase liquid chromatography: improved separation systems for polyamines in cerebro-spinal fluid, urine and tissue, *J. Chromatogr.*, 380, 19, 1986.
10. **Bontemps, J., Laschet, J., Dandrifosse, G., van Cutsen J-L, and Forget, P-Ph,** Analysis of dansyl derivatives of di-and polyamines in mouse brain, human serum and duodenal biopsy specimens by high-performance liquid chromatography on a standard reversed-phase column, *J. Chromatogr.*, 311, 59, 1984.
11. **Seiler, N. and Weichmann, M.,** *Progress In Thin-Layer Chromatography and Related Methods*, Vol. 1, Ann Arbor-Humphrey Science, Ann Arbor, MI, 1970, 95.
12. **Smith, M.A. and Davies, P.J.,** Separation and quantitation of polyamines in plant tissue by high performance liquid chromatography of their dansyl derivatives, *Plant Physiol.*, 78, 89, 1985.
13. **Gennaro, M.C., Mentasti, E., Sarzanni, C., and Porta, V.,** Aliphatic monoamines, diamines, polyamines: an HPLC method for their identification and separation by a dansylation reaction. The study of separation factors in homologous series, *Chromatographia*, 25, 117,1988.
14. **Heby, O. and Andersson, G.,** Simplified micro-method for the quantitative analysis of putrescine, spermidine and spermine in urine, *J. Chromatogr.*, 145, 73, 1978.
15. **Flores, H.E. and Galston, A.W.,** Osmotic stress-induced polyamine accumulation in cereal leaves, *Plant Physiol.*, 75, 102, 1984.
16. **Escribano, M.I. and Legaz, M.E.,** High performance liquid chromatography of the dansyl derivatives of putrescine, spermidine, and spermine, *Plant Physiol.*, 87, 519, 1988.
17. **Smith, T.A. and Best, G.R.,** Polyamines in barley seedlings, *Phytochemistry*, 16, 841, 1977.
18. **Birecka, H., DiNolfo, T.E., Martin, W.B., and Frohlich, M.W.,** Polyamines and leaf senescence in pyrrolizidine alkaloid-bearing *Heliotropium* plants, *Phytochemistry*, 23, 991, 1984.

19. **Kumar, P.P. and Thorpe, T.A.,** Putrescine metabolism in excised cotyledons of *Pinus radiata* cultured *in vitro*, *Physiol. Plant*, 76, 521, 1989.
20. **Brown, N.D., Strickler, M.P., and Whaun, J.M.,** Femtomolar ion-pair high-performance liquid chromatographic method for determining Dns-polyamine derivatives of red blood cell extracts utilizing an automated polyamine analyzer, *J. Chromatogr.*, 245, 101, 1982.
21. **Saeki, Y., Uehara, N., and Shirakawa, S.,** Sensitive fluorimetric method for the determination of putrescine, spermidine and spermine by high-performance liquid chromatography and its application to human blood, *J. Chromatogr.*, 145, 221, 1978.
22. **Bontemps, J., Etienne, A., Kadri, M., van Cutsem, J-L., Dandrifosse, G., and Forget, P. Ph.,** High-speed analysis of dansyl derivatives of polyamines, *J. Chromatographia*, 18, 525, 1984.
23. **Minocha, S.C., Minocha, R., and Robie, C.A.,** High-performance liquid chromatographic method for the determination of dansyl-polyamines, *J. Chromatogr.*, 511, 177, 1990.
24. **Walter, H.J-P. and Geuns, J.M.C.,** High speed HPLC analysis of polyamines in plant tissues, *Plant Physiol.*, 83, 232, 1987.
25. **Khan, A.J. and Minocha, C.C.,** Polyamine biosynthetic enzymes and the effects of their inhibition on the growth of some phytopathogenic fungi, *Plant Cell Physiol.*, 30, 655, 1989
26. **Hayman, A.R., Gray, D O., and Evans, S.V.,** New high-performance liquid chromatography system for the separation of biogenic amines as their Dns derivatives, *J. Chromatogr.*, 325, 462, 1985.
27. **Redmond, J.W. and Tseng, A.,** High-pressure liquid chromatographic determination of putrescine, cadaverine, spermidine and spermine, *J. Chromatogr.*, 170, 479, 1979.
28. **Corbin, J.L., March, B.H., and Peters, G.A.,** An improved method for analysis of polyamines in plant tissue by precolumn derivatization with *o*-phthalaldehyde and separation by high performance liquid chromatography, *Plant Physiol.*, 90, 434, 1989.
29. **Asotra, S., Mladenov, P.V., and Burke, R.D.,** Improved methods of benzoyl chloride derivatization of polyamines for high-performance liquid chromatography, *J. Chromatogr.*, 408, 227, 1987.
30. **Verkoelen, C.F., Romijn, J.C., Schroeder, F.H., Van Schalkwijk, W.P., and Splinter, T.A.W.,** Quantitation of polyamines in cultured cells and tissue homogenates by reversed-phase high-performance liquid chromatography of their benzoyl derivatives, *J. Chromatogr.*, 426, 41, 1988.
31. **Slocum, R.D., Flores, H.E., Galston, A.W., and Weinstein, L.H.,** Improved methods for HPLC analysis of polyamines, agmatine and aromatic monoamines in plant tissue, *Plant Physiol.*, 89, 512, 1989.
32. **Roberts, D.R., Walker, M.A., and Dumbroff, E.B.,** Mass spectral determination of benzamide derivatives of polyamine separated by HPLC, *Phytochemistry*, 24, 1089, 1985.
33. **Simons, S.S. and Johnson, D.F.,** The structure of the fluorescent adduct to the reaction of *o*-phthalaldehyde and thiols with amines, *J. Am. Chem. Soc.*, 98, 1976
34. **Skaaden. T. and Greibrokk, T.,** Determination of polyamines by pre-column derivatization with *o*-phthalaldehyde and ethanethiol in combination with reversed-phase high-performance liquid chromatography, *J. Chromatogr.*, 247, 111, 1982.
35. **Koski, P., Helander, I.M., Sarvas, M., and Vaara, M.,** Analysis of polyamines as their dabsyl derivatives by reversed-phase high-performance liquid chromatography, *Anal. Biochem.*, 164, 261, 1987.
36. **Samejima, K.,** Separation of fluorescamine derivatives of aliphatic diamines and polyamines by high-speed liquid chromatography, *J. Chromatogr.*, 96, 250, 1974.
37. **Sabri, M.I., Soiefer, A.I., Kisby, G.E., and Spencer, P.S.,** Determination of polyamines by precolumn derivatization with 9-fluorenylmethyl chloroformate and reverse-phase high-performance liquid chromatography, *J. Neurosci. Methods*, 29, 27, 1989.
38. **Wagner, J., Danzin, C., and Mamont, P.,** Reversed-phase ion-pair liquid chromatographic procedure for the simultaneous analysis of *S*-adenosylmethionine, its metabolites and the natural polyamines, *J. Chromatogr.*, 227, 349, 1982.
39. **Seiler, N.,** Ion-pair partition chromatographic separation of polyamines and their monoacetyl derivatives, *Meth. Enz.*, 94, 25, 1983.
40. **Seiler, N. and Knodgen, B.,** Determination of polyamines and related compounds by reversed-phase high-performance liquid chromatography: improved separation systems, *J. Chromatogr.*, 339, 45, 1985.
41. **Greenberg, M.L. and Cohen, S.S.,** Dicyclohexylamine-induced shift of biosynthesis from spermidine to spermine in plant protoplasts, *Plant Physiol.*, 78, 568, 1985.
42. **Cooke, N.H.C. and Olsen, K.,** Chemically bonded alkyl reversed-phase columns: properties and use, *Am. Lab.*, 11, 45, 1979.
43. **Ponchet, M., Martin-Tanguy, J., Poupet, A., Marais, A., and Beck, D.,** Separation and quantification of basic hydroxycinnamic amines and hydroxycinnamic acids by reversed-phase high-performance liquid chromatography, *J. Chromatogr.*, 240, 397, 1982.
44. **Villanueva, V.R., Adlakha, R.C., and Cantera-Soler, A.M.,** Changes in polyamine concentration during seed germination, *Phytochemistry*, 17, 1245, 1978.

45. **Aleksijevic, A., Grove, J., and Schuber, F.,** Studies on polyamine biosynthesis in *Euglena gracilis*, *Biochim. Biophys. Acta.*, 565, 199, 1979.
46. **Perdrizet, E. and Prevost, J.,** Aliphatic and aromatic amines during development of *Nicotiana tabacum*, *Phytochemistry*, 20, 2131, 1981.
47. **Tabor, C.W. and Tabor, H.,** Quantitative determination of naturally occurring aliphatic diamines and polyamines by an automated liquid chromatography procedure, *Meth. Enz.*, 94, 29, 1983.
48. **Villanueva, V.R., Adlakha, R.C., and Cantera-Soler, A.M.,** Détermination rapide de polyamines et de quelques mono-et diamines dans dês extraits végetaux, *J. Chromatogr.*, 139, 381, 1977.
49. **Villanueva, V.R., Simola, L.K., and Mardon, M.,** Polyamines in turions and young plants of *Hydrocharis morsus-ranae* and *Utricularia intermedia*, *Phytochemistry*, 24, 171, 1985.
50. **Villanueva, V.R., Mardon, M., and Le Goff, M.Th.,** A new high performance chromatographic method for polyamine analysis in *Picea* needles, without previous extract purification, *Inter. J. Environ. Anal. Chem.*, 25, 115, 1986.
51. **Villanueva, V.R., Mardon, M., and Le Goff, M.Th.,** Comparison of amino-compounds contained in the needles of healthy and damaged *Picea* trees in air polluted areas. Preliminary results, *Intern. J. Environ. Anal. Chem.*, 25, 127, 1986.

Chapter 17

ASSAY METHODS FOR ENZYMES OF POLYAMINE METABOLISM

Helena Birecka

TABLE OF CONTENTS

I. INTRODUCTION

Assays discussed in this chapter deal with (A) amino acid decarboxylases: namely, ornithine decarboxylase (ODC), arginine decarboxylase (ADC), lysine decarboxylase (LDC), and *S*-adenosylmethionine decarboxylase (SAMDC), which are involved in the biosynthesis of putrescine (Put), agmatine (Agm), cadaverine (Cad), and spermidine (Spd) and spermine (Spm), respectively; (B) two enzymes, Agm iminohydrolase and *N*-carbamoylputrescine amidohydrolase, which catalyze the transformation of Agm into Put; (C) Put and Spd aminopropyl transferases (Spd and Spm synthases), which are responsible for the transfer of the aminopropyl group from decarboxylated *S*-adenosylmethionine to Put and Spd, yielding Spd and Spm, respectively; and finally (D) di- and polyamine oxidases, which are responsible for oxidative degradation of di- and polyamines in plants. During the last ten years, most studies on the biosynthetic enzymes have employed assays using ^{14}C-labeled compounds as substrates.

II. BIOSYNTHETIC ENZYMES

A. AMINO ACID DECARBOXYLASES

In the overwhelming number of publications, the $^{14}CO_2$-trapping method, using crude enzyme extracts with or without dialysis, has estimated the activities of the plant decarboxylases, especially of ODC and ADC, on the basis of the released CO_2 from [1-^{14}C]- or [U-^{14}C]-labeled substrates; ammonium sulfate fractionation has rarely been applied. As has been shown, the CO_2 that is produced in such extracts from a variety of species, even after $(NH_4)_2SO_4$ fractionation, can be many-fold greater than the Put and/or Agm that is isolated from reaction mixtures containing Tris · HCl or phosphate buffer (pH 8.0), dithiothreitol (DTT), pyridoxal phosphate (PLP), and EDTA.[1,2]

Several factors can cause an artifactual CO_2 production. First, the presence of phenolic compounds that readily oxidize to quinones creates a danger of oxidizing both the enzyme and the substrates that are expected, in this case, to undergo nonoxidative decarboxylation;[3] however, addition of polyvinylpyrrolidone (PVP) to crude extracts may be quite effective in removing phenolics,[4] in particular when their levels are high.[5-7] Second, the presence of diamine oxidases in the extracts have a broad affinity to substrates (as discussed later), including not only di- and polyamines, but also some amino acids involved in amine synthesis. It has been shown that L-ornithine (Orn), L-arginine (Arg), L-lysine (Lys), and *S*-adenosylmethionine undergo decarboxylation due to the liberation of hydrogen peroxide (H_2O_2) during the amine oxidase-mediated catabolism of diamines like Put and Agm.[8,9] Thus, the presence of diamine oxidase in the extracts may affect the reliability of results based on both the CO_2 and the amine measurements.

Elimination of the oxidase interference is very important, especially in the case of dicots. In fact, aminoguanidine (AG), a potent diamine oxidase inhibitor, added at 0.1 to 0.2 m*M* to the reaction mixtures, prevented oxidation of Put and/or Agm. In many instances, the activities based on CO_2 release and Agm and/or Put recovery were similar. Moreover, 1m*M* DL-α-difluoromethylornithine (DFMO) or DL-α-difluoromethylarginine (DFMA), the most often used irreversible inhibitors of ODC and ADC, respectively, almost completely suppressed the release of CO_2.[1,2] However, in some instances, particularly in assays of ODC, production was much greater than the Put formed, even in the presence of AG the CO_2. Recently, with extracts from wheat and other plant species, an oxidative decarboxylation of Orn, Arg, and Lys (in which Put or Agm was not a mediator) has been detected. The enzyme in the presence of PLP oxidized both D- and L-Orn to 4-aminobutanamide with K_m^{Orn} of about 1 m*M* and optimum PLP concentration also at 1 m*M*. The oxidation of amino acids may be dependent on the Mn^{2+} present in the extracts. It has been concluded that (1) in order to reduce oxidative decarboxylation, EDTA (> 1 m*M*) and thiols (e.g., cysteine > 5 m*M*) or other reducing agents should be added and sufficient PLP should be included only to optimize the amine formation, not the CO_2 release; and (2) amino acid decarboxylase assays should not depend on $^{14}CO_2$ release from ^{14}C-labeled substrates unless stoichiometry with amine formation has been established.[12] The latter conclusion is of special importance, as has been shown,[1] in the case of [U-^{14}C]-labeled substrates, particularly of [U-^{14}C]Arg used for assaying ADC in extracts exhibiting high arginase activity; the release of [guanido-^{14}C] CO_2 due to arginase and urease activities was significantly, but not completely, suppressed by 20 m*M* Orn, which is a relatively effective inhibitor of arginase.[13-15]

1. Ornithine Decarboxylase (ODC; EC 4.1.1.17)

The enzyme has been found in the cytosol as well as in the nucleus, bound to the chromatin, at least in barley, corn, bean, and pea.[16,17] The barley cytosolic, as well as chromatin-bound, ODC have been purified; the latter was successfully extracted from the isolated chromatin with 1*M* NaCl followed by repeated freezing and thawing.[18] Both fractions require PLP and DTT for maximum activity and both are sensitive to DFMO; K_m for Orn was 0.36 m*M* and pH optimum, 8.5. Both the cytosolic and chromatin-bound ODC form a complex with antizyme, a protein noncompetitive inhibitor of ODC. The antizyme can be dissociated from the enzyme with 2 m*M* NaCl and 10% $(NH_4)_2SO_4$, respectively.[19] The ODC antizyme assay measures the inhibition of a partially purified preparation of ODC by extracts having antizyme activity.[20]

A pelletable form of ODC activity has been found in oat.[21] It was assayed in the supernatant of a sonicated suspension of a pellet which was obtained after the tissue homogenate in 0.1 m*M* buffer, pH 8, was centrifuged at 20,000xg for 10 min. A significant ODC, as well as ADC, activity has been found in isolated chloroplasts of *Pinus radiata* cotyledons and in mitochondria of *Helianthus tuberosus* tubers.[22]

Since the commonly used assay for ODC has taken into account only $^{14}CO_2$ released from D,L-[1-^{14}C]- or L-[1-^{14}C]-Orn, the reported characteristics of the enzyme tested in crude or semicrude extracts in many cases may not be fully reliable. However, in some studies the stoichiometry of Orn decarboxylation and the formation of Put was, in fact, tested. In addition to L- [1-^{14}C]-Orn, D,L-[5-^{14}C]-Orn was used for tobacco dialyzed ODC fraction precipitated with 0 to 50% saturation of ammonium sulfate;[23] Put was separated from Orn as dansyl derivatives on silica-G gel TLC plates with ethyl acetate and cyclohexane (2:3) as the solvent and as hydrochloride salts by descending paper chromatography with MeOH/pyridine/H_2O/ acetic acid (6:6:4:1).

A similar procedure was applied to tobacco and tomato ODC after the enzymes were, in addition, filtered through a Sephadex G-200 column.[24] ODC from both species was specific for L-Orn with an apparent K_m of 0.14 mM and an optimum pH of 8.0; it required PLP and thiol reagent (DTT) and was almost completely inhibited by canaline at 10 μM. A wheat ODC fraction, precipitated by 70% saturation of $(NH_4)_2SO_4$ and desalted over Sephadex G-25M column (apparent K_mOrn of 0.24 mM), showed an artifactual release of CO_2 amounting to about 25% of the total $^{14}CO_2$ from D,L-[1-^{14}C]-Orn when compared with the labeled Put from [5-^{14}C]-Orn after their separation as dansyl derivatives on silica TLC plates using chloroform and triethylamine (5:1) as solvents.[25] DFMO pre-incubated with the extract at 1 mM for 30 min inhibited the activity based on $^{14}CO_2$ by about 80%. In fact, the degree of DFMO-induced inhibition of decarboxylation is usually a good indicator of the extent of artifactual CO_2 production, except, of course, in cases of a genetically controlled ODC resistance to DFMO, as reported for a tobacco cell line.[26] It is worth mentioning that a jute embryo ODC preparation, purified over 700-fold, with K_m of 0.1 mM for Orn, pH optimum of 7.8, and requiring PLP (20 μM) and DTT (5 mM) for maximum activity, was not inhibited by 1-10 mM DFMO.[27] A tobacco ODC, in a crude extract passed through a Sephadex G-25 column, showed an optimum pH of 7.0 in Tris · HCl buffer and an apparent K_mOrn of 1 mM when assayed with nonlabeled substrate in a 2 ml reaction mixture containing 50 μM PLP, followed by estimation of Put formed in 50 to 200 μl aliquots using ion exchange chromatography and *o*-phthalaldehyde reaction in an amino acid analyzer.[28] The lowest reported enzyme activities were somewhat below 10 pkat/g fresh weight.

Phosphate or Tris · HCl buffers at 40 to 250 m*M*, pH 7.5 to 8.3, containing 50 to 100 μ*M* EDTA, 20 to 50 μ*M* PLP, and 5 to 10 m*M* DTT have been usually used for plant tissue homogenization. EDTA at 5 to 10 m*M* has been applied for tobacco tissues[7,26] and 2-mercaptoethanol at 2.5 to 10 m*M* and/or ascorbate at 0.5 to 10 m*M* has sometimes substituted for DTT; the salt concentration of the buffer used may affect the degree at which the chromatin-bound ODC is released into the solution. The assay mixtures (0.05 to 2.5 ml) usually contain lower concentrations of the buffer and DTT when semicrude or purified extracts are analyzed. Sodium fluoride and/or phenylmethylsulphonyl-fluoride, inhibitors of phosphatases and proteases, respectively, are sometimes included.[17,18,29,30] Incubation with labeled L-Orn at concentration ranging from 14 μ*M* to 5 m*M* at 30-37°C lasts 15 to 60 min.[14] The CO_2 reaction product is absorbed by KOH, protosol, or methylbenzethonium hydroxide on filter paper or in solution, and is removed from the reaction vials 30 to 60 min after the enzyme was inactivated with acid. Radioactivity in the collected $^{14}CO_2$ is then measured in a scintillation counter. The threshhold of CO_2 detectability depends, of course, on the specific radioactivity of the substrate used, its concentration, amount of enzyme, and the radioactivity of the blanks.

As previously noted, labeled Put formed in the reaction mixture from [5-^{14}C] Orn has been separated by paper or silica gel chromatography. A very good separation of [U-^{14}C]-labeled Put, Agm, Orn, and Arg with full recovery of their radioactivity has been obtained using TLE at pH 4.2 on silica gel 60 (250 μm) plates;[1] the marker spots were developed with ninhydrin, the corresponding silica gel bands scraped off, transferred into scintillation vials and radioactivity was measured.

A modification of an ion exchange method for the radiometric estimation of Put formed using [2,3-^{3}H]-Orn has been recently presented.[31] A 3 cm high column of Amberlite CG-50 ion exchange resin [H^+ form] in a pasteur pipet is washed with 0.2*M* ammonium acetate, pH 6.5. The 2 ml assay mixture, containing Orn with specific radioactivity of 20 μCi/μmol, is applied to the column. The unreacted substrate is washed out with the 0.2*M* ammonium acetate followed by 2*M* acetate; Put is eluted with 3 ml of 8*M* formic acid directly into scintillation vials and measured with a ^{3}H counting efficiency of 30%. The detection limit is about 50 pmoles [^{3}H]-Put.

A spectrophotometric estimation of ODC activity using 2,4,6-trinitrobenzene sulfonic acid, which reacts with Put to give a colored product soluble in 1-pentanol, is much less sensitive.[32] The molar extinction coefficient of the product at 420 nm is about 1.9×10^3 cm^{-1} and the detection limit is about 10 nmoles Put. The assay has been compared with HPLC measurements of Put as a dansyl derivative. Two assays for ODC activity using HPLC, tested on the bacterial and mammalian enzymes, have been recently published. In the first one,[33] Put is separated from its nonlabeled substrate and measured as an *o*-phthalaldehyde (OPA) derivative by HPLC equipped with a filter fluorescence detector with standard filters for OPA. The lower limit of detection was found to be about 0.5 pmole, the $t_{1/2}$ of the OPA derivative of Put being about 3 h. The second assay[34] uses isocratic reversed-phase HPLC without the necessity of generating a precolumn derivative. [1,2-^{3}H]-Put is separated from its substrate with a mobile phase containing 0. 01*M* Na dodecyl sulfate and 36% acetonitrile. Orn (R_t 4 to 5 min) and Put (R_t 7 to 8 min) is collected in 0.5 ml fractions directly in scintillation vials. The lower limit of detection is about 2.5 pmol Put.

2. Arginine Decarboxylase (ADC; EC 4.1.1.19)

The PLP- and DTT-stimulated enzyme yields Agm and CO_2 from L-Arg. It has been purified (partially or to homogeneity) from barley,[35] oat,[36] rice,[37] *Lathyrus sativus*,[38] *Vicia faba*,[39] and avocado.[5] The reported K_m values for Arg were 0.75, 0.03, 0.28, 1.75, 0.58, and 1.8 to 2.8 m*M*, respectively; pH optima, 7 to 9, 7.0 to 7.5, 8.0, 8.5, 8.5, and 8 to 9, respectively; and temperature optima, 37, 32, 45, 45, 45, and 60°C, respectively. Like ODC, ADC has been tested in most cases by the $^{14}CO_2$-trapping method and often using the same tissue extracts with a similar, broad range of L-Arg concentrations. Usually L-[U-^{14}C]-Arg has been used; rarely (e.g., with *Datura* or tobacco extracts), D,L-[1-^{14}C]-Arg was applied.[40,41] In order to account for $^{14}CO_2$ liberation from [U-^{14}C]-Arg due to arginase and urease activities, the enzyme from *Citrus* and *Poncirus* seedlings was additionally tested with [guanido-^{14}C]-Arg and the ADC activity was based on the difference between the two estimations.[42] The possible interference of arginase is not limited only to the artifactual release of $^{14}CO_2$. In cases of high arginase activity, arginine, labeled or nonlabeled, may be partially converted into Orn and, if the extract contains active ODC, the CO_2 released and/or amines formed may be due to the action of both decarboxylases. It should, however, be added that with crude or dialyzed extracts from a variety of plant species, in which ODC activity was negligible as compared with that of ADC, [U-^{14}C]-labeled Arg yielded not only Agm, but also Put,[2] thus indicating that under the assay conditions Agm as such may be converted to Put. Moreover, as has been shown, tobacco arginase can hydrolyze *in vitro* and *in vivo* DFMA to DFMO;[43] this fact supposedly explains the DFMA-induced inhibition of ODC observed in tobacco ovaries. DFMA-derived DFMO has been recently detected in roots and epicotyls of pea seedlings;[44] it did cause a partial inhibition of ODC activity.

New ADC inhbitors, such as α-monofluoromethylagmatine and [E]-α-monofluoromethyldehydroarginine might be of great interest if they do not serve as substrates for arginase or Agm metabolizing enzymes. When tested on barley and oat ADC, these inhibitors proved to be even more potent than DFMA.[45]

It is perhaps worth mentioning that a multifunctional enzyme, Put synthase, purified from cucumber seedlings, revealed ADC activity in addition to Agm iminohydrolase, Put transcarbamylase, Orn transcarbamylase, putrescine synthase, and carbamate kinase activities.[46,47] The ADC activity was assayed by measuring $^{14}CO_2$ liberated from [U-^{14}C]-Arg in a mixture containing 100 m*M* Tris · HCl, pH 8.2; 2 m*M* DTT, and 50 μ*M* PLP.

Two methods have been used for assaying plant ADC with nonlabeled Arg, namely (1) a manometric estimation of CO_2 released from L-Arg at 25 m*M* in a Warburg apparatus after incubation with the *Vicia faba* enzyme for 60 min at 30°C,[39] and (2) a spectrophotometric method coupling Agm production with pea diamine oxidase/peroxidase oxidation of guaiacol.[48] A sixfold more sensitive modification of this latter method has been achieved by substituting antipyrine and 3,5-dichloro-2-hydroxy-benzene sulfonic acid for guaiacol as chromogens.[49] This ADC assay is based on oxidation of Agm by the oxidase; the produced H_2O_2 is in turn used by peroxidase to oxidize the chromogens to a stable red product with a molar extinction coefficient, at 575 nm, of approximately 2.4×10^4 cm^{-1}. The reaction mixture is boiled after incubation; 0.1 ml is transferred to a cuvette containing peroxidase, the chromogens, and phosphate buffer up to 2.5 ml prior to adding 10 μl of purified diamine oxidase (10 nkat). The spectrophotometric estimation of Agm formed by oat ADC was comparable to that obtained with HPLC.

3. Lysine Decarboxylase (LDC; EC 4.1.1.18)

The enzyme has been isolated from *Lupinus polyphyllus* leaf chloroplasts.[50] When assayed (estimations based on ^{14}C-Cad production) with [U-^{14}C]-Lys in a reaction mixture containing 10 m*M* diethyldithiocarbamate (DIECA), a diamine oxidase inhibitor, the enzyme had a K_m of 0.76 m*M*, pH optimum of 8.0 and required PLP, dithioerythritol (DTE), and Fe^{2+}. LDC activity has been found in 46 different plant species.[51,52] Leaves and cell cultures of most of the species exhibited an activity below 7 nmol · h^{-1} g^{-1} acetone powder when tested in a reaction mixture, containing 0.1 *M* pyrophosphate pH 8.0, 3 m*M* DTE, 0.1 m*M* PLP, 2.5 m*M* $FeSO_4$, and 0.1 m*M* L-[U-^{14}C]-Lys, incubated for 5 h at 37°C (the activity was linear with time). When assayed under similar conditions, but without DIECA, and using the $^{14}CO_2$-trapping method, 40 to 70% of the total LDC activity of *Nicotiana glauca* roots and aerial parts has been found in the pellet after centrifugation of the tissue homogenates at 20,000 g for 20 min.[53] In contrast to the results with lupin plants, 2.5 m*M* $FeSO_4$, 5 to 10 m*M* DIECA, and 5 to 10 m*M* DTE, caused a significant inhibition of LDC activity in extracts (acetone powder) from *Heimia salicifolia* cultures assayed in reaction mixtures, containing 100 m*M* phosphate, pH 8.0, 0.1 m*M* lysine, 0.3 μ*M* PLP, incubated at 37°C for 60 min.[54] The activity was measured by $^{14}CO_2$-trapping and sometimes verified by ^{14}C-Cad formed (data not shown). An inhibition of 40% by Fe^{2+} has been reported for pea LDC assayed in crude extracts using also the $^{14}CO_2$-trapping method (occasionally verified with ^{14}C-Put);[55] the K_m, pH, and temperature optima were 20 m*M*, 8.0, and 45°C, respectively, and the activities, at 5 m*M* Lys, reached 70 to 160 and even over 400 nmol $CO_2 \cdot h^{-1} \cdot mg^{-1}$ protein in ethylene treated tissues.[56]

An enzyme purified over 100-fold from *Lathyrus sativus* catalyzed decarboxylation of Lys as well as of homoArg, leading to Cad formation;[57] its K_m for Lys and homoArg were 0.88 and 3.33 m*M*, respectively; pH and temperature optima for both substrates 8.4 and 41°C, respectively. The enzyme had an absolute requirement for Mn^{2+} or Fe^{2+} when incubated with 100 m*M* phosphate, pH 8.4, 2 m*M* DTT, 50 μ*M* PLP, and 4 m*M* substrate for 4 to 6 h; its activity was measured by $^{14}CO_2$-trapping.

A spectrophotometric assay, developed for bacterial LDC and based on the reaction of nonlabeled Cad with 2,4,6-trinitrobenzene sulfonic acid to give a product soluble in toluene and measured at 340 nm, has a relatively low sensitivity, the threshhold of detectability being about 7 nmol Cad.[58] However, it has been successfully used with purified LDC from *Hafnia alvei*, exhibiting a specific activity of 0.5 to 6.4 mmol $\cdot h^{-1} \cdot mg^{-1}$ protein.[59]

A continuous assay for decarboxylase activity using a CO_2-selective electrode, developed for bacterial LDC, requiring tens and hundreds nmoles of CO_2 released per minute in the reaction mixture at pH 6,[60] is not sensitive enough for plant LDC. A continuous LDC assay by the pH-Stat method[61] cannot be applied to plant ODC since the most favorable pH range for determining activity is pH 3.1 to 5.4. A similar objection may be raised in the case of an indicator assay, that is applicable to amino acid decarboxylases active at acid pH.[62] A continuous spectrophotometric assay, which couples the formation of CO_2 to decreases in NADH absorbance at 340 nm by phosphoenolpyruvate carboxylase and malate dehydrogenase, has been tested on bacterial LDC, ADC, and diaminopimelate decarboxylase and in principle, can be applied to any decarboxylase which is active in the pH range between 6.6 and 7.7.[63] The sensitivity limit is 1.5 nmol product $\cdot min^{-1}$, which corresponds to a change of 0.008 OD $\cdot min^{-1}$. An almost identical assay has been tested on LDC, orotidine-5′-phosphate decarboxylase and 4′-phosphopantothenoyl-L-cysteine decarboxylase;[64] 1 nmol $CO_2 \cdot min^{-1}$ is indicated as the lower limit of detectability in this assay. A rapid assay for aliphatic amino acid decarboxylases, with a detection limit for the product of 40 pmol, has been tested on Orn, Arg, Lys, aspartate, 2,6-diaminopimelate, histidine, and glutamate decarboxylases; it is based on separation of the product from the substrate by ion-pairing reversed-phase HPLC and subsequent fluorometric detection.[65]

4. *S*-Adenosylmethionine Decarboxylase (SAMDC; EC 4.1.1.50)

This enzyme is basically a cytosolic enzyme although some of its activity has been detected in chloroplasts and mitochondria.[66] It has been partially purified from corn,[67] tobacco,[68] *Lathyrus sativus*,[8] and Chinese cabbage.[69] No stimulation by Put was observed, especially after removal of diamine oxidase from the extract;[8] Mg^{2+} stimulated the activity of *L. sativus* SAMDC, but had no effect on the corn, pea,[70] or cabbage enzyme. The latter was stimulated by PLP.

Carbonyl group reagents, such as methylglyoxal *bis*-(guanylhydrazone) (MGBG), hydroxylamine, or semicarbazide, and sulfhydryl reagents, such as *p*-hydroxymercuriphenylsulphonate or *N*-ethylmaleimide have proved to be very effective SAMDC inhibitors.[8,25,67,70] However, SAMDC isolated from a mutant of *Nicotiana tabacum* cells displayed a high resistance to MGBG *in vitro*.[71] Isonicotinic acid hydrazide, a specific inhibitor of some pyridoxal-dependent enzymes, did not affect the corn SAMDC; added PLP had no effect on the potato tuber SAMDC activity.[72] Optimal pH of 7.4 to 7.6, 8.0, 8.3, and 8.6 have been reported for the enzyme from *L. sativus*, pea, and carrot;[73] tobacco and wheat;[25] *Helianthus tuberosus*;[63] and rice[70] and corn, respectively. K_m values of 5, 17, and 200 μ*M* SAM were reported for the corn, wheat, and pea SAMDC, respectively.

SAMDC has been assayed, mostly in crude extracts, using *S*-adenosyl-L-[carboxy-^{14}C]-methionine and the $^{14}CO_2$-trapping method. It has been extracted with phosphate, Tris · HCl or HEPES buffers containing, in most cases, 2.5 to 10 m*M* DTT or 1 to 2 m*M* 2-mercaptoethanol,[67,74] and 0.05 to 10 m*M* EDTA; PLP at 20 to 40 μ*M* is sometimes added.[25,66,71] The reaction mixtures (0.1 to 0.5 ml) usually include 50 to 100 μ*M* EDTA and 2 to 5 m*M* DTT; occasionally PLP or 5 m*M* Mg^{2+} [66,67,75,76] is also added. The substrate concentrations ranged from 40 μ*M* to 2.7 m*M*. Incubation at 37°C (KOH, hyamine hydroxide, or methylbenzethonium hydroxide used to absorb CO_2) lasted from 15 to 60 min and the reaction was stopped by adding acid or KH_2PO_4. In the case of a dialyzed SAMDC extract from tobacco, the observed nonenzymic decarboxylation amounted to 5 to 10% of the total activity.[77] The reported SAMDC activities range from 0.2 to 600 pmol/h · mg^{-1} protein; activities of 1 to 4 nmol/h · mg^{-1} protein were recorded for barley leaves[76] and pea epicotyls.[70]

B. AGMATINE IMINOHYDROLASE

Agmatine iminohydrolase (AIH; EC 3.5.3.12) converts Agm to *N*-carbamylputrescine (NCP) with the formation of ammonia. The cytosolic enzyme, inhibited by thiol agents and arcaine, has been purified from corn;[79] groundnut;[80] and rice;[81] its K_m values for Agm were 0.19; 0.75; and 15 m*M*, respectively and pH optima 6.5 to 7.5., 5.5 to 8.5, and 6.0, respectively. It has also been found in extracts from oat, sunflower, cabbage, brussel sprouts,[78] and *Passiflora*.[82] AIH has been assayed in reaction mixtures containing 25 to 250 m*M* phosphate buffer and 0.15 to 15 m*M* Agm, incubated for 1 to 18 h. Its activity has been estimated either by measuring NCP formation, using the method of Archibald,[83] or by measuring ammonia released with Nessler's Reagent (Conway microdiffusion units). In the latter case, the stoichiometry of the reaction was determined by separating NCP using high voltage paper electrophoresis, followed by its colorimetric estimation.[79]

C. *N*-CARBAMOYLPUTRESCINE AMIDOHYDROLASE

N-Carbamoylputrescine amidohydrolase (EC 3.5.1.-) degrades NCP to Put, releasing CO_2 and NH_3. It has been detected in wheat, rye, oat, corn, pea, radish, sunflower, and barley[84] and purified 70-fold from corn seedlings.[85] The purified enzyme has a K_m for NCP of 0.09 m*M* and pH optimum 6.5 to 7.0; the optimum pH for the barley enzyme ranges between 7 and 8. Both enzymes were inhibited by thiol reagents. The activity was estimated on the basis of NH_3 released, as in the case of AIH. Stoichiometry of the reaction was studied by determining CO_2 manometrically and Put, separated by paper electrophoresis, colorimetrically.

D. PUTRESCINE AND SPERMIDINE AMINOPROPYLTRANSFERASES

Putrescine aminopropyltransferase (Spd synthase; EC 2.5.1.16) and spermidine aminopropyltransferase (Spm synthase; EC 2.5.1.22) catalyze the transfer of the propylamine moiety from *S*-adenosyl-(5′)-3-methylpropylamine (dSAM) to Put and Spd, yielding Spd and Spm, respectively, and 5′-methylthioadenosine. The enzyme activities can be determined using labeled Put (Spd) or dSAM labeled in the propylamine moiety with subsequent isolation of labeled Spd (Spm) and measurement of its radioactivity.[86-88] Also, the enzyme activities can be based on the recovery of labeled 5′-methylthioadenosine from methyl-labeled dSAM.[86,89] Finally, Spd synthase activity can be tested by a fluorometric method which uses a polyamine oxidase/peroxidase/homovanillic acid couple and unlabeled substrates.[90] All of these methods have been used in assaying the synthases in crude or purified extracts from animal and bacterial cells, and the choice of the method may depend on the interference by substrate- and/or product-degrading enzymes present in the extract. Assays employing ^{14}C-labeled Put (Spd) have usually been used for plant systems.

Spermidine synthase activity in crude extract form *Lathyrus sativus*[8] was determined following the removal of DAO activity through an immunoaffinity column prepared by coupling to Sepharose antibodies against highly purified DAO from the same plant species. The second substrate, dSAM, was generated from SAM by SAMDC present in the extract. SAM as the source of dSAM was also used in assaying Spd synthase from wheat seedlings, purified by ammonium sulfate (70% saturation) precipitation followed by desalting over a Sephadex G-25 column.[25] In assays of Spd and Spm synthase activities in crude or partially purified extracts from Chinese cabbage, dSAM has been used as the propylamine donor;[91,92] 150 m*M* glycine-NaOH buffer, pH 8.8 (optimum) was used for tissue homogenization. In the case of wheat seedlings crushed in liquid N_2, tissues were suspended in 200 m*M* phosphate buffer, pH 8, used also for assaying Orn, Arg, and SAM decarboxylases.

For the *L. sativus* enzyme, the reaction mixture contained 100 m*M* Tris · HCl, pH 7.6, 100 μ*M* SAM, 0.5 m*M* ^{14}C-labeled Put, 5 m*M* Mg^{2+}, and 5 m*M* DTT (2 h incubation at 37°C). For the wheat enzyme, it contained 100 m*M* phosphate buffer, pH 8.0, 2 m*M* DTT, 50 μ*M* EDTA, 50 μ*M* PLP, 0.5 m*M* SAM, and 0.1 m*M* ^{14}C-labeled Put (40 min incubation at 37°C). And, for the cabbage synthases, it contained 150 m*M* glycine-NaOH buffer (pH 8.8), 37 to 100 μ*M* labeled Put (Spd) and 25 μ*M* dSAM (60 min incubation at 37°C). The labeled Spd and Spm formed in the reaction mixture were separated, after addition of carrier polyamines, by paper chromatography[8] or on silica TLC plates as Dns-polyamines.[25] A necessity of removing the labeled substrate from the labeled product(s) before dansylation has been stressed.[91] A Dowex-50 [H^+ form] column has been used for this purpose; 93% of residual Put and 90% of Spd is removed by elution with 2.3 *M* and 3.3 *M* HCl, respectively, leaving 93% of Spm on the column to be eluted by 6 *M* HCl. The recovery of Spd and Spm from the TLC plates was about 80 to 90%. With crude extracts, the necessity of taking into account their own Put (Spd) content for determining the specific radioactivity of the substrate was emphasized.

A partially purified (about 160-fold) cabbage Spd synthase was tested using a fluorometric method, which is reliable within the range between 1 and 10 nmoles of Spd. The method is based on the oxidation of Spd by polyamine oxidase (PAO) yielding H_2O_2 that is subsequently used by peroxidase to oxidize homovanillic acid.[90] The reaction mixture (1.2 ml) contained 0.1 *M* phosphate buffer, pH 7.4, 0.42 m*M* Put, 50 μ*M* dSAM, 0.01 mg fluorogenic homovanillic acid, 0.01 mg horseradish peroxidase, and 0.1 mg of oat PAO (K_m for Spd less than 5 μ*M*), devoid of activity to Put and also devoid of Spd synthase activity. After 60 min incubation with the Spd synthase extract at 37°C, the reaction was measured with excitation at 323 nm and emission at 426 nm. Although the Spd synthase activity at pH 7.4 was about 55% of that at the optimum pH of 8.8, the simplicity of the method permitted analysis of many samples. The use of the fluorogenic 3-(*p*-hydroxyphenyl)propionic acid (excitation at 320 nm and emission at 404 nm) may increase the sensitivity of the method to 0.1 nmol.[93] The purified cabbage enzyme was inhibited by dicyclohexylamine, cyclohexylamine, and 3-adenosyl-3-thio-1,8-diaminooctane. However, the wheat Spd synthase was not inhibited by dicyclohexylamine at 1 m*M*.[25] It has been revealed that the Sigma preparation, named "dicyclohexylamine," was, in fact, *bis*-cyclohexylammonium sulfate.[94] Thus, the previously observed inhibition of Spd synthase by this preparation might have been due to the action of cyclohexylamine at double the concentration quoted.

A second pathway of Spd biosynthesis involving carboxy-Spd synthase has been found in *L. sativus*.[95] The NADPH-dependent enzyme catalyzes the formation of carboxy-Spd from Put and aspartate semialdehyde; the product is then decarboxylated to give rise to Spd. The enzyme activity was tested in a reaction mixture (1 ml) containing P_i buffer, pH 7.5, 5 m*M* L-threonine,

3 m*M* DTT, 0.25 m*M* ^{14}C-Put, 10 m*M* aspartic β-semialdehyde, and 0.5 m*M* NADPH. After 60 min incubation at 37°C, the labeled product was fractionated on phosphocellulose and purified by paper electrophoresis.

A thiol enzyme, *sym*-homo Spd synthase with NAD^+ as a coenzyme, has also been isolated from *L. sativus* and purified; it catalyzes synthesis of the polyamine from two molecules of Put.[96] The enzyme activity was assayed using a reaction mixture (1 ml) containing 50 m*M* Tris · HCl, pH 8.4, 2 m*M* DTT, 10 m*M* KCl, 0.5 m*M* ^{14}C-Put and 1 m*M* NAD^+. After terminating the reaction, the amines were purified on a Dowex-50 [H^+-form] column, separated by paper chromatography or electrophoresis and the radioactivity of the product measured.

III. DIAMINE OXIDASES AND POLYAMINE OXIDASES

Diamine oxidases (DAO; EC 1.4.3.6) oxidize primary amines according to the following equation: $RCH_2NH_2 + O_2 + H_2O \rightarrow RCHO + NH_3 + H_2O_2$. Generally, the oxidases are most active with Put and Cad, producing 1-pyrroline and piperideine, which are formed by the spontaneous cyclization of 4-aminobutyraldehyde and 5-aminopentanal, respectively. However, DAO exhibit a broad substrate specificity and may oxidize, among others, Agm, Spd, Spm, Lys, Arg, and/or Orn[49,97-109] although the K_m values for those substrates may be higher than corresponding values for Put and Cad.

DAO extracted from the alkaloid-producing roots of *Nicotiana tabacum*, *Datura stramonium*, and *Hyoscyamus niger* have a greater affinity for *N*-methylputrescine than for Put or Cad, in contrast to the pea DAO, whose K_m for this substrate is higher and/or V_{max} lower than for Put.[110-114]

The lowest K_m values for Put (about 10 to 40 μ*M* at optimum pH) have been reported for purified DAO from rice,[98] cucumber,[100] clover,[102] and pea,[115] and the highest values (200 to 400 μ*M*) for DAO from *Cicer*,[97] *Lens esculenta*,[101] *Lathyrus cicera*,[109] and *Euphorbia* latex.[103] Extremely high K_m values for Put of about 1.2 and 7.0 m*M* have been recently reported for the purified pea DAO at pH 6.3[114] and for *Phaseolus vulgaris* DAO.[109] The optimum pH for DAO activity ranges between 7 and 8; most often it is 7.5. However, it may also depend on the substrate; in the case of a purified DAO from *Vicia faba*, it was 7.0 for Cad and Agm, 7.2 for Put and Spd, and 8.6 for Lys.[104] An optimum pH of 8.4 was found for the *Lathyrus sativus* DAO, regardless of the substrate used.[107]

Heavy metal chelators, as well as carbonyl group reagents, have proven to be potent inhibitors of DAO. Copper chelators, such as diethyldithiocarbamide, *o*-phenanthraline, or α,α-dipyridyl, and carbonyl group agents, such as semicarbazide, aminoguanidine, or phenylhydrazine, showed the strongest inhibitory effects;[101,104-107,109,116] at 2×10^{-5} *M* concentration, the two latter compounds completely inactivated the soybean enzyme.[105] However, it has been recently reported that *Helianthus tuberosus* DAO reacted to aminoguanidine at 5×10^{-4} *M* with a 65% inhibition only, and to hydrazine at the same concentration with a 100% inhibition.[117] The *N*-methylputrescine oxidase from *N. tabacum* and *H. niger* was partially or completely inactivated by phenylhydrazine and 2,4-dinitrophenylhydrazine at 10^{-5} to 10^{-4} *M*.[113,118]

Sulfhydryl reagents at 10^{-4} to 10^{-3} *M* concentrations had no effect, or a very weak one, on the activities of DAO isolated from various plant species. However, the enzyme activity from rice embryos was inhibited 80% by 10 m*M* Hg^{2+} and exhibited a 40% decline in activity in the presence of Ca^{2+} and Mg^{2+} at the same concentration.[98] The presence of an endogenous DAO inhibitor has been reported for *Arachis hypogea* and this nondialyzable compound could be removed with Triton X-100.[119]

In most cases, DAO has been extracted from plant tissues with 50 to 100 m*M* phosphate buffers. It has been shown that in lentil and pea seedlings, most of the enzyme activity is located in the cell wall and can be extracted by repeated infiltration with 10 m*M* K P_i buffer.[120,121]

Helianthus tuberosus DAO was extracted with phosphate buffer containing 2 m*M* DTT[117] and the *N*-methylputrescine oxidase from tobacco roots with 50 m*M* HEPES, pH 8.0, containing 2.5 m*M* 2-mercaptoethanol, 1 m*M* EDTA, and 0.1 *M* phenylmethanesulfonyl fluoride.[110]

Several methods have been applied to assay DAO, based on estimation of (a) O_2 consumption using an oxygen electrode, (b) H_2O_2 production using a coupled peroxidase system, or (c) 1-pyrroline production using either spectrophotometry or radiometry. The assays have been performed in crude or dialyzed extracts or after enzyme purification.

Oxygen consumption has been measured in most cases by Gibson oxygraph equipped with a Clark electrode using air as the gaseous phase. The reaction mixture (total volume between 1 and 2.5 ml) is preincubated at 30°C to 38°C for 2 to 10 min in the presence or absence of added catalase, and the reaction is started with the addition of Put at the final concentration of 1 or 5 m*M*.[97,101-103,106,122,123] The coupled peroxidase system has been used to assay DAO from rice,[98] pea,[116] and cucumber.[100] In the case of the rice enzyme, the 4 ml assay mixture contained 0.2 mg peroxidase, 0.5 mg hydroquinone (as the electron donor for H_2O_2) and 0.25 m*M* Put. The oxidation product was measured spectrophotometrically at 470 nm. In the case of the pea DAO, scopoletin was used as the electron donor. The activity was determined spectrophotofluorometrically as a decrease in scopoletin fluorescence at 465 nm with excitation wavelength of 370 nm. An often used DAO assay,[100,119,121,124-127] is based on the interaction of 1-pyrroline or piperideine with *o*-aminobenzaldehyde to form a yellow adduct with A_{max} at 430 nm and $E^{1M}_{1cm} = 1.86 \times 10^3$. The method is reliable between 0.1 and 1 μmol 1-pyrroline and is not affected by the presence of catalase or peroxidase in the mixture.[128] A typical reaction mixture of 2.5 to 4.0 ml containing 50 to 100 m*M* P_i buffer and 1 to 10 m*M* Put, is incubated with the enzyme extract for 10 to 30 min at 37°C; the reaction is stopped with 0.5 ml of 10% TCA followed by addition of 0.05 to 0.10 ml of 1% *o*-aminobenzaldehyde. After 60 min at room temperature in the dark to allow the color to develop,[146] the proteins are removed by centrifugation and the absorbance of the supernatant is measured. *o*-Aminobenzaldehyde can also be added prior to incubation with the enzyme, as in the original method.[128] The formed adduct can be measured by gas chromatography on oxidation with CrO_3.[129]

A method in which *o*-aminobenzaldehyde was replaced by a ninhydrin reagent, used for Orn quantitation, proved to be more sensitive and is reliable in the range between 15 and 150 nmol.[130] After the incubation is terminated by adding 0.5 ml 10% TCA, an aliquot of the centrifuged supernatant made up to 1 ml is heated with 1 ml of the reagent followed by 1.5 ml acetic acid and kept in a boiling water bath for 30 min. The red colored complex is measured at 570 nm. The presence of Put or Spd does not interfere with the determination.

DAO activity can be assayed using a radiometric method in which ^{14}C-labeled piperideine or pyrroline produced from ^{14}C-labeled Cad or Put, respectively, is extracted with 10 ml toluene from 2 ml of the reaction mixture after addition of 200 mg sodium bicarbonate to bring the pH to 8;[131] a reaction mixture preincubated with aminoguanidine before addition of the substrate serves as a blank. The recovery of the product with one extraction was about 70 to 80%. Only trace activities of labeled Cad or Put were found in the toluene extract. The radiometric method, with some modification, has been used to assay pea DAO[132] and *H. tuberosus* DAO.[117] In both cases, catalase, which enhanced the enzyme activity, was added to the reaction mixture. Without added catalase, the method was applied to DAO, extracted from wheat and alfalfa.[133] With *N*-methylputrescine as a substrate, the method has been modified as follows:[111] [CH_3-^{14}C]-labeled substrate is added to the reaction mixture of 0.2 ml total volume. After 30 min incubation with the enzyme at 30°C in the presence of added catalase, the mixture is treated with 0.2 ml of 1% KCN which reacts quickly and completely with *N*-methylpyrroline to give 1-methyl-2-cyanopyrrolidine-$^{14}CH_3$, which is extracted with 5 ml toluene. The recovery of the product was over 95% and the labeled substrate was not extracted by toluene. Unlabeled *N*-methylpyrrolinium can also be detected without derivatization by UV absorbance (210 nm) after elution from an HPLC cation exchange column at pH 3.5.[112]

Polyamine oxidases (PAO; EC 1.5.3.3) oxidize polyamines at the secondary amino group producing H_2O_2 and 1-pyrroline and diaminopropane, in the case of Spd, or diazabicyclononane and diaminopropane, in the case of Spm. The enzyme has been found in the Gramineae in oat, barley, corn, wheat, rye, millet, and rice[49,110,129,134-137] and, in contrast to DAO occurring mostly within the Leguminosae, it shows a very high substrate specificity for polyamines, Spd and Spm versus diamines. The enzymes from oat, corn, millet, and barley have been purified and partially characterized.

The optimum pH for PAO activity ranges between 5.5 and 6.6. However, in the case of the barley PAO, the pH optimum depends on the substrate used; it is about 7.5 and 4.5 for Spd and Spm, respectively.[115,136] The K_m values for Spd and Spm range between 5×10^{-7} to 8×10^{-5} *M*.[123,126] In oat, $K_m = 2 \times 10^{-6}$ for Spd and 8×10^{-6} *M* for Spm;[136] in millet,[137] 5×10^{-7} and 6×10^{-5}; and in barley,[115] 2×10^{-5} and 3×10^{-5} *M*, respectively, for Spd and Spm at optimum pHs. K_m values similar to those of the barley enzyme have been found with corn PAO at pH 6.0 and 5.5 in the case of Spd and Spm, respectively.[135]

Homologs of diguanidines inhibit oat PAO; the enzyme activity was enhanced by chlorides of K^+, NH_4^+, Na^+, and Li^+, while Ca^{2+} and Mg^{2+} had an inhibitory effect.[123,138] Phosphate ions decreased the rate of Spd oxidation by barley PAO but not that of Spm oxidation.[115] Neither carbonyl group nor sulfhydryl group reagents affected the activity of corn PAO.[135]

It has been shown that in oat leaves, PAO is mainly located in cell walls[139] and over 90% of total enzyme activity can be extracted by vacuum infiltration with 200 m*M* NaCl in 5 m*M* phosphate buffer, pH 6.5.[140,141] In contrast to the apoplastic oat PAO, barley PAO proved to be symplastic and none of its activity could be extracted by a similar vacuum infiltration.[141] In most cases, PAO has been extracted with 100 m*M* phosphate or citrate buffer containing 0.5 to 1 *M* NaCl.

The DAO assay in which 1-pyrroline reacts with the ninhydrin reagent,[130] has been also used for measuring PAO activities in barley substituting 3 m*M* Spd for Put as substrate. The enzyme from rice has been assayed by the peroxidase/hydroquinone couple[134] identical to that used for DAO[98] with 0.25 m*M* Spd or Spm substituting for Put. However, methods based on a peroxidase/guaiacol couple[136,138-142] and on O_2 consumption[123,124,135-137,142,143] have been used more often. In a typical assay with guaiacol, the reaction mixture (1 to 2.5 ml), preincubated at 30°C, contains 100 m*M* buffer, 1.25 m*M* guaiacol, 50 μg horseradish peroxidase (10 to 20 Units), and 0.5 to 1.25 m*M* Spd or Spm. Since the activity declines rapidly during the assay, especially with Spm, Spd as a substrate is preferred. After incubation with the substrate for 2 min at 30°C, A_{470} is measured.

In the oxygen electrode assay, the reaction mixture similar to that used for DAO activity determination, usually contains 100 m*M* citrate or phosphate buffer at the appropriate pH, 1 *M* NaCl, substrate, and catalase to remove H_2O_2. However, addition of catalase did not prevent the rapid decrease in PAO activity during the assay; the activity was assessed on the basis of the slope of the line representing O_2 consumption during the first 2 min.[123]

A histochemical staining method has been recently used to study the distribution of the amine oxidases in *Canavalia*.[144]

Methods of purification of DAO and PAO are presented in a series of publications.[97,98,100-104,106,111,116,124,126,135,136,142,145]

ACKNOWLEDGMENTS

I am most grateful to Mr. Sreekumar Subramanian for his extremely effective technical help in preparation of this manuscript.

REFERENCES

1. **Birecka, H., Bitonti, A.J., and McCann, P.P.,** Assaying ornithine and arginine decarboxylases in some plant species, *Plant Physiol.*, 79, 509, 1985.
2. **Birecka, H., Bitonti, A.J., and McCann, P.P.,** Activities of arginine and ornithine decarboxylases in various plant species, *Plant Physiol.*, 79, 515, 1985.
3. **Loomis, W.D. and Battaile, J.,** Plant phenolic compounds and the isolation of plant enzymes, *Phytochemistry*, 5, 423, 1966.
4. **Loomis, W.D.,** Overcoming problems of phenolics and quinones in the isolation of plant enzymes and organelles, *Meth. Enz.*, 31, 528, 1974.
5. **Winer, L., Winkler, C., and Apelbaum, A.,** Partial purification and characterization of arginine decarboxylase from avocado fruit, a thermostable enzyme, *Plant Physiol.*, 76, 233, 1984.
6. **Tiburcio, A.F., Kaur-Sawhney, R., Ingersoll, R.B., and Galston, A.W.,** Correlation between polyamines and pyrrolidine alkaloids in developing tobacco callus, *Plant Physiol.*, 78, 323, 1985.
7. **Slocum, R.D. and Galston, A.W.,** Changes in polyamine biosynthesis associated with postfertilization growth and development in tobacco ovary tissues, *Plant Physiol.*, 79, 336, 1985.
8. **Suresh, M.R. and Adiga, P.R.,** Putrescine-sensitive (artifactual) and insensitive (biosynthetic) *S*-adenosyl-L-methionine decarboxylase activities of *Lathyrus sativus* seedlings, *Eur. J. Biochem.*, 79, 511, 1977.
9. **Adiga, P.R. and Prasad, G.L.,** Biosynthesis and regulation of polyamines in higher plants, *Plant Growth Regul.*, 3, 205, 1985.
10. **Smith, T.A. and Marshall, J.H.A.,** The oxidative decarboxylation of ornithine by extracts of higher plants, *Phytochemistry*, 27, 703, 1988.
11. **Smith, T.A. and Marshall, J.H.A.,** Oxidation of amino acids by manganous ions and pyridoxal phosphate, *Phytochemistry*, 27, 1611, 1988.
12. **Smith, T.A. and Marshall, J.H.A.,** Oxidative decarboxylation of amino acids by plant extracts, *Biochem. Soc. Trans.*, 16, 972, 1988.
13. **Wright, L.C., Brady, C.J., and Hinde, R.W.,** Purification and properties of the arginase from Jerusalem artichoke tubers, *Phytochemistry*, 20, 2641, 1981.
14. **Villanueva, V.R. and Giret, M.,** Human platelet arginase, *Mol. Cell Biochem.*, 33, 97, 1980.
15. **Carvajal, N., Acoria, M., Rodriguez, J. P., Fernandez, M., and Martinez, J.,** Evidence for cooperative effect in human liver arginase, *Biochem. Biophys. Acta*, 701, 146, 1982.
16. **Kyriakidis, D.A., Panagiotidis, C.A., and Georgetsos, J.G.,** Ornithine decarboxylase (germinated barley seeds), *Meth. Enz.*, 94, 162, 1983.
17. **Foudouli, A.C. and Kyriakidis, D.A.,** Chromatin-associated ornithine decarboxylase at early stages of growth and differentiation of etiolated mono- and dicotyledonous plants, *Plant Growth Regul.*, 8, 233, 1989.
18. **Koromilas, A.E. and Kyriakidis, D.A.,** Purification and properties of ornithine decarboxylase from germinated barley seeds, *Phytochemistry*, 27, 989, 1988.
19. **Koromilas, A.E. and Kyriakidis, D.A.,** The existence of ornithine decarboxylase-antizyme complex in germinated barley seeds, *Physiol. Plant.*, 72, 718, 1988.
20. **Kyriakidis, D.A., Heller, J.S., and Canellakis, E.S.,** Purification of ornithine decarboxylase antizymes, *Meth. Enz.*, 94, 193, 1983.
21. **Young, N.D. and Galston, A.W.,** Physiological control of arginine decarboxylase activity in K-deficient oat shoots, *Plant Physiol.*, 76, 331, 1984.
22. **Torrigiani, P., Serafini-Fracassini, D., Biondi, S., and Bagni, N.,** Evidence for the subcellular localization of polyamines and their biosynthetic enzymes in plant cells, *J. Plant Physiol.*, 124, 23, 1986.
23. **Heimer, Y.M., Mizrahi, Y., and Bachrach, U.,** Ornithine decarboxylase activity in rapidly proliferating plant cells, *FEBS Lett.*, 104, 146, 1979.
24. **Heimer, Y.M. and Mizrahi, Y.,** Characterization of ornithine decarboxylase of tobacco cells and tomato ovaries, *Biochem. J.*, 201, 373, 1982.
25. **Christ, M., Felix, H., and Harr, J.,** Inhibitors influencing plant enzymes of the polyamine biosynthetic pathway, *Ztschr. Naturforsch.*, 44c, 49, 1989.
26. **Hiatt, A. and Malmberg, R.L.,** Utilization of putrescine in tobacco cell lines resistant to inhibitors of polyamine synthesis, *Plant Physiol.*, 86, 441, 1988.
27. **Pandit, M. and Ghosh, B.,** Ornithine decarboxylase from embryos of jute seeds, *Phytochemistry*, 27, 1609, 1988.
28. **Negrel, J., Vallee, J.-C., and Martin, C.,** Ornithine decarboxylase activity and the hypersensitive reaction to tobacco mosaic virus in *Nicotiana tabacum*, *Phytochemistry*, 23, 2747, 1984.
29. **Altman, A., Friedman, R., and Levin, N.,** Alternative metabolic pathways for polyamine biosynthesis in plant development, in *Advances in Polyamine Research*, Vol. 4, Bachrach, U., Kaye, A., and Chayen, R., Eds., Raven Press, New York, 1983, 395.

30. **Orazi, D.D'. and Bagni, N.,** Ornithine decarboxylase activity in *Helianthus tuberosus*, *Physiol. Plant.*, 71, 177, 1987.
31. **Weber, L.W.D.,** A practical variant of the ion exchange method for the radiometric estimation of ornithine decarboxylase activity, *Experientia*, 43, 176, 1987.
32. **Ngo, T.T., Brillhart, K.L., Davis, R.H., Wong, R.C., Bovaird, J.H., Digangi, J.J., Ristow, J.L., Marsh, J.L., Phan, A.P.H., and Lenhoff, H.H.,** Spectrophotometric assay for ornithine decarboxylase, *Anal. Biochem.*, 160, 290, 1987.
33. **Kvannes, J. and Flatmark, T.,** Rapid and sensitive assay of ornithine decarboxylase activity by high performance liquid chromatography of the *o*-phthalaldehyde derivative of putrescine, *J. Chromatogr.*, 419, 291, 1987.
34. **Beeman, C.S. and Rossomando, E.F.,** Assay of ornithine decarboxylase activity by reversed-phase high performance liquid chromatography, *J. Chromatogr.*, 496, 101, 1989.
35. **Smith, T.A.,** L-arginine de carboxylase of higher plants and its relation to potassium nutrition, *Phytochemistry*, 4, 241, 1963.
36. **Smith, T.A.,** Arginine decarboxylase of oat seedlings, *Phytochemistry*, 18, 1447, 1979.
37. **Choudhuri, M.M. and Ghosh, B.,** Purification and partial characterization of arginine decarboxylase from rice embryos (*Oryza sativa* L.), *Agric. Biol. Chem.*, 46, 739, 1982.
38. **Ramakrishna, S. and Adiga, P.R.,** Arginine decarboxylase from *Lathyrus sativus* seedlings: purification and properties, *Eur. J. Biochem.*, 59, 377, 1975.
39. **Matsuda, H.,** Some properties of the arginine decarboxylase in *Vicia faba* leaves, *Plant Cell Physiol.*, 25, 523, 1984.
40. **Chiriqui, D., Orazi, D.D., and Bagni, N.,** Ornithine and arginine decarboxylases and polyamine involvement during *in vivo* differentiation and *in vitro* differentiation of *Datura innoxia* leaf extracts, *Physiol. Plant.*, 68, 589, 1986.
41. **Pinol, T., Palazon, J., Altabella, T., and Serrano, M.,** Correlation between K^+ content, activities of arginine and ornithine decarboxylase and levels of putrescine and nicotine in cultured tobacco collus, *Physiol. Plant.*, 69, 221, 1987.
42. **Rabe, E. and Lovatt, C.J.,** *De novo* arginine biosynthesis in leaves of phosphorus deficient *Citrus* and Poncirus species, *Plant Physiol.*, 76, 747, 1984.
43. **Slocum, R.D., Bitonti, A.J., McCann, P.P., and Feifer, R.P.,** DL-α-difluoromethyl [3,4-3H] arginine metabolism in tobacco and mammalian cells. Inhibition of ornithine decarboxylase activity after arginase-mediated hydrolysis of DL-α-difluoromethylarginine to DL-α-difluoromethylornithine, *Biochem. J.*, 255, 197, 1988.
44. **Birecka, H., Birecki, M., Bitonti, A.J., and McCann, P.P.,** Inhibition of arginine, ornithine, and lysine decarboxylases in pea seedlings, *Phytochemistry*, 30, 99, 1991.
45. **Bitonti, A.J., Casara, P.J., McCann, P.P., and Bey, P.,** Catalytic irreversible inhibition of bacterial and plant arginine decarboxylase activities by novel substrate and product analogues, *Biochem. J.*, 242, 69, 1987.
46. **Prasad, G.L. and Adiga, P.R.,** Arginine decarboxylase is a component activity of the multifunctional enzyme putrescine synthase in cucumber seedlings, *J. Biosci.*, 11, 571, 1987.
47. **Prasad, G.L. and Adiga, P.R.,** Purification and characterization of arginine decarboxylase from cucumber (*Cucumis sativus*) seedlings, *J. Biosci.*, 7, 331, 1985.
48. **Smith, T.A.,** Spectrophotometric method for the estimation of arginine decarboxylase, *Anal. Biochem.*, 92, 331, 1979.
49. **Smith, T.A. and Barker, J.H.A.,** The di- and polyamine oxidases of plants, *Adv. Exp. Med. Biol.*, 250, 573, 1988.
50. **Hartmann, T., Schoofs, G., and Wink, M.,** A chloroplast-localized lysine decarboxylase of *Lupinus polyphyllus*, *FEBS Lett.*, 115, 35, 1980.
51. **Wink, M. and Hartmann, T.,** Sites of enzymatic synthesis of quinolizidine alkaloids and their accumulation in *Lupinus polyphyllus*, *Z. Pflanzenphysiol.*, 102, 337, 1981.
52. **Schoofs, G., Teichman, S., Hartmann, T., and Wink, M.,** Lysine decarboxylase in plants and its integration in quinolizidine alkaloid biosynthesis, *Phytochemistry*, 22, 65, 1983.
53. **Bagni, N., Creus, J., and Pistocchi, R.,** Distribution of cadaverine and lysine decarboxylase activity in *Nicotiana glauca* plants, *J. Plant Physiol.*, 125, 9, 1986.
54. **Pelosi, L.R., Rother, A., and Edwards, J.M.,** Lysine decarboxylase activity and alkaloid production in *Heimia salicifolia* cultures, *Phytochemistry*, 25, 2315, 1986.
55. **Bakhanashvili, M., Icekson, I., and Apelbaum, A.,** Cadaverine formation by specific lysine decarboxylation in *Pisum sativum* seedlings, *Plant Cell Rept.*, 4, 297, 1985.
56. **Icekson, I., Bakhanashvili, M., and Apelbaum, A.,** Inhibition by ethylene of polyamine biosynthetic enzymes enhanced lysine decarboxylase activity and cadaverine accumulation in pea seedlings, *Plant Physiol.*, 82, 607, 1985.

57. **Ramakrishna, S. and Adiga, P.R.,** Decarboxylation of homoarginine and lysine by an enzyme from *Lathyrus sativus* seedlings, *Phytochemistry*, 15, 83, 1976.
58. **Phan, A.P.H., Ngo, T.T., and Lenhoff, H.M.,** Spectrophotometric assay for lysine decarboxylase, *Anal. Biochem.*, 120, 193, 1982.
59. **Beier, H., Fecker, L.F., and Berlin, J.,** Lysine decarboxylase from *Hafnia alvei*: purification and molecular data, and preparation of polyclonal antibodies, *Z. Naturforsch.*, 42c, 1307, 1987.
60. **Tonelli, D., Budini, R., Gattavecchia, E., and Girotti, S.,** A simple and continuous assay for decarboxylase activity using a carbon dioxide-selective electrode, *Anal. Biochem.*, 111, 189, 1981.
61. **Vienozinskiene', R. Januseciute, Pauliukonis, A., and Kazlauskas, K.,** Lysine decarboxylase assay by the pH-Stat method, *Anal. Biochem.*, 146, 180, 1985.
62. **Rosenberg, R.M., Herreid, R.M., Plazza, G.J., and O'Leary, M.H.,** Indicator assay for amino acid decarboxylases, *Anal. Biochem.*, 181, 59, 1989.
63. **Scriven, F., Wlasichuk, K.B., and Palcic, M.M.,** A continual spectrophotometric assay for amino acid decarboxylases, *Anal. Biochem.*, 170, 367, 1988.
64. **Burns, D.H. and Aberhart, D.J.,** A general coupled spectrophotometric assay for decarboxylases, *Anal. Biochem.*, 171, 339, 1988.
65. **Kochhar, S., Mehta, P.K., and Christen, P.,** Assay for aliphatic amino acid decarboxylases by high-performance liquid chromatography, *Anal. Biochem.*, 179, 182, 1989.
66. **Torrigiani, P., Serafini-Fracassini, D., Biondi, S., and Bagni, N.,** Evidence for the subcellular localization of polyamines and their biosynthetic enzymes in plant cells, *J. Plant Physiol.*, 124, 23, 1986.
67. **Suzuki, Y. and Hirasawa, E.,** *S*-adenosylmethionine decarboxylase of corn seedlings, *Plant Physiol.*, 66, 1091, 1980.
68. **Hiatt, A.C., McIndoo, J., and Malmberg, R.L.,** Regulation of polyamine biosynthesis in tobacco. Effects of inhibitors and exogenous polyamines on arginine decarboxylase, ornithine decarboxylase, and *S*-adenosylmethionine decarboxylase, *J. Biol. Chem.*, 261, 1293, 1986.
69. **Cohen, S.S., Balint, R., and Sindhu, R.K.,** The synthesis of polyamines from methionine in intact and disrupted leaf protoplasts of virus-infected Chinese cabbage, *Plant Physiol.*, 68, 1150, 1981.
70. **Icekson, I., Goldlust, A., and Apelbaum, A.,** Influence of ethylene on *S*-adenosylmethionine decarboxylase activity in etiolated pea seedlings, *J. Plant Physiol.*, 119, 335, 1985.
71. **Malmberg, R.L. and Rose, D.J.,** Biochemical genetics of resistance to MGBG in tobacco: mutants that alter SAM decarboxylase or polyamine ratios, and floral morphology, *Mol. Gen. Genet.*, 207, 9, 1987.
72. **Kaur-Sawhney, R., Shih, L.M., and Galston, A.W.,** Relation of polyamine biosynthesis to the initiation of sprouting in potato tuber, *Plant Physiol.*, 69, 411, 1982.
73. **Montague, M.J., Armstrong, T.A., and Jaworski, E.G.,** Polyamine metabolism in embryogenic cells of *Daucus carota*, *Plant Physiol.*, 63, 341, 1979.
74. **Mukhopadhyay, A. and Ghosh, B.,** Polyamine metabolism in relation to viability of rice seeds, *Seed Sci. Tech.*, 737, 1986.
75. **Bagni, N., Barbieri, P., and Torrigiani, P.,** Polyamine titer and biosynthetic enzymes during tuber formation of *Helianthus tuberosus*, *J. Plant Growth Regul.*, 2, 177, 1983.
76. **Walters, D.R., Wilson, P.W.F., and Shuttleton, M.A.,** Relative changes in levels of polyamines and activities of their biosynthetic enzymes in barley infected with the powdery mildew fungus, *Erysophe graminis* D. C. ex Merat f. sp. hordei marchal, *New Phytol.*, 101, 695, 1985.
77. **Hiatt, A. and Malmberg, R.L.,** Utilization of putrescine in tobacco cell lines resistant to inhibitors of polyamine synthesis, *Plant Physiol.*, 86, 441, 1988.
78. **Smith, T.A.,** Agmatine iminohydrolase in maize, *Phytochemistry*, 8, 2111, 1969.
79. **Yanagisawa, H. and Suzuki, Y.,** Corn agmatine iminohydrolase, *Plant Physiol.*, 67, 697, 1981.
80. **Sindhu, R.K. and Desai, H.V.,** Purification and properties of agmatine iminohydrolase from groundnut cotyledons, *Phytochemistry*, 18, 1937, 1979.
81. **Choudhuri, M.M. and Ghosh, B.,** Agmatine deiminase in rice seedlings, *Phytochemistry*, 24, 2433, 1985.
82. **Desai, H.V. and Mehta, A.R.,** Changes in polyamine levels during shoot formation, root formation, and callus induction in cultured *Passiflora* leaf discs, *J. Plant Physiol.*, 119, 45, 1985.
83. **Archibald, R.M.,** Determination of citrulline and allantoine and demonstration of citrulline in blood plasma, *J. Biol. Chem.*, 156, 121, 1944.
84. **Smith, T.A.,** *N*-carbamylputrescine amidohydrolase of higher plants and its relation to potassium nutrition, *Phytochemistry*, 4, 599, 1965.
85. **Yanagisawa, H. and Suzuki, Y.,** Purification and properties of *N*-carbamylputrescine amidohydrolase from maize shoots, *Phytochemistry*, 21, 2201, 1982.
86. **Raina, A., Eloranta, T., and Pajula, K.L.,** Rapid assays for putrescine aminopropyltransferase (spermidine synthase) and spermidine aminopropyltransferase (spermine synthase), *Meth. Enz.*, 94, 257, 1983.

87. **Anton, D.L.**, A rapid nonchromatographic assay for aminopropyltransferases, *Anal. Biochem.*, 156, 43, 1986.
88. **White Tabor, C. and Tabor, H.**, Putrescine aminopropyl transferase (*Escherichia coli*), *Meth. Enz.*, 94, 265, 1983.
89. **Pegg, A. E.**, Assay of aminopropyltransferases, *Meth. Enz.*, 94, 260, 1983.
90. **Suzuki, O., Matsumoto, T., Oya, M., Katsumata, Y., and Samejima, K.**, A new fluorometric assay for spermidine synthase, *Anal. Biochem.*, 115, 72, 1981.
91. **Sindhu, R.K. and Cohen, S.S.**, Some properties of the spermidine synthase of Chinese cabbage, in *Advances in Polyamine Research*, Vol. 1, Bachrach, U., Kaye, A., and Chayen, R., Eds., Raven Press, New York, 1983, 371.
92. **Sindhu, R.K. and Cohen, S.S.**, Putrescine aminopropyltransferase (spermidine synthase) of Chinese cabbage, *Meth. Enz.*, 94, 279, 1983.
93. **Zaitsu, K. and Ohkura, Y.**, New fluorogenic substrates for horseradish peroxidase: rapid and sensitive assays for hydrogen peroxide and the peroxidase, *Anal. Biochem.*, 109, 109, 1980.
94. **Batchelor, K.W., Smith, R.A., and Watson, N.S.**, Dicyclohexylamine is not an inhibitor of spermidine synthase, *Biochem. J.*, 233, 307, 1985.
95. **Srivenugopal, K.S. and Adiga, P.R.**, Coexistence of two pathways of spermidine biosynthesis in *Lathyrus sativus* seedlings, *FEBS Lett.*, 112, 260, 1980.
96. **Srivenugopal, K.S. and Adiga, P.R.**, Enzyme synthesis of *sym*-homospermidine in *Lathyrus sativus* (grass pea) seedlings, *Biochem. J.*, 190, 461, 1980.
97. **Angelini, R., DiLisi, F., and Federico, R.**, Immunoaffinity, purification and characterization of diamine oxidase from *Cicer*, *Phytochemistry*, 24, 2511, 1985.
98. **Choudhuri, M.M. and Ghosh, B.**, Purification and characterization of diamine oxidase from rice embryos, *Phytochemistry*, 23, 241, 1984.
99. **Hoon, K.J.**, Purification and properties of diamine oxidase from soybean (Glycine Max), *Chem. Abstr.*, 1990, 112, 3241, 1989.
100. **Percival, F.W. and Purves, W.K.**, Multiple amine oxidases in cucumber seedlings, *Plant Physiol.*, 54, 601, 1974.
101. **Floris, G., Gartosio, A., and Rinaldi, A.**, Diamine oxidase from *Lens esculenta* seedlings: purification and properties, *Phytochemistry*, 22, 1871, 1983.
102. **Delhaize, E. and Webb, J.**, Purification and characterization of diamine oxidase from clover leaves, *Phytochemistry*, 26, 641, 1987.
103. **Rinaldi, A., Floris, G., and Finazzi-Agro, A.**, Purification and properties of diamine oxidase from *Euphorbia* latex, *Eur. J. Biochem.*, 127, 417, 1982.
104. **Matsuda, H. and Suzuki, Y.**, Purification and Properties of the diamine oxidase from *Vicia faba*, *Plant Cell Physiol.*, 22, 737, 1981.
105. **Suzuki, Y.**, Some properties of the amine oxidase of soybean seedlings, *Plant Cell Physiol.*, 14, 413, 1973.
106. **Cogoni, A., Piras, C., Farci, R., Melis, A. and Floris, G.**, *Hordeum vulgare* seedlings amine oxidase, *Plant Physiol.*, 93, 818, 1990.
107. **Suresh, M.R., Ramakrishna, S., and Adiga, P.R.**, Diamine oxidase of *Lathyrus sativus* seedlings, *Phytochemistry*, 15, 483, 1976.
108. **Suresh, M.R. and Adiga, P.R.**, Diamine oxidase of *Lathyrus sativus* seedlings: purification and properties, *J. Biosci.*, 1, 109, 1979.
109. **Cogoni, A., Farci, R., Medda, R., Rinaldi, A., and Floris, G.**, Amine oxidase from *Lathyrus cicera* and *Phaseolus vulgaris*: purification and properties, *Preparative Biochem.*, 19, 95, 1989.
110. **Walton, N.J. and McLauchlan, W.R.**, Diamine oxidation and alkaloid production in transformed root cultures of *Nicotiana tabacum*, *Phytochemistry*, 29, 1455, 1990.
111. **Mizusaki, S., Tanabe, Y., Noguchi, M., and Tamaki, E.**, *N*-methylputrescine oxidase from tobacco roots, *Phytochemistry*, 11, 2757, 1972.
112. **Feth, F., Wray, V., and Wagner, K.G.**, Determination of methylputrescine oxidase by high performance liquid chromatography, *Phytochemistry*, 24, 1653, 1985.
113. **Hashimoto, T., Mitani, A., and Yamada, Y.**, Diamine oxidase from cultured roots of *Hyoscyamus niger*, *Plant Physiol.*, 93, 216, 1990.
114. **Cooper, A., Equi, A.M., Ner, S.K., Watson, A.B., and Robins, D.J.**, Oxidation of *N*-alkylputrescines by diamine oxidases, *Tetrahedron Lett.*, 38, 5167, 1989.
115. **Smith, T.A.**, Polyamine oxidation by enzymes from *Hordeum vulgare* and *Pisum sativum* seedlings, *Phytochemistry*, 13, 1075, 1974.
116. **McGowan, R.E. and Muir, R.M.**, Purification and properties of amine oxidase from epicotyls of *Pisum sativum*, *Plant Physiol.*, 47, 644, 1971.
117. **Torrigiani, P., Serafini-Fracassini, D., and Fara, A.**, Diamine oxidase activity in different physiological stages of *Helianthus tuberosus* tuber, *Plant Physiol.*, 89, 69, 1989.

118. **Davies, H.M., Hawkins, D.J., and Smith, L.A.,** Quinoprotein characteristics of *N*-methylputrescine oxidase from tobacco roots, *Phytochemistry*, 28, 1573, 1989.
119. **Sindhu, R.K. and Desai, H.V.,** Inhibitor of diamine oxidase in cotyledons of groundnut seedlings, *Phytochemistry*, 19, 317, 1980.
120. **Federico, R., Angelini, R., Cesta, A., and Pini, C.,** Determination of diamine oxidase in lentil seedlings by enzymic activity and immunoreactivity, *Plant Physiol.*, 79, 62, 1985.
121. **Federico, R. and Angelini, R.,** Occurrence of diamine oxidase in the apoplast of pea epicotyls, *Planta*, 300, 1986.
122. **Federico, R. and Angelini, R.,** Distribution of polyamines and their related catabolic enzyme in etiolated and light grown leguminosae seedlings, *Planta*, 173, 317, 1988.
123. **Smith, T.A.,** The inhibition and activation of polyamine oxidase from oat seedlings, *Plant Growth Regul.*, 3, 269, 1985.
124. **Yanagisawa, H., Hirasawa, E., and Suzuki, Y.,** Purification and properties of diamine oxidase from pea epicotyls, *Phytochemistry*, 20, 2105, 1981.
125. **Shrivastava, S.K., Prakash, V., and Naik, B.I.,** Regulation of diamine oxidase activity in germinating pea seedlings, *Phytochemistry*, 15, 185, 1977.
126. **Shrivastava, S.K. and Prakash, V.,** Purification and properties of pea cotyledon and embryo diamine oxidase, *Phytochemistry*, 16, 189, 1977.
127. **Hirosawa, E.,** Diamine oxidase in cotyledons of *Pisum sativum* develops as a result of the supply of oxygen through the embryonic axis during germination, *Plant Physiol.*, 88, 441, 1988.
128. **Holmstedt, B., Larsson, L., and Tham, R.,** Further studies of a spectrophotometric method for the determination of diamine oxidase, *Biochim. Biophys. Acta*, 48, 182, 1961.
129. **Smith, T.A., Croker, S.J., and Loeffler, R.S.T.,** Occurrence in higher plants of 1-(3-aminopropyl)-pyrrolinium and pyrroline: products of polyamine oxidation, *Phytochemistry*, 25, 683, 1986.
130. **Naik, B.I., Goswami, R.G., and Srivastava, S.K.,** A rapid and sensitive colorimetric assay of amine oxidase, *Analyt. Biochem.*, 111, 146, 1981.
131. **Okuyama, T. and Kubayashi, Y.,** Determination of diamine oxidase activity by liquid scintillation counting, *Arch. Biochem. Biophys.*, 95, 242, 1961.
132. **Yamasaki, E.F., Swindell, R., and Reed, D.J.,** Some aspects of catalysis by the amine oxidase of pea seedlings, *Biochemistry*, 9, 1206, 1970.
133. **Nadeau, P., Delaney, S., and Chouinard, L.,** Effects of cold hardening on the regulation of polyamine levels in wheat (*Triticum aestivum* L.) and alfalfa (*Medicago sativa* L.), *Plant Physiol.*, 84, 73, 1987.
134. **Mukhopadhyay, A., Choudhuri, M.M., Sen, K., and Ghosh, B.,** Changes in polyamines and related enzymes with loss of viability in rice seeds, *Phytochemistry*, 22, 1547, 1983.
135. **Suzuki, Y. and Yanagisawa, H.,** Purification and properties of maize polyamine oxidase: a flavoprotein, *Plant & Cell Physiol.*, 21, 1085, 1980.
136. **Smith, T. S.,** Further properties of the polyamine oxidase from oat seedlings, *Phytochemistry*, 16, 1647, 1977.
137. **Hirasawa, E., Watanabe, H., and Suzuki, Y.,** Polyamine oxidase of millet shoots, *Phytochemistry*, 25, 1739, 1986.
138. **Smith, T.A.,** Polyamine oxidase from barley and oats, *Phytochemistry*, 15, 633, 1976.
139. **Kaur-Sawhney, R., Flores, H.E., and Galston, A.W.,** Polyamine oxidase in oat leaves: a cell wall localized enzyme, *Plant Physiol.*, 68, 494, 1981.
140. **Li, Z.C. and McClure, J.W.,** Contamination of oat mesophyll protoplasts by apoplastic polyamine oxidase, *Physiol. Plant.*, 77, 347, 1989.
141. **Li, Z.C. and McClure, J.W.,** Polyamine oxidase of primary leaves is apoplastic in oats but symplastic in barley, *Phytochemistry*, 26, 2255, 1989.
142. **Smith, T.A.,** Purification and properties of the polyamine oxidase of barley plants, *Phytochemistry*, 11, 899, 1972.
143. **Smith, T.A. and Bickley, D.A.,** Further properties of the polyamine oxidase of barley leaves, *Phytochemistry*, 13, 2437, 1974.
144. **Scorer, K.N., Coamal, R.M., Oropeza, C., and Loyola-Vargas, V.M.,** Amine oxidase activity in *Canavalia ensiformis* during early growth. A histochemical study., *J. Plant Physiol.*, 135, 346, 1989.
145. **Glatz, Z., Kovar, J., Marcholan, L., and Pec, P.,** Pea (*Pisum sativum*) diamine oxidase contains pyrroloquinoline quinone as a cofactor, *Biochem. J.*, 242, 603, 1987.
146. **Slocum, R.,** personal communication.

ABBREVIATIONS

ABA, abscisic acid
ACC, 1-aminocyclopropane-1-carboxylic acid
ADC/ARGdc, arginine decarboxylase
AdoDATAD, *S*-adenosyl-1,12-diamino-3-thio-9-azadodecane
AdoDATO, *S*-adenosyl-1,8,-diamino-3-thiooctane
AG, aminoguanidine
Agm, agmatine
AIH, agmatine iminohydrolase
AO, amine oxidase
AOA, aminooxyacetic acid
AOAP, 1-aminooxy-3-aminopropane
Arg, arginine
ARGase, arginase
AVG, aminoethoxyvinylglycine
Cad/CD, cadaverine
CAL, canaline
CAV, canavanine
CCCP, carbonylcyanide-*m*-chlorophenylhydrazone
CHA/DCHA, cyclohexylamine
CitDC, citrulline decarboxylase
2,4-D, 2,4-dichlorophenoxyacetic acid
DAO, diamine oxidase
Dap, 1,3-diaminopropane
DCCD, dicyclohexylcarbodiimide
DCHA, *N,N*-dimethylcyclohexylamine
DES, diethylstilbestrol
DFMA, DL-a-difluoromethylarginine
DFMO, DL-a-difluoromethylornithine
DMSO, dimethylsulfoxide
2,4-DNP, 2,4-dinitrophenol
dSAM/dAdoMet, decarboxylated *S*-adenosylmethionine
EFE, ethylene forming enzyme
EGF, epidermal growth factor
FCCP, carbonylcyanide-*p*-trifluoromethoxyphenylhydrazone
FMA, a-monofluoromethylagmatine
FMOC, 9-fluorenylmethyl chloroformate
GA_3, gibberellic acid
GABA, 4-aminobutyric acid
H_2O_2, hydrogen peroxide
IAA, indole-3-acetic acid
IBA, indolebutyric acid
LDC, lysine decarboxylase
Lys, lysine
MGBG, methylglyoxal-*bis*-(guanylhydrazone)
MTA, 5′-methylthioadenosine
MTR, 5′-methylthioribose
NAA, napthaleneacetic acid

NaSCN, sodium thiocyanate
NCPut/NCP, *N*-carbamoylputrescine
NEM, *N*-ethylmaleimide
OCT, ornithine carbamoyltransferase
ODC/ORNdc, ornithine decarboxylase
Orn, ornithine
OPA, *o*-phthalaldehyde
PA, polyamine
PAO, polyamine oxidase
PCA, perchloric acid
PLP, pyridoxal phosphate
POD, peroxidase
PQQ, pyrroloquinolinequinone
Put/PU, putrescine
SAH, *S*-adenosylhomocysteine
SAM/AdoMet, *S*-adenosylmethionine
SAMDC, *S*-adenosylmethionine decarboxylase
Spd/SD, spermidine
Spm/SM, spermine
TGase, transglutaminase

INDEX

A

B

C

D

T

U